# Synthesis, Study and Utilization of Natural Products

# Synthesis, Study and Utilization of Natural Products

Special Issue Editors

**Pavel B. Drasar**
**Vladimir A. Khripach**

MDPI • Basel • Beijing • Wuhan • Barcelona • Belgrade

MDPI

*Special Issue Editors*

Pavel B. Drasar
Department of Chemistry of
Natural Compounds, University
of Chemistry and Technology
Czech Republic

Vladimir A. Khripach
The Institute of Bioorganic
Chemistry, The National
Academy of Sciences of Belarus
Republic of Belarus

*Editorial Office*
MDPI
St. Alban-Anlage 66
4052 Basel, Switzerland

This is a reprint of articles from the Special Issue published online in the open access journal *Molecules* (ISSN 1420-3049) from 2018 to 2019 (available at: https://www.mdpi.com/journal/molecules/ special_issues/molecules_SSUoNP).

For citation purposes, cite each article independently as indicated on the article page online and as indicated below:

LastName, A.A.; LastName, B.B.; LastName, C.C. Article Title. *Journal Name* **Year**, *Article Number*, Page Range.

**ISBN 978-3-03928-152-7 (Pbk)**
**ISBN 978-3-03928-153-4 (PDF)**

# Contents

# About the Special Issue Editors

**Pavel B. Drasar**, prof., RNDr. DSc. Education and Recognition of professional experience: 2008 Chartered Scientist, 2004 Full professor of organic chemistry, 2004 DSc in organic chemistry, 2002 associated professor (docent), 1997 EurChem, 1993 CChem, FRSC, 1972–77 PhD study, Institute of Organic Chemistry CAS, Prague, 1972 RNDr. (Rerum Naturalium Doctor), 1966–71 Charles University Prague, Faculty of Natural Sciences. Scientific Activity and Professional Positions: 2002– UCT Praha, educator and research worker, 1972–2002, Institute of Organic Chemistry and Biochemistry (IOCB), CAS, PhD student, later research worker, 1971–1972 Charles University, assistant. Board and Committee Membership: 2002–3, 2007–8 President Assoc. Czech Chemical Societies; 2004– member, 2006–08 vice-chairman, 2008–2013 chairman (2013–14 past chair) 2019– chairman ECTN Label Committee; 2015–2018 ECTN president, 2018–19 past-president; 2004–2019 European Chemical Society (EuChemS, formerly FECS) ExComm member; 1997– ECRB member; 1996– vice president of the Czech Chemical Society; 1990– member of the Czech committee for organic chemistry nomenclature; 1987– Bulletin of the Czech(oslovak) Chemical Society, editor. 1997– Chemicke Listy, editor; 2015– Steroids (Elsevier) editorial board member, managing guest editor; 2018– Molecules, guest editor; 1994– Alfred Bader Prize Committees, member; 2019– Scientific Secretary of the Czech Association of Scientific and Technical Societies. 2014– Isoprenoid Society General Secretary. Areas of the main scientific interest: Synthesis and biological and physicochemical evaluation of steroids and their conjugates, steroidal and terpene lactones, alkaloids, brassinosteroids, carbohydrate and their conjugates, ion channel modifiers, synthesis of natural products with fluorescent labels for bioimaging, targeting of biologically active compounds by peptide vectors, i.a. Publication Activity: 256 documents and 1289/886 citations in WoS, h-index 16, over 160 conferences, 16 books, 38 patents.

**Vladimir A. Khripach**, prof., DSc., member of the Academy of Sciences Education: M. Sc. (Chemistry), 1971, Byelorussian State University, Minsk, Ph.D. (Organic Chemistry, with Prof. A.A. Akhrem and F.A. Lakhvich), 1978, Institute of Physical Organic Chemistry, Byelorussian SSR Acad. Sci., Minsk, Dr. Sc. (Chemistry), 1990, N.D. Zelinsky Inst. of Org. Chem., USSR Acad. Sci., Moscow. Career/Employment: Probationer–researcher, Laboratory of cortico-steroids, N.D. Zelinsky Inst. of Org. (position and date) Chem., USSR Acad. Sci., Moscow (1970–1971). Research worker, Inst. of Bioorganic Chemistry, Belarus Acad. Sci. (1971–1982). Head of the Laboratory of Steroid Chemistry, Inst. of Bioorganic Chemistry, Belarus Acad. Sci. (1982–date). Specialization: Organic and Bioorganic Chemistry. Main Field: Synthesis of biologically important natural substances and structure-activity relationships. Steroid chemistry and biochemistry. Other Fields: Bioactivity and practical application of natural bioregulators. Plant physiology. Immunochemistry. Medicinal chemistry. Current Research Interests: Synthesis, biosynthesis and analysis of steroid plant hormones and related compounds, their biological study and application. Synthetic and biomedical aspects of natural polyhydroxysteroids and their analogues. New pharmacologically important steroids. Supramolecular chemistry and complex chemical systems. Honours, Awards, Fellowships: Academic rank of Senior Researcher (1984). Membership of professional Prize of D.I. Mendeleev Scientific Chemical Society (1985). Gold Medals of All-Russian Exhibition Center, Moscow (1993, 1995, 1996). State Prize Winner (1996). Jubilee Medal of the Academy of Sciences of Belarus (2009). Hanuš Medal of Czech Chemical Society (2009). Medal of the Ukrainian State Foundation for Fundamental

Research (2009). D.I. Mendeleev Scientific Chemical Society Fellowship. Publications, Patents: More than 500 publications, including five books and more than 50 patents.

*molecules*   MDPI

*Editorial*

# Growing Importance of Natural Products Research

**Pavel B. Drasar** [1,*] and **Vladimir A. Khripach** [2]

1    Department of Chemistry of Natural Compounds, University of Chemistry and Technology, Technicka 5, 166 28 Prague, Czech Republic
2    Institute of Bioorganic Chemistry, National Academy of Sciences of Belarus, 5/2 Academician V. F. Kuprevich Street, BY-220141 Minsk, Belarus; khripach@iboch.bas-net.by
*    Correspondence: Pavel.Drasar@vscht.cz

Received: 16 December 2019; Accepted: 17 December 2019; Published: 18 December 2019

Natural products and preparations based on them play a stable and ever-increasing role in human and veterinary medicine, agriculture, in food and the cosmetic industry, and in other increasing numbers of fields. Their importance is based on the fact that they are mostly bound to renewable sources, which in fact makes them valuable within a circular economy, inter alia. At the same time, natural products give the origin of stereochemistry, optical activity, regioselectivity, chirality, and many other concepts and directions within science, development, and industry in a scope, which is indispensable. They serve as a constant powerful stimulus and model that inspires researchers to create new effective tools, similar to natural ones for controlling bioregulation mechanisms and solving practical problems. This was the reason for organizing this Special Issue aimed to underline current developments in all fields connected to natural products.

Hence, the Molecules Special Issue "Synthesis, Study and Utilization of Natural Products" brought in 15 papers, four reviews, and 11 full research communications.

The scope of the selected topics was rather broad, it showed the importance of the pegylated purpurin 18 for photodynamic therapy of cancer [1], it presented the anti-platelet aggregation activity study of ginkgolide-1,2,3-triazole derivatives [2], it showed that the overexpression of the melatonin synthesis-related gene *SLCOMT1* improves the resistance of tomato to salt stress [3]. Another study revealed the effects of isosorbide incorporation into flexible polyurethane foams: reversible urethane linkages and antioxidant activity [4]. The synthesis and in vitro evaluation of caffeoylquinic acid derivatives as potential hypolipidemic agents [5] and the first total synthesis of varioxiranol A [6] were also presented. Another study introduced to the readers the preparation of polysaccharides from *Ramulus mori*, and their antioxidant, anti-inflammatory, and antibacterial activities [7]. Studied were also the effect of enzymolysis on the performance of soy protein-based adhesive [8] and the study of new octadecanoid enantiomers from the whole plants of *Plantago depressa* [9]. Connected studies that combined the biological properties of another type of secondary metabolite described the biosynthesis of fluorescent β subunits of C-phycocyanin from *Spirulina subsalsa* in *Escherichia coli*, and their antioxidant properties [10] and presented the synthesis of the sex pheromone of the tea tussock moth based on a resource chemistry strategy [11].

Review articles described well the synthesis and anticancer activity of CDDO and CDDO-Me, two derivatives of natural triterpenoids [12], the advances in biosynthesis, pharmacology, and pharmacokinetics of pinocembrin, a promising natural small-molecule drug [13], as well as recent advances in the discovery and biosynthetic study of eukaryotic RiPP natural products [14]. Another review article addressed the issue of whether polyphenols could help in the control of rheumatoid arthritis [15].

Summing up, the current development in the chemistry of natural products proved to be so exciting that now Molecules itself organized recently several special issues oriented to this unfinished and fruitful field of the activity of the world chemical community. It is important to wish chemists and their friends in connected fields much enthusiasm and success in their work as it brings so many useful fruits and tools for all humankind.

**Acknowledgments:** The Guest Editor wish to thank all the authors for their contributions to this Special Issue, all the Reviewers for their work in evaluating the submitted articles and the editorial staff of Molecules for their kind assistance.

**Conflicts of Interest:** The author declares no conflict of interest.

## References

1. Pavlíčková, V.; Rimpelová, S.; Jurášek, M.; Záruba, K.; Fähnrich, J.; Křížová, I.; Bejček, J.; Rottnerová, Z.; Spiwok, V.; Drašar, P.; et al. PEGylated Purpurin 18 with Improved Solubility: Potent Compounds for Photodynamic Therapy of Cancer. *Molecules* **2019**, *24*, 4477. [CrossRef] [PubMed]
2. Cui, J.; Hu, L.; Shi, W.; Cui, G.; Zhang, X.; Zhang, Q.-W. Design, Synthesis and Anti-Platelet Aggregation Activity Study of Ginkgolide-1,2,3-triazole Derivatives. *Molecules* **2019**, *24*, 2156. [CrossRef] [PubMed]
3. Liu, D.-D.; Sun, X.-S.; Liu, L.; Shi, H.-D.; Chen, S.-Y.; Zhao, D.-K. Overexpression of the Melatonin Synthesis-Related Gene SlCOMT1 Improves the Resistance of Tomato to Salt Stress. *Molecules* **2019**, *24*, 1514. [CrossRef] [PubMed]
4. Shin, S.-R.; Liang, J.-Y.; Ryu, H.; Song, G.-S.; Lee, D.-S. Effects of Isosorbide Incorporation into Flexible Polyurethane Foams: Reversible Urethane Linkages and Antioxidant Activity. *Molecules* **2019**, *24*, 1347. [CrossRef] [PubMed]
5. Tian, Y.; Cao, X.-X.; Shang, H.; Wu, C.-M.; Zhang, X.; Guo, P.; Zhang, X.-P.; Xu, X.-D. Synthesis and In Vitro Evaluation of Caffeoylquinic Acid Derivatives as Potential Hypolipidemic Agents. *Molecules* **2019**, *24*, 964. [CrossRef] [PubMed]
6. Lásiková, A.; Doháňošová, J.; Štiblariková, M.; Parák, M.; Moncol, J.; Gracza, T. First Total Synthesis of Varioxiranol A. *Molecules* **2019**, *24*, 862. [CrossRef] [PubMed]
7. Yu, W.; Chen, H.; Xiang, Z.; He, N. Preparation of Polysaccharides from *Ramulus mori*, and Their Antioxidant, Anti-Inflammatory and Antibacterial Activities. *Molecules* **2019**, *24*, 856. [CrossRef] [PubMed]
8. Xu, Y.; Xu, Y.; Han, Y.; Chen, M.; Zhang, W.; Gao, Q.; Li, J. The Effect of Enzymolysis on Performance of Soy Protein-Based Adhesive. *Molecules* **2018**, *23*, 2752. [CrossRef] [PubMed]
9. Song, X.-Q.; Zhu, K.; Yu, J.-H.; Zhang, Q.; Zhang, Y.; He, F.; Cheng, Z.-Q.; Jiang, C.-S.; Bao, J.; Zhang, H. New Octadecanoid Enantiomers from the Whole Plants of *Plantago depressa*. *Molecules* **2018**, *23*, 1723. [CrossRef] [PubMed]
10. Wu, X.-J.; Yang, H.; Chen, Y.-T.; Li, P.-P. Biosynthesis of Fluorescent β Subunits of C-Phycocyanin from *Spirulina subsalsa* in *Escherichia coli*, and Their Antioxidant Properties. *Molecules* **2018**, *23*, 1369. [CrossRef] [PubMed]
11. Zhang, H.-L.; Sun, Z.-F.; Zhou, L.-N.; Liu, L.; Zhang, T.; Du, Z.-T. Synthesis of the Sex Pheromone of the Tea Tussock Moth Based on a Resource Chemistry Strategy. *Molecules* **2018**, *23*, 1347. [CrossRef] [PubMed]
12. Borella, R.; Forti, L.; Gibellini, L.; De Gaetano, A.; De Biasi, S.; Nasi, M.; Cossarizza, A.; Pinti, M. Synthesis and Anticancer Activity of CDDO and CDDO-Me, Two Derivatives of Natural Triterpenoids. *Molecules* **2019**, *24*, 4097. [CrossRef] [PubMed]
13. Shen, X.; Liu, Y.; Luo, X.; Yang, Z. Advances in Biosynthesis, Pharmacology, and Pharmacokinetics of Pinocembrin, a Promising Natural Small-Molecule Drug. *Molecules* **2019**, *24*, 2323. [CrossRef] [PubMed]
14. Luo, S.; Dong, S.-H. Recent Advances in the Discovery and Biosynthetic Study of Eukaryotic RiPP Natural Products. *Molecules* **2019**, *24*, 1541. [CrossRef] [PubMed]
15. Sung, S.; Kwon, D.; Um, E.; Kim, B. Could Polyphenols Help in the Control of Rheumatoid Arthritis? *Molecules* **2019**, *24*, 1589. [CrossRef] [PubMed]

![molecules logo] *molecules*

MDPI

*Article*

# PEGylated Purpurin 18 with Improved Solubility: Potent Compounds for Photodynamic Therapy of Cancer

Vladimíra Pavlíčková [1], Silvie Rimpelová [1,*] , Michal Jurášek [2] , Kamil Záruba [3] ,
Jan Fähnrich [3], Ivana Křížová [4], Jiří Bejček [1], Zdeňka Rottnerová [5], Vojtěch Spiwok [1] ,
Pavel Drašar [2,*] and Tomáš Ruml [1,*]

[1]  Department of Biochemistry and Microbiology, University of Chemistry and Technology in Prague,
    Technická 3, 166 28 Prague 6, Czech Republic; vladimira.pavlickova@vscht.cz (V.P.);
    jiri.bejcek@vscht.cz (J.B.); vojtech.spiwok@vscht.cz (V.S.)
[2]  Department of Chemistry of Natural Compounds, University of Chemistry and Technology in Prague,
    Technická 5, 166 28 Prague 6, Czech Republic; michal.jurasek@vscht.cz
[3]  Department of Analytical Chemistry, University of Chemistry and Technology in Prague, Technická 5,
    166 28 Prague 6, Czech Republic; kamil.zaruba@vscht.cz (K.Z.); jan.fahnrich@vscht.cz (J.F.)
[4]  Department of Biotechnology, University of Chemistry and Technology in Prague, Technická 5,
    166 28 Prague 6, Czech Republic; ivana.krizova@vscht.cz
[5]  Central laboratories, University of Chemistry and Technology in Prague, Technická 5, 166 28 Prague 6,
    Czech Republic; zdenka.rottnerova@vscht.cz
*   Correspondence: silvie.rimpelova@vscht.cz (S.R.); pavel.drasar@vscht.cz (P.D.); tomas.ruml@vscht.cz (T.R.);
    Tel.: +420-220-44-4360 (S.R.)

Received: 30 September 2019; Accepted: 1 December 2019; Published: 6 December 2019

check for
updates

**Abstract:** Purpurin 18 derivatives with a polyethylene glycol (PEG) linker were synthesized as novel photosensitizers (PSs) with the goal of using them in photodynamic therapy (PDT) for cancer. These compounds, derived from a second-generation PS, exhibit absorption at long wavelengths; considerable singlet oxygen generation and, in contrast to purpurin 18, have higher hydrophilicity due to decreased logP. Together, these properties make them potentially ideal PSs. To verify this, we screened the developed compounds for cell uptake, intracellular localization, antitumor activity and induced cell death type. All of the tested compounds were taken up into cancer cells of various origin and localized in organelles known to be important PDT targets, specifically, mitochondria and the endoplasmic reticulum. The incorporation of a zinc ion and PEGylation significantly enhanced the photosensitizing efficacy, decreasing $IC_{50}$ (half maximal inhibitory compound concentration) in HeLa cells by up to 170 times compared with the parental purpurin 18. At effective PDT concentrations, the predominant type of induced cell death was apoptosis. Overall, our results show that the PEGylated derivatives presented have significant potential as novel PSs with substantially augmented phototoxicity for application in the PDT of cervical, prostate, pancreatic and breast cancer.

**Keywords:** apoptosis; cancer cells; cytotoxicity; flow cytometry; live-cell fluorescence microscopy; PEGylated purpurin 18; photodynamic therapy; photosensitizer; phototoxicity; singlet oxygen

## 1. Introduction

Chlorins are natural photosensitive chlorophyll derivatives containing twenty $\pi$ electrons in the aromatic ring. Various modified substructures derived from their basic core have been discovered within the plant kingdom [1–4]. Owing to their strong absorption between 650–700 nm, wavelengths that penetrate tissue effectively, chlorins have been investigated as photosensitizers (PSs) for use in the photodynamic therapy (PDT) of cancerous and noncancerous diseases [5–8].

During the PDT treatment of cancer, ubiquitous oxygen in the triplet state turns into highly reactive singlet oxygen [9] that triggers cell death via oxidative damage to proteins, lipids and other cellular content, resulting in apoptosis [7,10], necrosis [11] and/or autophagy [12]. In addition to these direct mechanisms of tumor elimination, PDT leads to microvascular damage [13], which is a significant advantage over traditionally used treatments, such as chemo- and radiotherapy. Moreover, PDT also induces immunogenic cell death by stimulating the immune system response to the tumor [13]. PS-induced phototoxic damage initiates the release of anti-inflammatory mediators that attract neutrophils and other immune cells [14]. Indeed, a PS can even trigger adaptive immunity leading to long-term immune response [15].

Chlorins possess optimal properties for use in PDT but are rather hydrophobic and, thus, aggregate in aqueous media, limiting their application. Consequently, various chemical modifications of chlorin-based PSs have been investigated with the aim of improving their physico-chemical characteristics: core metalation [16]; PEGylation [17–20]; conjugation with peptides [21–24], amino acids [1,25–27], sugars [28–31], choline [7,32] and gold nanoparticles [32].

A chlorin worth further derivatization is purpurin 18 (compound **1**, Scheme 1), which comprises a fused anhydride and an aliphatic side chain terminated with a carboxylic group. With its strong absorption at 700 nm and good singlet oxygen quantum yield (0.7) [33], this PS has been previously evaluated as a highly potent inductor of PDT-mediated cell death [6,34,35]. Nevertheless, in its natural form, its hydrophobicity causes aggregation at physiological pH and, thus, preferential localization in compartments undesirable for PDT, such as lipid vesicles and lysosomes. Moreover, under the in vivo conditions of PDT, the anhydride ring moiety is readily hydrolyzed into another PS chlorin, p6 [36], which is less effective than compound **1** [34].

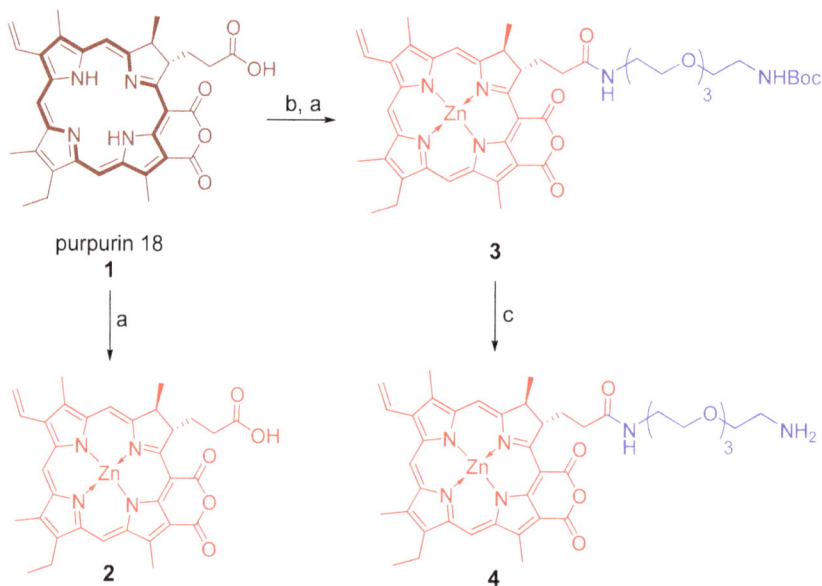

**Scheme 1.** Synthesis of derivatives of PEGylated purpurin 18 (compound **1**). Reagents and conditions: (**a**) Zn(OAc)$_2$·2H$_2$O, MeOH, CHCl$_3$, 50 °C, 13 h; yield of compound **2** was 61%; (**b**) DIC, EDIPA, THF, HOBt, 24 h, RT (22 °C); yield of compound **3** was 41% over two steps; (**c**) TFA, wet DCM, 1 h, RT (22 °C); yield of compound **4** was 56%.

However, despite the drawbacks associated with compound **1**, the natural advantages of purpurins makes it worthwhile to investigate the modification of this chlorin. Therefore, we here synthesize and evaluate PEGylated derivatives of compound **1** as novel PDT agents. We show that the attachment of short PEG$_3$ moieties terminated by Boc (**3**) or an amino group (**4**) via an amide bond to the zinc chelate of purpurin 18 (**2**) does not hamper its ability to generate singlet oxygen in cell culture media in vitro; in fact, it actually enhances singlet oxygen generation, and photodynamic efficiency, by a factor of at least two. Furthermore, live-cell imaging showed that the PEGylation of compound **1** improves PS accumulation in the mitochondria and endoplasmic reticulum, the preferred targets for PDT drugs; in the case of compound 4, it also improves PS accumulation in lysosomes. Moreover, compound phototoxicity and dark toxicity were compared in six cancerous cell lines using WST-1 assay. These tests confirmed the increased PDT efficacy of the PEGylated analogues of compound **1**; these analogues also augmented the proportion of apoptotic cells when photoactivated. In addition, we show that these novel compounds have enhanced hydrophilicity (calculated) and are weaker binders of the prevalent transport protein, human serum albumin (HSA), than the parental compound.

## 2. Results and Discussion

### 2.1. Synthesis of Purpurin 18 Derivatives

The single carboxylic moiety of compound **1** was chosen as the site of synthetic modifications to its structure. A purpurin zinc complex (**2**) was prepared as described by Olshevskaya et al. [37]. Purpurin-18-PEG$_3$-amine conjugates **3** and **4** were synthesized in three steps (see Scheme 1). The conjugation of Boc-protected PEG$_3$-diamine to **1** was performed using carbodiimide chemistry. *N*, *N*-diisopropylcarbodiimide (DIC) with *N*-hydroxybenzotriazole (HOBt) and Hünig's base (EDIPA) were used as the coupling conditions. PEGylated compound **1** was only filtered through a silica plug and the first dark band collected as crude product. Zinc was inserted into the chlorin core using zinc(II) acetate as a metal donor and product **2** was purified by two-step column chromatography with a yield of 41%. The Boc protecting group was cleaved by an excess of trifluoroacetic acid (TFA) in wet dichloromethane (DCM) to obtain amine **4** with a yield of 56%. The obtained products were lyophilized from aqueous dioxane and stored in a fridge in the dark. The acquired spectra are shown in Supplementary Information (SI, Figures S1–S6-2), Section 1.

### 2.2. Singlet Oxygen Generation

The quantum yield of singlet oxygen production by the PSs was evaluated using absorption spectrometry, with 9,10-anthracenediyl-bis(methylene)dimalonic acid (**AB**) as the probe. The PS-mediated singlet oxygen production was monitored by decreases in the absorbance of **AB** at 381 and 403 nm, which were due to the formation of the corresponding endoperoxide [38,39]. There was a negligible decrease in AB absorption without PS (SI, Figure S6-3). The rate of a decrease in **AB** relative absorbance was considered to be proportional to singlet oxygen production.

The singlet oxygen quantum yield for a tested compound ($\phi_x$) was compared with the known quantum yield ($\phi_s$) of a standard

$$\phi_x = \phi_s \, \gamma_x/\gamma_s$$

where $\gamma_x$ and $\gamma_s$ are chemical photodynamic efficiencies of the tested and standard compounds, respectively, evaluated from the **AB** absorbance decrease plotted against relative light exposure ($I_A$) (Figure 1). Using the singlet oxygen quantum yield of Rose Bengal (**RB**) in phosphate buffered saline $\phi_s = 0.75$ [40,41], quantum yields for **RB** and studied compounds **1–4** were evaluated in Dulbecco's Modified Eagle Medium with fetal bovine serum (DMEM+FBS) (Table 1, SI Figure S6-4).

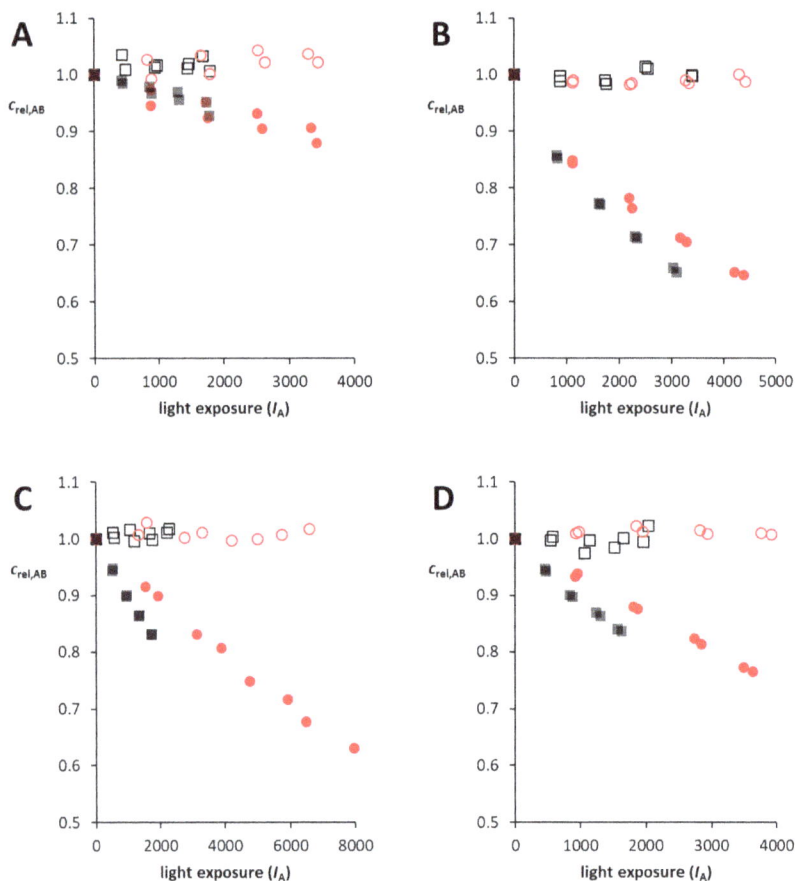

**Figure 1.** Depletion of 9,10-anthracenediyl-bis(methylene)dimalonic acid (**AB**, $7 \times 10^{-5}$ M) with photosensitizer-generated singlet oxygen in Dulbecco's Modified Eagle Medium with 10% fetal bovine serum. Photosensitizers: (**A**) compound **1** ($7.7 \times 10^{-6}$ M, $1.5 \times 10^{-5}$ M), (**B**) compound **2** ($8.0 \times 10^{-6}$ M, $1.6 \times 10^{-5}$ M), (**C**) compound **3** ($7.2 \times 10^{-6}$ M, $1.4 \times 10^{-5}$ M), (**D**) compound **4** ($7.5 \times 10^{-6}$ M, $1.5 \times 10^{-5}$ M). The experiments were duplicated. ■ ●—Solution exposed to light, □ ○—Solution kept in dark. $c_{rel,AB}$—Relative concentration of **AB** (actual concentration with respect to concentration at experiment start).

**Table 1.** Estimated chemical photodynamic efficiencies $\gamma$ and singlet oxygen quantum yields $\phi$ for compounds **1–4**. Values were measured in phosphate buffered saline (PBS) (except $\phi_s = 0.75$ of **RB** in PBS, as reported by Gottfried et al. [40]) and cell culture media supplemented with 10% fetal bovine serum (DMEM+FBS). Standard deviations of all calculated values were less than 10%.

| Compound | Solvent | $\gamma \times 10^4$ | $\phi$ |
|----------|---------|----------------------|--------|
| RB | PBS | 16.5 | 0.75 [1] |
|  | DMEM+FBS | 2.68 | 0.122 |
| 1 | DMEM+FBS | 0.34 | 0.015 |
| 2 | DMEM+FBS | 1.23 | 0.056 |
| 3 | DMEM+FBS | 0.63 | 0.029 |
| 4 | DMEM+FBS | 0.81 | 0.037 |

[1] Reference value according to Gottfried et al. [40].

## 2.3. Uptake and Intracellular Localization of the Compounds

The ability of a PS to cross the plasma membrane is the initial prerequisite for good PDT efficacy [42,43]. Therefore, using live-cell fluorescence microscopy, we determined the ability of compound **1** and its derivatives **2–4** (0.2 to 2 μM) to accumulate in human cells of various origin after 3, 16 and 24 h. Cell lines derived from breast (MCF-7), prostate (PC-3, LNCaP) and cervical (HeLa) carcinoma, as well as from pancreatic adenocarcinoma (MiaPaCa-2) and immortalized human keratinocytes (HaCaT), were used. Based on the microscopic images (Figure 2, SI, Figure S7), it is clear that the efficacy of the cell uptake of the individual compounds varied. At the same concentration and incubation time, the fluorescence emission intensities (Table 2; SI, Figure S16) of compounds **1** and **2** were weaker than those of PEGylated derivatives **3** and **4** (data in Table 2 for PC-3 cells). Compared with compounds **3** and **4**, the low fluorescence emission intensities of compounds **1** and **2** might be caused by their less efficient penetration through the plasma membrane and/or by faster efflux. In turn, this could be due to their distinct molecule sizes as well as to differences in their lipophilicity, which is one of the key factors for compound penetration through cell membranes. The lipophilicity of compounds may be enhanced at the lower pH of cancer cells. This has been documented by the increased uptake of hematoporphyrin at lower than physiological pH [44,45], though it was not observed for mTHPP, mTHPC and TPPS2a [45]. Similarly, Sharma et al. reported the augmented cell uptake of chlorin p6 at decreased pH for Colo-205 cells, but not for MCF-7 cells [46]. Therefore, apart from being pH dependent, the cell uptake of a PS is also cell line specific. This corresponds with the uptake and intracellular localization of compounds **1–4** differing both among the tested compounds and evaluated cell lines.

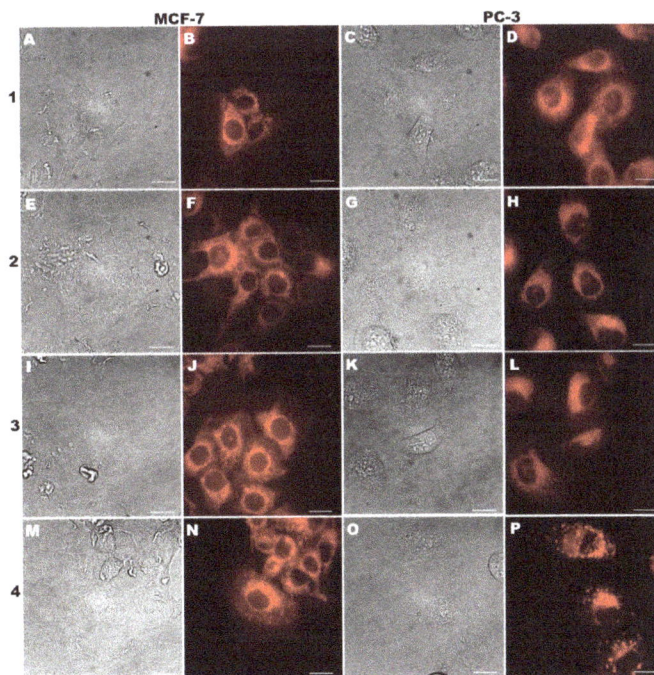

**Figure 2.** Fluorescence microscopy images of intracellular localization of purpurin 18 (compound **1**) and its derivatives (compounds **2–4**) at 0.5 μM concentration in human cancer cell lines of MCF-7 (breast carcinoma) and PC-3 (prostate carcinoma) after 24 h incubation. In the first and third columns, there are bright field images; the second and fourth columns show compound localization. The scale bars represent 20 μm.

**Table 2.** Corrected total cell fluorescence (CTCF) of compounds **1**–**4** (1 μM, 24 h) localized in PC-3 cells (see Figure S16 for raw data).

| Compound | CTCF $\times 10^3$ |
|----------|-----------|
| 1 | $1.404 \pm 0.134$ |
| 2 | $1.593 \pm 0.208$ |
| 3 | $5.042 \pm 0.263$ |
| 4 | $6.643 \pm 0.405$ |

Sharma et al. [36] reported that the aggregation of compound **1** (6 μM) led to its limited availability. Nevertheless, probably due to the lower concentration used (0.5 μM), we did not observe any aggregation of this compound but, rather, homogenous localization in the intracellular space of the HaCaT, LNCaP and PC-3 cells (Figure 2 and SI, Figure S7) after 3 h. In the MCF-7 cells (Figure 2), compounds **1**–**3** localized in organelles visible as a network-like structure. Compound **4** localized in the HaCaT and PC-3 cells, preferentially in small vesicles with high fluorescence intensity. Regarding the MCF-7 cell line, compound **4** localized in both a network-like structure and in small vesicles with high fluorescence intensity.

### 2.4. Colocalization Study

To determine the exact intracellular localization of the tested compounds, commercial markers of cell organelles were used. Colocalization with the endoplasmic reticulum marker, ER-Tracker Blue-White DPX, was detected for all tested compounds in the PC-3 (Figure 3), MCF-7, LNCaP and HaCaT cells (SI, Figures S8–S10). Moreover, compounds **1**–**3** colocalized with mitochondrial sensors (MitoTracker Green and/or our patented green-emitting dimethinium salt [47]) in the PC-3 (Figure 4), MCF-7, LNCaP and HaCaT cells (SI, Figures S11–S13). Compound **4** also localized in the endoplasmic reticulum, but not in the mitochondria of the PC-3, MCF-7, LNCaP and HaCaT cells. Because another fluorescent signal not originating from the endoplasmic reticulum was surprisingly detected, further colocalization studies were performed. Using fluorescent markers of the Golgi apparatus (CellLight Golgi-GFP) and lysosomes (LysoTracker Green DND-26), lysosomal localization was confirmed, except in the Golgi apparatus (SI, Figure S14) of compound **4** in the HaCaT (SI, Figure S15), PC-3 (Figure 5) and MCF-7 cells.

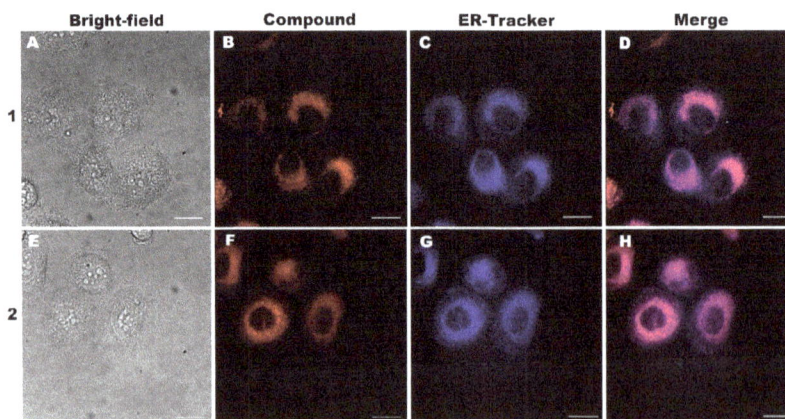

**Figure 3.** *Cont.*

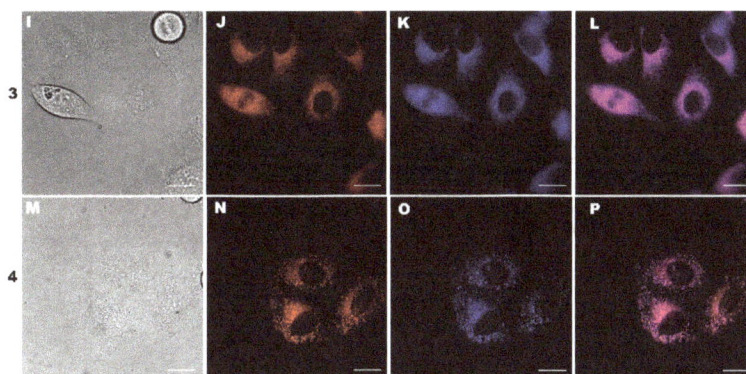

**Figure 3.** Fluorescence microscopy images of localization of purpurin 18 (compound **1**) and its derivatives (compounds **2–4**) in the endoplasmic reticulum of human PC-3 cells derived from prostate carcinoma. Colocalization of compounds **1–2** (0.5 μM, 24 h) or compounds **3–4** (0.5 μM, 24 h) with ER-Tracker™ Blue-White DPX (70 nM, 30 min). (**A,E,I,M**) Bright-field images; (**B,F,J,N**) localization of the tested compounds; (**C,G,K,O**) ER-Tracker™ Blue-White DPX; (**D,H,L,P**) merged fluorescent images. The scale bars represent 20 μm.

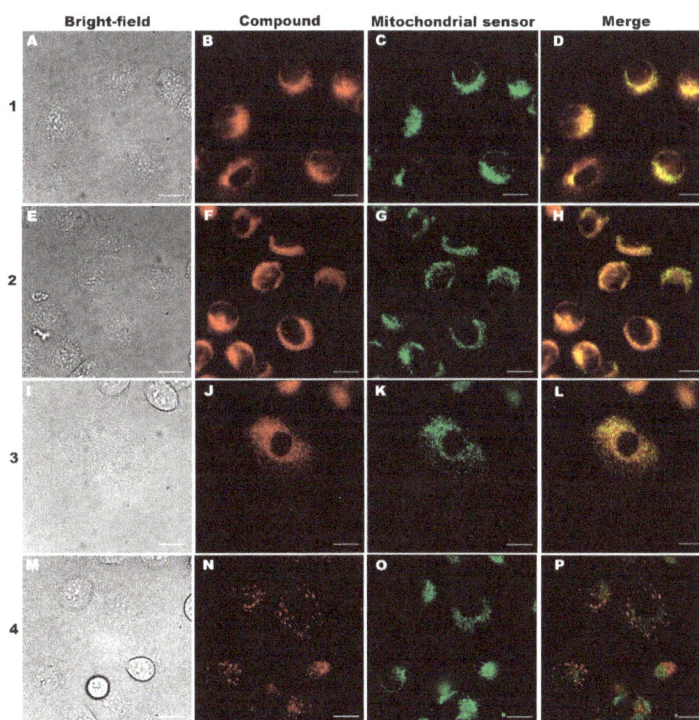

**Figure 4.** Fluorescence microscopy images of localization of purpurin 18 (compound **1**) and its derivatives (compounds **2–4**) in the mitochondria of human PC-3 cells derived from prostate carcinoma. Colocalization of compounds **1–2** (0.5 μM, 3 h) or compounds **3–4** (0.5 μM, 3 h) with a mitosensor (70 nM, 10 min) based on our patented dimethinium salt [47]. (**A,E,I,M**) Bright-field images; (**B,F,J,N**) localization of the tested compounds; (**C,G,K,O**) mitosensor; (**D,H,L,P**) merge of the fluorescent images. The scale bars represent 20 μm.

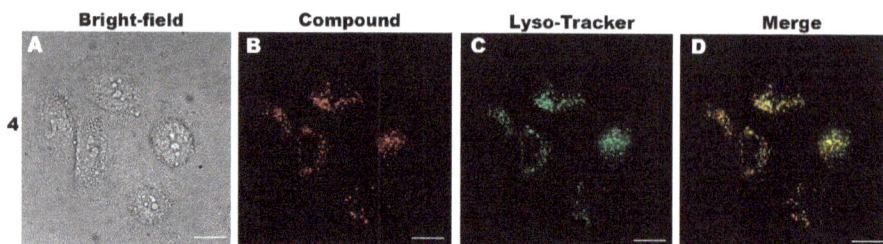

**Figure 5.** Fluorescence microscopy images of compound **4** localization in lysosomes of human PC-3 cells derived from prostate carcinoma. Colocalization of compound **4** (0.5 μM, 24 h) with LysoTracker Green DND-26 (70 nM, 20 min). (**A**) Bright-field images; (**B**) localization of compound **4**; (**C**) LysoTracker Green DND-26; (**D**) merge of the fluorescent images. The scale bars represent 20 μm.

The results for compound **1** correspond to those reported by other research groups focused on chlorophyll-derived PS photochemistry. The localization of purpurin 18 and its derivative chlorin p6 has been detected in the mitochondria, lysosomes and endoplasmic reticulum [19,42,48–52]. The localization of any PS in such organelles is key to high PDT efficacy.

### 2.5. Photo- and Dark Toxicity of the Compounds In Vitro

PSs **2–4** not only exhibited localization in preferable cell organelles (meaning that high PDT efficacy can be expected) but also produced good quantum yields (Table 1) exceeding those of compound **1** (Table 1). Therefore, we investigated their phototoxicity in human cancer cells. LNCaP, PC-3, MCF-7, U-2 OS (osteosarcoma), MIA PaCa-2 and HeLa cells were treated with compounds **1–4** (0.5–10 μM) for 24 h followed by light activation (light dose of 4 J·cm$^{-2}$, 13 min) and incubation (a further 24 h). Dark toxicity (without photoactivation) was also evaluated for all compounds. Compound toxicity is expressed as a decrease in cell viability (SI, Figures S17 and 18) and by half maximal inhibitory compound concentration (IC$_{50}$) values (Table 3).

**Table 3.** Photo- and dark toxicity of compounds **1–4** in human cancer cell lines in vitro 24 h after photoactivation (48 h after compound treatment).

| | IC50 (μM) [1] | | | | | | | |
|---|---|---|---|---|---|---|---|---|
| Compound | 1 | | 2 | | 3 | | 4 | |
| Cell Line | Light | Dark | Light | Dark | Light | Dark | Light | Dark |
| LNCAP | 0.34 ± 0.02 | >10 | 0.47 ± 0.03 | >10 | 0.04 ± 0.03 | 7.20 ± 0.08 | 0.02 ± 0.00 | >10 |
| PC-3 | 0.16 ± 0.01 | >10 | 0.21 ± 0.01 | >10 | 2.33 ± 0.03 | >10 | 0.65 ± 0.00 | >10 |
| U-2OS | 1.96 ± 0.01 | >10 | 7.01 ± 0.05 | >10 | 3.17 ± 0.05 | >10 | 1.83 ± 0.01 | >10 |
| MIA PACA-2 | 1.51 ± 0.03 | >10 | 1.04 ± 0.03 | >10 | 1.12 ± 0.01 | >10 | 0.45 ± 0.05 | >10 |
| MCF-7 | 1.62 ± 0.02 | >10 | 2.95 ± 0.01 | >10 | 2.00 ± 0.02 | >10 | 0.59 ± 0.03 | >10 |
| HELA | 3.40 ± 0.02 | >10 | >10 | >10 | 0.06 ± 0.05 | 7.95 ± 0.06 | 0.02 ± 0.01 | >10 |

[1] IC$_{50}$—Half maximal inhibitory compound concentration.

Up to a concentration of 10 μM, compound **1** did not induce any dark toxicity in the MCF-7, PC-3, MIA PaCa-2 and U-2 OS cells. This corresponds to the assumption that compound **1** (and thereby potentially also its derivatives) has low dark toxicity, as reported for a number of human cancer cell lines (HL-60 [8], Colo-205 [36], Hep-G2 [42], A549 [53,54], MCF-7 [55]). Darmostuk et al. [6] determined that the $IC_{50}$ (dark toxicity) of compound **1** exceeded 100 μM in HaCaT and VH10 cells and was 54 μM for NIH 3T3 cells. In our case, compound **1** did not exhibit dark toxicity up to 10 μM (the highest concentration tested), thus fulfilling a basic criterion for use in PDT. Likewise, the novel derivatives **2**–**4** did not display any dark toxicity (up to 10 μM), except in the case of compound **3** in the LNCaP and HeLa cells, whose $IC_{50}$ values were 7.20 and 7.95 μM, respectively.

After light activation, a significant decrease in cell viability was observed, especially for compounds **3** and **4**. Compound **1** exhibited the highest phototoxicity in the prostatic cancer cell lines with $IC_{50}$ values of 0.16 and 0.34 μM for the PC-3 and LNCaP cells, respectively. Regarding U-2 OS, MIA PaCa-2 and MCF-7, the phototoxic effect of compound **1** corresponded to $IC_{50}$ values below 2 μM. Compound **2**, which contained a zinc ion but no $PEG_3$ spacers, exhibited higher phototoxicity than compound **1** ($IC_{50} = 1.04$ μM) in the MIA PaCa-2 cells and slightly increased $IC_{50}$ values for the prostatic cancer cell lines: 0.21 and 0.47 μM for PC-3 and LNCaP, respectively. Interestingly, PEGylated derivatives **3** and **4** of compound **1** manifested extraordinary phototoxicity in the LNCaP cells; $IC_{50}$ values of 0.02 and 0.04 μM for compounds **4** and **3** were 18 and 9 times lower, respectively, than those for parental compound **1**. An even bigger difference in phototoxicity between the parental compound and its PEGylated derivatives was detected in the HeLa cells, for which there was an approximately 170- and 57-fold decrease in the $IC_{50}$ values of compounds **4** and **3**, respectively. In contrast, up to 10 μM, compound **2** did not reach $IC_{50}$ in the HeLa cells. The slightly increased phototoxicity of compounds **3** and **4** was also determined in the MIA PaCa-2 cells and, in the case of compound **4**, in the MCF-7 cells. Overall, the HeLa, LNCaP and MIA PaCa-2 cell lines were most sensitive to the PEGylated derivatives of compound **1**.

These results correspond to the compound localization determined by live-cell fluorescence microscopy, during which the highest fluorescence emission intensities were detected for compounds **3** and **4**. As previously reported [17,19,20,56,57], derivatization by a PEG spacer leads to increased compound hydrophilicity and, thus, to improved aqueous solubility, which in the case of porphyrins leads reduced aggregate formation. Thus, here the incorporation of a $PEG_3$ spacer probably improved the bioavailability of the compounds, resulting in the augmented cell uptake of compounds **3** and **4** compared with compound **1** and its zinc derivative. Moreover, the presence of a zinc ion in the structure of compound **1** enhances absorption in the red region of the visible spectra, which facilitates deeper tissue penetration.

Similar to purpurinimides [58], compounds **1**–**4** passively diffused into the cells, but their intracellular localization differed after light treatment, upon which compounds **3** and **4** may have translocated to and/or become better sequestered in sensitive organelles, probably the mitochondria, thereby enhancing PDT efficacy. Another possible explanation for the enhanced efficiency of compounds **3** and **4** is that they induce different mechanisms of action and cell death type than compounds **1** and **2**. We investigate the latter below.

*2.6. Evaluation of Cell Death*

To date, three different mechanisms of PDT action in cancer have been proposed [59]: direct cell damage, vascular shutdown and immune response activation. The photoactivation of a PS results in an acute stress response that leads to changes in calcium ion concentration and lipid metabolism, as well as the production of cytokines and stress response mediators [60]. These responses hamper mitochondrial processes, resulting in reactive oxygen species production and, consequently, in damage to the mitochondrial membrane; such damage induces cytochrome c release into cytosol [7], which, in turn, causes apoptosis. In addition to apoptosis, at excessive PS concentrations, PDT can also lead to cell death via necrosis; sometimes, a combination of both apoptosis and necrosis is involved. What is not yet clear is the role of autophagy, which also can occur under certain PDT conditions. There is considerable disagreement between researchers regarding the effect of autophagy on PDT outcome, with some suggesting it enhances outcome and others that it inhibits efficacy; this debate is comprehensively reviewed in Mroz et al. [60] Generally, the prevalent cell death type is dependent not only on the structure, intracellular localization and concentration of a PS, but also on the light dose applied and on cell origin.

To verify the mechanism of cell death induced by the tested PSs, MCF-7 cells were treated with compounds **1–4** (0.1–5 µM) for 24 h and photoactivated (light dose of 4 J·cm$^{-2}$). After a further 24 h of incubation, the cells were stained with Annexin V and propidium iodide (PI) and cell death type was determined by flow cytometry. The controls (untreated photoactivated and nonphotoactivated cells) displayed a physiological level of ca. 10% of apoptotic cells, corresponding with [61,62], but exhibited no necrosis (Figure 6, SI Table S1). Similar results were detected for compound **1** at 0.1–1 µM concentration without photoactivation. However, at the highest tested concentration (5 µM, no photoactivation) while the proportion of apoptotic cells remained ca. 10%, necrotic cells accounted for ca. 11% of all cells (Figure 6, SI Table S1). At 5 µM, very similar results were observed for compounds **2–4**. At this highest concentration, the mechanism of cell death was probably governed by excessive PS concentration, corresponding to Stefano et al. [8]. At lower concentrations (0.1–1 µM) of compounds **2–4**, the proportion of apoptotic cells [ca. 10% to 22% compared with almost no necrotic cells (0% to 0.1%)] increased in direct dependence to concentration.

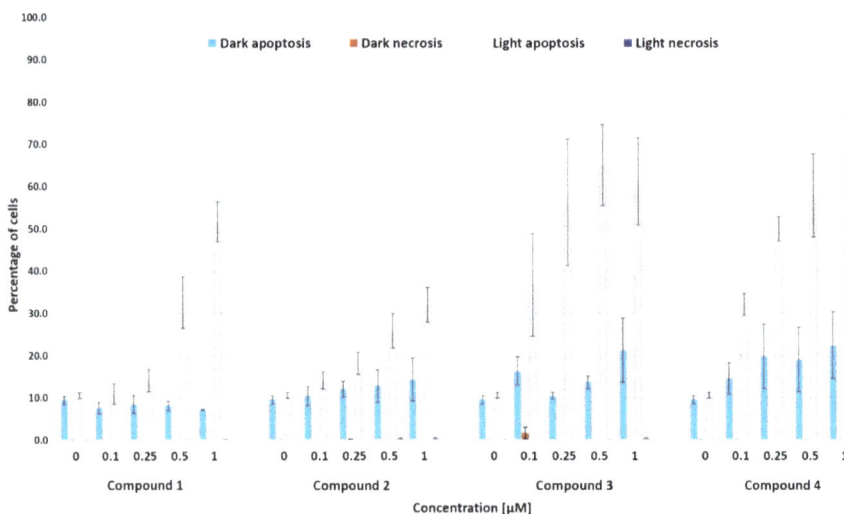

**Figure 6.** Dose-dependent mechanisms of cell death in MCF-7 cells induced by compounds **1–4** after 24-h treatment and light induction (Light) measured by flow cytometry. Control represents untreated cells and cells incubated with the same compounds without illumination (Dark). Total light dose was 4 J·cm$^{-2}$. The data values and the errors are stated in Table S1 in Supplementary information.

More interestingly, after photoactivation, the proportion of apoptotic cells among the MCF-7 cells treated with compound **1** rose to 52% as the concentration rose to 1 µM (Figure 6, SI Table S1). Under the same conditions, compound **2** induced apoptosis in 32% of cells. More promisingly, after photoactivation of the PEGylated derivatives of purpurin 18, compounds **3** and **4** (1 µM) triggered apoptosis in 61% and 68% of cells, respectively. Figure 6 and Table S1 in SI make it clear that the level of necrotic cells did not exceed 1.7% (mostly only 0.1%) after the photoactivation of compounds **1–4** (1 µM).

Researchers who have tested other PSs have reported similar results. For example, Stefano et al. [8] reported that HL-60 cells treated with a low concentration (0.2 µM) of photoactivated compound **1** (light dose of 1 J·cm$^{-2}$) predominantly underwent apoptosis, while at higher concentrations (>2 µM) necrosis was prevalent. Tsai et al. showed that not only the concentration and the structure of a PS influences cell death type but also the light dose applied; for instance, at a light dose of 8 J·cm$^{-2}$, 5-aminolevulic acid (ALA, 1 mM, 3 h; a precursor of protoporphyrin IX) induced the apoptosis of MCF-7 cells while a doubled light dose induced necrosis [63]. Light dose also affected cell death type in Sharma et al. [36], in which the 5 min light treatment (10 W·m$^{-2}$) of compound **1** in liposomes caused the apoptosis of Colo-205 cells while 40 min treatment caused necrosis. These findings indicate that manipulating the desired type of cell death in PDT involves achieving a careful balance between PS type, PS concentration, light dose and cell line.

In summary, it is probable that both the site of PS localization and the initial location of PDT-related damage determine which cell death pathway is activated. It is possible that autophagy is initially activated to rescue the cells, but that later, when the PDT effect is sufficient and the cells are damaged beyond repair, apoptosis occurs [60]. Following this, at high PS doses, necrosis takes place, as the proteins participating in autophagy and apoptosis are destroyed and cellular integrity is lost.

## 2.7. Molecular Docking of Purpurin 18 Derivatives with Human Serum Albumin

To ensure that a PS is efficiently delivered to a pathological site in the body, it should interact with transport proteins, particularly albumins such as human serum albumin (HSA). HSA, one of the most abundant plasma proteins, is the key endogenous vehicle for the biodistribution of molecules by blood plasma [64]. Thus, we studied the association constants of compounds **1–4** by performing molecular docking with HSA.

The docked ligands, purpurin 18 and its three derivatives (Scheme 1), differed in zinc ion coordination (**2–4**) and PEG spacers (**3–4**). Moreover, the carboxyl group of compound **3** contained poly (ethylene glycol) diamine (PEGDA) with a *tert*-butyloxycarbonyl protecting group on the nitrogen atom; compound **4** was derivatized with PEGDA without a protecting group. The binding pocket for porphyrins in HSA is localized in its 1B domain [65,66]; similarly our ligands were docked to this domain, albeit with a different orientation. Regarding compound **1**, the following were present: hydrophobic interactions for LEU$_{135}$, LEU$_{139}$, ALA$_{158}$, LEU$_{115}$ and PHE$_{149}$; $\pi$–$\pi$ interactions for TYR$_{161}$ and TYR$_{138}$ with pyrrole cycles; a hydrogen bond between the carboxylic group and ARG$_{186}$. Compounds **2–4** were stabilized in their HSA binding sites by similar hydrophobic interactions, as well as by the $\pi$–$\pi$ interaction of TYR$_{161}$ and TYR$_{138}$ with pyrrole cycles and by the interaction of the hydroxyl group of TYR$_{161}$ with Zn$^{2+}$. Furthermore, compound **2** formed a hydrogen bond between the carboxyl group and ARG$_{117}$ in HSA. The PEGDA spacer in compounds **3** and **4** was partially exposed to the solvent. The only interaction observed for compound **4**, with no PEGDA protecting group, was that of the terminal nitrogen with ASP$_{187}$.

Figure 7 shows the positions of docked compounds **1–4** with the lowest binding energies. The ligand-HSA binding energies are summarized in Table 4. The presence of a zinc ion and/or PEGDA spacer affected the binding mode by which the ligands docked to the HSA. Compound **1** docked with the lowest binding energy and ligand **4** with the highest (Table 5). Compound **2** was rotated by ca. 45°, resulting in the polar part of the glutaric acid anhydride being partially docked to the nonpolar part of the HSA binding pocket. This is probably the reason why compound **2** had a higher binding energy to HSA than compound **1**; a similar phenomenon was observed by Akimova et al. [67]. The zinc ion in compounds **2–4** interacted with the hydroxyl group of TYR$_{161}$ in HSA; a comparable interaction was described for a complex of HSA with heme and hemin [65,68]. Other authors have also reported the increased binding affinity of HSA to a PS after a metal ion was introduced to the PS [69–74]. The PEGDA spacer in compounds **3** and **4** was oriented out of the cavity and led to molecule rotation (Figure 7). In some cases, the scoring function can assign a better score to bigger, but worse, ligands based on the higher number of interactions, but this is not the case of compounds **3** and **4**. In these compounds, despite the porphyrin core being to some extent symmetric, the binding modes with higher binding energies were not as favorable as the binding modes for compounds **1** and **2**.

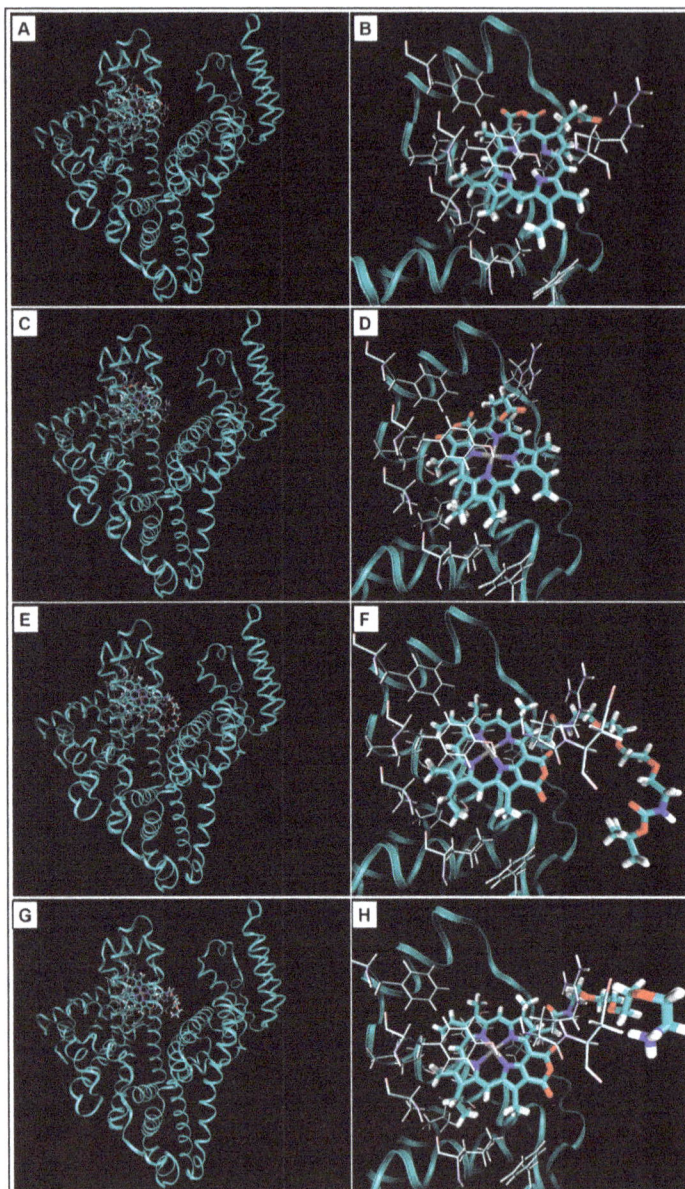

**Figure 7.** Best positions of docked ligands **1–4** shown in licorice representation. Carbon atoms are shown in light blue, hydrogen atoms in white, nitrogen atoms in dark blue, oxygen atoms in red and zinc atoms in grey. Human serum albumin is depicted as a ribbon. The images were captured by VMD software (Theoretical and Computational Biophysics Group, NIH Center for Macromolecular Modeling and Bioinformatics at the Beckman Institute, University of Illinois at Urbana-Champaign, USA, version 1.9.2).

**Table 4.** Calculated binding energy of docked ligands (compounds **1**–**4**) with human serum albumin.

| Ligands | Calculated Binding Energy (kcal/mol) |
| --- | --- |
| 1 | −13.37 |
| 2 | −11.58 |
| 3 | −9.32 |
| 4 | −8.52 |

**Table 5.** Calculated logarithm of partition coefficients between *n*-octanol and water (logP) for compounds **1**–**4**.

| Compounds | Calculated logP |
| --- | --- |
| 1 | 4.55 |
| 2 | 3.78 |
| 3 | 4.05 |
| 4 | 2.88 |

Since HSA binds a wide range of compounds and, thus, is able to reduce their active concentration in blood plasma, it reduces their bioavailability, thereby leading to decreased activity [75–77]. Because the HSA complex with compound **1** exhibited the lowest binding energy, it is less vulnerable to dissociation than compounds **2**–**4**, meaning that a higher free concentration of the novel derivatives and, consequently, higher activity can be expected.

### 2.8. Logarithm of Partition Coefficients

The lipophilicity of xenobiotics affects their binding to blood proteins and plays a key role in molecular discovery [78]. The level of their affinity to these proteins is expressed by the quantitative descriptor of lipophilicity, logP (partition coefficient between *n*-octanol and water): the more lipophilic the compound, the stronger its binding to these proteins [79]. Moreover, compound lipophilicity also has a direct impact on its other pharmacological parameters, such as distribution volume and biological half-life. Because logP is related to the cell uptake level of a compound, logP was calculated for our purpurin 18 derivatives. The data presented in Table 5 show that all compounds had logP of less than 5, which indicates that they should yield a good PDT response [54]. The derivatization of compound **1** with a metal ion and/or a PEG spacer led to lower logP values; the corresponding increase in hydrophilicity is an important factor for the use of such a compound in PDT. Gryshuk et al. described an inverse relationship between the lipophilicity and photosensing efficacy of purpurinimides, compounds structurally related to ours. At the same light doses, the purpurinimide concentrations effective for PDT were 30× lower for hydrophilic derivatives than for hydrophobic ones [58].

The logP results confirm those obtained by our docking study. Probably due to the absence of a PEGDA protecting group, the most hydrophilic compound was **4**, whose complex with HSA also displayed the highest binding energy. Conversely, compound **1** docked to HSA with the lowest binding energy and exhibited the highest logP. The presence of a zinc ion decreased the lipophilicity of compound **2** while correspondingly increasing its binding energy. The PEGDA spacer in compound **3** resulted in it having higher lipophilicity, but a less favorable binding energy, than compound **2**.

Our experimental data showed that compounds **3** and **4** exhibited the highest PDT efficacy, suggesting that the PDT activity of the studied ligands increased as their lipophilicity decreased. However, this is in contrast with Akimova et al. [67], who reported that the increased lipophilicity of structurally related compounds resulted in enhanced PDT. Furthermore, Henderson et al. [80], showing the dependence of PDT activity on lipophilicity for various pyropheophorbides, found that the best PDT outcome was achieved at a logP of between 5.6 and 6.6; at higher and lower values, PDT efficacy decreased. Based on these studies, we assume that the same is valid for our tested compounds; namely, as lipophilicity reduced, logP shifted closer to the optimal.

Indeed, when evaluating the pharmacological properties of compounds, it is necessary to consider the fact that their ability to cross cell membranes and, thus, accumulate in cells is improved with augmented lipophilicity [81]. Although compound **1** theoretically had the highest lipophilicity, it exhibited the lowest binding energy in our docking study with HSA. It seems that its high lipophilicity enhances not only its ability to pass through biological membranes but also its binding affinity to HSA, which, in turn, lowers its active concentration and leads to reduced PDT activity. This corresponds with the results of our in vitro experiments. The docking study also showed that compounds **3** and **4** exhibited higher binding energies than compounds **1** and **2**. From this, we conclude that lower binding energy and higher lipophilicity negatively influenced the active concentrations of the compounds, thereby reducing their PDT efficacy. Other researchers have reached similar conclusions about the significant impact of PS lipophilicity on PDT outcome [80,82,83].

## 3. Materials and Methods

### 3.1. Chemistry

#### 3.1.1. General Methods and Materials

Boc-protected PEG-3 amine was purchased from Iris Biotech GMBH (Marktredwitz, Germany), purpurin 18 (compound **1**) from Frontier Scientific, Inc. (Utah, USA). These chemicals were used as supplied. Other common chemicals were purchased from Sigma-Aldrich (Missouri, USA). NMR spectra were recorded by Agilent 400-MR DDR2 (Varian, Palo Alto, USA) spectrometer ($^1$H 400 MHz), solvent $CD_3OD$ was used as calibrator. Chemical shifts are given in δ (ppm). HRMS spectra were measured by Micro Q-TOF with ESI ionization (Thermo Scientific, Waltham, USA). For thin-layer chromatograms, aluminum TLC sheets for detection in UV light (TLC Silica gel 60 $F_{254}$, Merck, Darmstadt, Germany) were used. For column chromatography, silica gel (30-60 μm, SiliTech, MP Biomedicals, Eschwege, Germany) was used.

#### 3.1.2. Synthesis of Purpurin Zinc Complex—Compound 2

{3-[(22S,23S)-17-Ethenyl-12-ethyl-13,18,22,27-tetramethyl-3,5-dioxo-4-oxa-8,24,25,26-tetraazahexacyclo[19.2.1.1$^{6,9}$.1$^{11,14}$.1$^{16,19}$.0$^{2,7}$]heptacosa-1(24),2(7),6(27),9,11(26),12,14,16,18,20-decaen-23-yl-κ$^4N^8$,$N^{24}$, $N^{25}$,$N^{26}$]propanoato(2-)}zinc$^{36}$

To a solution of compound **1** (30 mg, 53 μmol) in chloroform (7 mL), Zn(OAc)$_2$·2H$_2$O (117 mg, 0.53 mmol; in 3 mL of MeOH) was added. This mixture was stirred for 12 h at 45 °C. The solvents were removed under reduced pressure and the residue was chromatographed (CHCl$_3$-MeOH, 40/1). The obtained product was redissolved in a small amount of chloroform and precipitated by the addition of hexanes. Product **2** (20 mg, 32 μmol; Figure 8) was obtained as a dark green solid in 61% yield. After analyses, the product was lyophilized from 1,4-dioxane. $R_F$ = 0.7 in DCM-MeOH, 10/1. $^1$H NMR (400 MHz, CD$_3$OD) δ: 1.42 (t, *J* = 7.2 Hz, 3 H), 1.77 (d, *J* = 7.0 Hz, 3 H), 1.98–2.08 (m, 1 H), 2.39–2.59 (m, 2 H), 2.72 (dt, *J* = 10.0, 4.9 Hz, 1 H), 2.76–2.82 (m, 2 H), 3.18 (s, 3 H), 4.33 (q, *J* = 7.3 Hz, 1 H), 5.04 (dd, *J* = 9.0, 2.0 Hz, 1 H), 6.01 (dd, *J* = 11.5, 1.0 Hz, 1 H), 6.09 (dd, *J* = 17.8, 1.0 Hz, 1 H), 7.78 (dd, *J* = 17.8, 11.5 Hz, 1 H), 8.38 (s, 1 H), 8.78 (br. s., 1 H), 8.84 (br. s., 1 H); Figure S1 in SI. HRMS-ESI: *monoisotopic mass* 626.15077 Da, found *m/z* 625.14392 [M-H]$^-$; Figure S4 in SI.

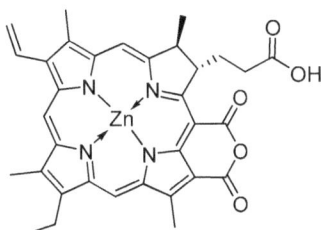

**Figure 8.** Structure of compound **2**.

### 3.1.3. Synthesis of Purpurin-PEG3-Boc Zinc Complex—Compound **3**

[*Tert*-butyl{15-[(22*S*,23*S*)-17-ethenyl-12-ethyl-13,18,22,27-tetramethyl-3,5-dioxo-4-oxa-8,24,25,26-tetra azahexacyclo[19.2.1.1$^{6,9}$.1$^{11,14}$.1$^{16,19}$.0$^{2,7}$]heptacosa-1(24),2(7),6(27),9,11(26),12,14,16,18,20-decaen-23-yl-κ$^4$$N^8$,$N^{24}$,$N^{25}$,$N^{26}$]-13-oxo-3,6,9-trioxa-12-azapentadecan-1-yl}carbamatato(2-)]zinc

To a solution of compound **1** (150 mg, 0.27 mmol) and Boc-PEG$_3$-diamine (124 mg, 0.43 mmol) in THF (5 mL), EDIPA (70 mg, 0.54 mmol) and HOBt (37 mg, 0.27 mmol) were added. The mixture was stirred for 5 min, after which DIC (50 mg, 0.4 mmol in 1 mL of THF) was added and the mixture was stirred for 24 h. Then, the solvent was evaporated and the residue was filtered through a short pad of silica (DCM-MeOH, 25/1) to obtain the crude product (250 mg), R$_F$ = 0.5 (DCM-MeOH, 20/1). This material was redissolved in chloroform (7 mL) and the solution of Zn(OAc)$_2$·2H$_2$O (658 mg, 3 mmol) in MeOH (3 mL) was added. The mixture was stirred overnight at 60 °C. Thereafter, the mixture was diluted with chloroform (90 mL) and washed with brine (1 × 100 mL) and water (1 × 100 mL), dried over Na$_2$SO$_4$, filtered and the solvent was evaporated under reduced pressure. The residue was chromatographed using AcOEt-MeOH, 20/1 as an eluent to obtain compound **3** (99 mg, 0.11 mmol; Figure 9) as a green solid in 41% yield. R$_F$ = 0.48 in DCM-MeOH, 10/1. $^1$H NMR (400 MHz, CD$_3$OD) δ: 1.25 (t, *J* = 7. 6 Hz, 3 H), 1.37 (s, 8 H), 1.80 (d, *J* = 7.4 Hz, 3 H), 1.94–2.06 (m, 1 H), 2.31–2.43 (m, 2 H), 2.54 (s, 3 H), 2.55–2.62 (m, 1 H), 2.84 (tt, *J* = 14.3, 7.2 Hz, 2 H), 3.09 (q, *J* = 5.5 Hz, 2 H), 3.15 (d, *J* = 2.7 Hz, 5 H), 3.32–3.36 (m, 2 H), 3.40–3.55 (m, 9 H), 4.31 (q, *J* = 7.3 Hz, 1 H), 5.05 (d, *J* = 8.2 Hz, 1 H), 5.96 (dd, *J* = 11.4, 1.6 Hz, 1 H), 6.05 (dd, *J* = 18.0, 1.6 Hz, 1 H), 7.70 (dd, *J* = 17.8, 11.4 Hz, 1 H), 8.01 (t, *J* = 5.3 Hz, N*H*), 8.34 (s, 1 H), 8.44 (s, 1 H), 8.55 (s, 1 H); Figure S2 in SI. HRMS-ESI: *monoisotopic mass* 900.34002 Da, found *m/z* 923.32928 [M+Na]$^+$; Figure S5 in SI.

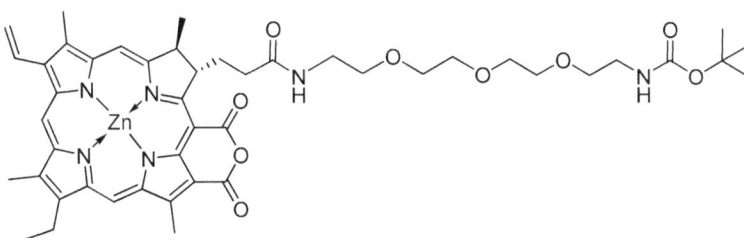

**Figure 9.** Structure of compound **3**.

### 3.1.4. Synthesis of Purpurin-PEG3-Amine Zinc Complex—Compound **4**

[*N*-(2-{2-[2-(2-Aminoethoxy)ethoxy]ethoxy}ethyl)-3-[(22*S*,23*S*)-17-ethenyl-12-ethyl-13,18,22,27-tetrame thyl-3,5-dioxo-4-oxa-8,24,25,26-tetraazahexacyclo[19.2.1.1$^{6,9}$.1$^{11,14}$.1$^{16,19}$.0$^{2,7}$]heptacosa-1(24),2(7),6(27), 9,11(26),12,14,16,18,20-decaen-23-yl-κ$^4$$N^8$,$N^{24}$,$N^{25}$,$N^{26}$]propanamidato(2-)]zinc

Purpurin derivative **3** (99 mg, 0.11 mmol) was dissolved in DCM (5 mL). Five drops of water were added and TFA (1 mL) was added dropwise via a syringe. The mixture was stirred for 1 h, after which it was repetitively evaporated with toluene. The residue was chromatographed (triethylamine-deactivated silica) CHCl$_3$-MeOH (30/1→10/1) to obtain product **4** (66 mg, 0.08 mmol;

Figure 10) as a dark green solid in 56% yield. $R_F = 0.2$ in DCM-MeOH, 10/1. $^1$H NMR (400 MHz, CD$_3$OD) δ: 1.30 (t, *J* = 7.6 Hz, 3 H), 1.76 (d, *J* = 7.4 Hz, 3 H), 1.80–2.09 (m, 4 H), 2.14–2.26 (m, 2 H), 2.28–2.43 (m, 2 H), 2.65 (s, 3 H), 2.70–2.85 (m, 3 H), 2.89–3.03 (m, 3 H), 3.05 (s, 3 H), 3.13 (s, 3 H), 3.15–3.19 (m, 2 H), 3.21–3.25 (m, 2 H), 4.25 (q, *J* = 7.3 Hz, 1 H), 5.07 (dd, *J* = 7, 2.4 Hz, 1 H), 5.95 (dd, *J* = 11.4, 1.2 Hz, 1 H), 6.03 (dd, *J* = 17.8, 1.2 Hz, 1 H), 7.72 (dd, *J* = 17.8, 11.5 Hz, 1 H), 8.31 (s, 1 H), 8.54 (s, 1 H), 8.69 (s, 1 H); Figure S3 in SI. HRMS-ESI: *monoisotopic mass* 800.28759 Da, found *m/z* 801.29590 [M+H]$^+$, 823.27637 [M+Na]$^+$; Figure S6 in SI.

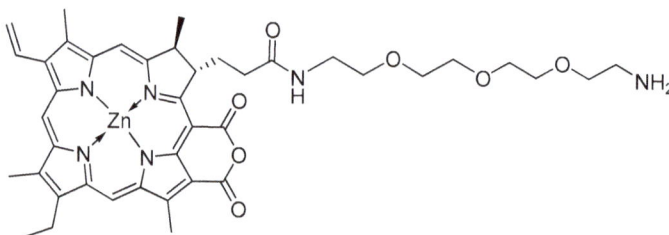

**Figure 10.** Structure of compound 4.

*3.2. Indirect Spectrophotometric Measurement of Singlet Oxygen Production*

3.2.1. Data Measurement

Singlet oxygen production by compounds **1**–**4** in DMEM + FBS was evaluated using 9,10-anthracenediyl-bis(methylene)dimalonic acid (**AB**, Sigma Aldrich, Saint Loui, MO, USA) as a compound reactive with singlet oxygen. A photosensitizer Rose Bengal (**RB**, 95%, Sigma Aldrich, Saint Loui, MO, USA) was used as a reference. A stock solution of **RB** was prepared in water. Stock solutions of compounds **1**–**4** were prepared by dissolving a solid in DMSO. Working solutions were prepared by diluting the stock solutions with air-saturated solvent which was either phosphate buffered saline (PBS, pH 7.4) or DMEM + FBS. Concentrations of the working solutions of **RB** and compounds **1**–**4** were $1.34 \times 10^{-5}$, $3.19 \times 10^{-4}$, $3.34 \times 10^{-4}$, $2.99 \times 10^{-4}$ and $3.12 \times 10^{-4}$ M, respectively. A stock solution of **AB** was prepared by dissolving solid substance in DMSO ($4.87 \times 10^{-3}$ M). Two different amounts of **RB** and compounds **1**–**4** (25 and 50 µL of stock solution) and 15 µL of **AB** stock solution were mixed with 1 mL of solvent in a plastic cuvette (1.000 cm, PMMA, Kartell, Milan, Italy). Two replicates of the same PS concentration were prepared. The first was kept in the dark, the second was illuminated. Absorption spectra of both solutions (illuminated and kept in the dark) were collected against the recorded baseline (Cintra 404, GBC Scientific, 270–800 nm, step 0.2 nm, slit 2 nm) after 0, 10, 20, 30 and 40 min All experiments were done in duplicates.

A 150 W halogen lamp with an edge-pass filter (Panchromar filter (58 mm), VEB Glastechnik Lommatzsch, Lommatzch, Germany) that transmitted light at wavelengths longer than 500 nm was used for illumination of solutions. The fluency rate at the cuvette was 5 mW·cm$^{-2}$. Relative spectral emission intensity of the illuminationg source was measured using spectrofluorometer Fluoromax 2 (Horiba Scientific, Horiba Ltd., Kyoto, Japan). The lamp illuminated a plate made of barium sulfate positioned on front surface accessory of the spectrofluorometer and its spectrum was recorded as reported by Pavlíčková et al. [5]

Emission correction function supplied by manufacturer was applied to the measured spectra. The same correction factor was used to correct spectra of quinine sulfate solution in sulfuric acid and results corresponded well with standard spectra by Velapoldi and Tønnesen [84].

3.2.2. Data Evaluation

Data were evaluated using MS Excel 2010 (Microsoft, Redmond, WA, USA) with a procedure described in detail by Pavlíčková et al. [5] Briefly, from concentrations and a spectrum of a tested compound, its absorbance $A(\lambda,t)$ in solutions was estimated, and then, relative exposure $I_A(t)$ for solution illuminated for time $t$ was calculated as

$$I_A(t) = \int\limits_0^t \int\limits_{\lambda_1}^{\lambda_2} \frac{I(\lambda)}{I_0} \cdot \left(1 - 10^{-A(\lambda,\tau)}\right) \cdot d\lambda d\tau$$

where $I(\lambda)/I_0$ is source relative emission intensity in photons per wavelength $\lambda$ and $A(\lambda,t)$ is absorbance of a tested compound at wavelength $\lambda$ and time $\tau$ from the beginning of illumination. Wavelength integration limits were $\lambda_1 = 450$ nm and $\lambda_2 = 800$ nm, and trapezoidal rule for integration was used for absorbance in intervals between spectral measurements. Chemical photodynamical efficiency $\gamma_x$ of a photosensitizer was evaluated supposing single exponential dependence of relative **AB** concentration on light exposure

$$c_{rel,AB} = \exp[-\gamma_x \, I_A(t)]$$

Then singlet oxygen quantum yields of photosensitizers were calculated using **RB** in PBS and DMEM+FBS, respectively, as a standard ($\phi = 0.75$ in PBS, according to Gottfried et al. [40,41]) using equation

$$\phi_x = \phi_s \, \gamma_x/\gamma_s$$

where $\phi_x$ is estimated quantum yield; $\phi_s$ is quantum yield for the standard and $\gamma_x$ and $\gamma_s$, respectively, are chemical photodynamical efficiencies of an evaluated compound and a standard, respectively.

### 3.3. Biology

#### 3.3.1. Cell Lines and Cultivation Conditions

In our study, we used the following human cell lines: MCF-7 (breast carcinoma), LNCaP (prostate carcinoma, PSMA$^+$), PC-3 (prostate carcinoma), U-2 OS (osteosarcoma), MIA PaCa-2 (pancreatic adenocarcinoma), HeLa (cervical carcinoma) and HaCaT (keratinocytes). Unless otherwise specified, the cells were cultured in DMEM medium GlutaMAX (Merck, Kenilworth, NJ, USA) supplemented with 10% FBS (Thermo Fisher Scientific, Waltham, MA, USA). Cells were maintained at exponential phase of growth under standard physiological conditions at 37 °C in humidified atmosphere with 5% $CO_2$.

#### 3.3.2. Cell Uptake of Purpurin 18 Derivatives

The number of $1 \times 10^5$ cells was seeded on individual 35-mm glass bottom (1.5#) dishes for live-cell imaging (MatTek Corporation, Ashland, MA, USA) and left to adhere for 16 h. Then, the cells were washed with PBS and incubated with compounds **1–4** (0.2, 0.5 and 1 µM) dissolved in complete cell cultivation medium without phenol red at 37 °C for 3 and 24 h. After that, the cells were washed twice with PBS and the medium was exchanged for phenol-red free DMEM. Stock solutions of compounds **1–4** were prepared in dimethylsulfoxide (DMSO) fresh before the experiments. The final concentration of the vehicle (DMSO) in cell culture medium did not exceed 0.02%.

#### 3.3.3. Determination of Intracellular Localization of Purpurin 18 Derivatives

In order to determine the intracellular localization of compounds **1–4**, the cells were seeded and treated with the tested compounds as described in Section 3.3.2. Then, the cells were incubated with a marker for staining of endoplasmic reticulum (ER-TrackerTM Blue-White DPX, 70 nM, 30 min), mitochondria (a green-emitting dimethinium salt from Bříza et al. [47], 70 nM, 10 min and MitoTracker™ Green FM, 70 nM, 15 min), lysosomes (LysoTracker Green DND-26, 70 nM, 15 min) and Golgi apparatus

(CellLight™ Golgi-GFP, BacMam 2.0, 24 h, $2 \times 10^4$ particles per cell). All the markers used from stock solutions as supplied by the manufacturer (Thermo Fisher Scientific, Waltham, MA, USA).

### 3.3.4. Fluorescence Microscopy

The intracellular localization of purpurin 18 derivatives was studied by real-time live-cell fluorescence microscopy at 37 °C and in 5% $CO_2$ atmosphere. The images were acquired by an inverse fluorescence microscope Olympus IX-81 operated by xCellence System (Olympus, Tokyo, Japan) using a high-stability 150 W xenon arc burner and EM-CCD camera C9100-02 (Hamamatsu, Herrsching am Ammersee, Germany). Living cells were analyzed under physiological conditions (37 °C, 5% $CO_2$) by a 60× oil immersion objective (Olympus, Tokyo, Japan) with the numerical aperture of 1.4. All images were deconvolved using xCellence 2D deconvolution module and background-corrected.

### 3.3.5. Corrected Total Cell Fluorescence

To calculate corrected total cell fluorescence (CTCF), the number of $1 \times 10^5$ PC-3 cells was seeded in DMEM+FBS onto glass-bottom MatTek dishes (35 mm, 1.5#). After 16 h, compounds **1–4** at 1 µM concentration were added to the cells in DMEM with 10% FBS and incubated for another 24 h. Then, the medium was removed, the cells were washed with PBS, which was replaced with FluoroBrite DMEM media (Thermo Fisher Scientific, Waltham, MA, USA) and subjected to fluorescence microscopy. The fluorescence emission intensity was measured in cells in at least 10 view fields at 600× magnification. The images were taken at the same exposition time (600 ms) and light intensity (100%). Then, the data were evaluated using ImageJ 1.52a software by an equation:

CTCF = Integrated Density − (Area of selected cell × Mean fluorescence of background readings).

### 3.3.6. Cell Lines and Cultivation Conditions

Photo- and dark toxicity of compounds **1–4** was evaluated in vitro by WST-1 viability assay (Sigma, Saint Loui, MO, USA) similarly as in Rimpelová et al. [85]. The WST-1 assay is based on reduction of a tetrazolium salt (WST-1 substrate) into soluble formazan in metabolically active cells. The following cell lines were used: MCF-7, LNCaP, PC-3, U-2 OS, MIA PaCa-2 and HeLa. The cells were seeded into individual wells of 96-well plates (3500 cells per well; except of LNCaP, for which 7000 cells per well was seeded) in 100 µL of DMEM media supplemented with 10% FBS. After 16 h of incubation, the cells were treated with the tested compounds (0.5, 1, 5 and 10 µM) in 100 µL of DMEM with 10% FBS. Then, after 24 h, the cells were washed with PBS, and 100 µL of phenol red-free DMEM was added. One half of the samples was illuminated by 150-W halogen lamp for 13 min with an edge-pass filter (Panchromar, filter (58 mm), VEB Glastechnik Lommatzsch, Lommatzch, Germany) that transmitted wavelengths longer than 500 nm (the total light dose of 4 J·cm$^{-2}$). The second half of the samples was kept in the dark. Next, 24 h after illumination, the cell culture medium was removed and 100 µL of fresh phenol red-free DMEM with 4 µL of WST-1 was added. After 2-h incubation, the absorbance of formed formazan was measured spectrophotometrically at 450 nm (the reference wavelength of 650 nm) using UV–Vis spectrometer (Tecan). The absorbance is directly proportional to the oxidoreductase activity, and thus to the number of metabolically active cells. Cells treated only with cell culture medium and cells treated with a vehicle (DMSO) served as controls. All samples were tested in quadruplicates. The IC$_{50}$ values were determined (GraphPad Prism 6) as the concentration necessary to kill 50% of cells.

### 3.3.7. Cell Death Evaluation by Flow Cytometry

Evaluation of the proportion of dead cells was adapted from Vermes et al. [86] MCF-7 cells were plated in a 6-well plate (25,000 cells per well) and treated with compounds **1–4** (0.1–5 µM) for 24 h. Then, the cells were illuminated for 13 min by a 150-W halogen lamp with an edge-pass filter transmitting wavelengths longer than 500 nm (the total light dose of 4 J·cm$^{-2}$). The cells were incubated at standard cultivation conditions for next 24 h. Afterwards, the cells were harvested by trypsinization, washed in cold PBS and resuspended in annexin-binding buffer followed by labeling with Alexa Fluor® 488 Annexin V and propidium iodide according to the manufacturer's protocol (Dead Cell Apoptosis Kit, Thermo Fisher Scientific, Waltham, MA, USA) as described in Kirakci et al. [87] and Rumlová et al. [88]. The stained cells were then analyzed by flow cytometer BD FACSAria III, by which live and dead (necrotic and apoptotic) cells were determined using BD FACSDiva 8. The experiments were done in triplicates.

*3.4. Theoretical Studies*

3.4.1. Docking Into Human Serum Albumin

Three-dimensional structures of compounds **1–4** (ligands) were created by CORINA Classic (v. 4.2.0; Molecular Networks GmbH, Nuremberg, Germany) software. Other modifications of ligands and proteins were done by Maestro (v. 2018-4, NY, USA) software. To minimize the ligand energy and addition of missing hydrogen atoms, a function "ligand preparation" with the force field OPLS3e was used. The structure of HSA protein with the code of 1N5U was obtained from Protein Data Bank (PDB; https://www.rcsb.org/structure/1N5U) database. From the structure of HAS 1N5U, molecules of myristic acid, protoporphyrin IX and water were removed. In order to define the binding pocket, we chose four amino acids (TYR$_{161}$, MET$_{123}$, HIS$_{146}$, LYS$_{190}$) present in the binding pocket of HSA and using the function "receptor grid generator", a small and big box was created with the edge length depending on the docked ligand (Table 6). The molecule center was not enabled to leave the small box and the molecule as whole could not overhang the big box. In addition to minimization and optimization of the HSA structure, all amino acids present in this protein were assigned a protonation state corresponding to pH 7 using a function PROPKA. OPLS3e was used as force field. Ligand docking was done in the "extra precisions" mode.

**Table 6.** Edge size of a box used for ligand docking.

| Ligand | Box Edge Size (Å) | |
|:---:|:---:|:---:|
| | Small | Big |
| 1 | 14 | 22 |
| 2 | 14 | 22 |
| 3 | 15 | 30 |
| 4 | 15 | 28 |

3.4.2. Calculation of the Logarithm of a Partition Coefficient

For compounds **1–4**, logP between *n*-octanol and water was calculated as the arithmetic average of results obtained by algorithms XLOGP3, WLOGP, MLOGP, SILICO-IT in SwissADME software (Molecular Modeling Group, Swiss Institute of Bioinformatics, Lausanne, Switzerland) [89,90].

**4. Conclusions**

We have designed, synthesized and presented the biological activity of two PEGylated derivatives of purpurin 18 (compound **1**). Compared to their parent compound, both derivatives (compounds **3** and **4**) exhibited improved accumulation in all of the tested cell lines (PC-3, LNCaP, MCF-7 and HaCaT). Live-cell fluorescence microscopy showed that compounds **3** and **4** localized predominantly in the endoplasmic reticulum and mitochondria, which are desired targets for PDT; moreover, compound **4** also localized in lysosomes. Upon illumination, both compounds efficiently generated singlet oxygen production. This corresponded with their good photodynamic activity at nanomolar to micromolar concentrations in all tested cell lines, with the strongest effect detected for compounds **3** and **4** in the LNCaP and HeLa cell lines, respectively. At submicromolar concentrations, the photoactivated compounds **1–4** prevalently induced apoptosis with negligible necrosis. The most efficient apoptosis inducers (61% to 68% of cells in apoptosis) were compounds **3** and **4**. In terms of photodynamic therapy, we believe that our PEGylated derivatives have the ability to outperform their parent photosensitizer purpurin 18, application of which is limited by its aggregation. Furthermore, their enhanced water solubility can overcome the high hydrophobicity and, thus, limited bioavailability associated with photosensitizers, such as chlorin, currently used in photodynamic therapy.

**Supplementary Materials:** The following are available online at http://www.mdpi.com/1420-3049/24/24/4477/s1, Chemical analysis (NMR spectra, HRMS spectra), Biological analysis (compound uptake by human cells, compound colocalization with organelle markers, cell death, photo- and cytotoxicity graphs). Figure S1: $^1$H-NMR spectra of compound **2**, Figure S2: $^1$H-NMR spectra of compound **3**, Figure S3-1: $^1$H-NMR spectra of compound **4**, Figure S3-2: $^{13}$C-NMR spectra of compound **3**, Figure S4: HRMS spectra of compound **2**, Figure S5: HRMS spectra of compound **3**, Figure S6-1: HRMS spectra of compound **4**, Figure S6-2: Absorption, excitation and emission spectra of compounds **3** and **4**, Figure S6-3: Depletion of 9,10-anthracenediyl-bis(methylene)dimalonic acid without presence of photosensitizer-generated singlet oxygen, Figure S6-4: Depletion of 9,10-anthracenediyl-bis(methylene)dimalonic acid with RB-generated singlet oxygen, Figure S7: Fluorescence microscopy images of the intracellular localization of purpurin 18 (compound **1**) and its derivatives (compounds **2–4**), Figure S8: Fluorescence microscopy images of purpurin 18 (compound **1**) and its derivatives (compounds **2–4**) localization in the endoplasmic reticulum of human MCF-7 cells (breast carcinoma), Figure S9: Fluorescence microscopy images of purpurin 18 (compound **1**) and its derivatives (compounds **2–4**) localization in the endoplasmic reticulum of human immortalized keratinocytes (HaCaT cells), Figure S10: Fluorescence microscopy images of purpurin 18 (compound **1**) and its derivatives (compounds **2–4**) localization in the endoplasmic reticulum of human LNCaP cells (prostate carcinoma), Figure S11: Fluorescence microscopy images of purpurin 18 (compound **1**) and its derivatives (compounds **2–4**) localization in the mitochondria of human MCF-7 cells (breast carcinoma), Figure S12: Fluorescence microscopy images of purpurin 18 (compound **1**) and its derivatives (compounds **2–4**) localization in the mitochondria of human LNCaP cells (prostate carcinoma), Figure S13: Fluorescence microscopy images of purpurin 18 (compound **1**) and its derivatives (compounds **2–4**) localization in the mitochondria of human keratinocytes HaCaT, Figure S14: Fluorescence microscopy images of compound **4** localization in the Golgi apparatus in human PC-3 and MCF-7 cells, Figure S15: Fluorescence microscopy images of compound **4** localization in lysosomes of human immortalized keratinocytes (HaCaT cells), Figure S16: Corrected total cell fluorescence (CTCF) of compounds **1–4** (1 μM, 24 h) localized in PC-3 cells, Table S1: Dose-dependent mechanisms of cell death in MCF-7 cells induced by compounds **1–4**, Figure S17: Photo- and dark toxicity of compounds **1–4** in vitro, Figure S18: Photo- and dark toxicity of compounds **1–4** in vitro.

**Author Contributions:** Conceptualization, S.R., M.J., K.Z., T.R., V.S. and P.D.; methodology, S.R., M.J., I.K., K.Z., J.F. and Z.R.; software, J.B. and V.S.; formal analysis, Z.R., J.F. and I.K.; resources, T.R. and P.D.; writing—original draft preparation, S.R., M.J. and V.P.; writing—review and editing, T.R., V.S. and P.D.; supervision, S.R., V.S., T.R. and P.D.

**Funding:** This research was funded by the Ministry of Education, Youth and Sports, grant numbers LO1220, LO1601, LM2015063, by the CZ.02.1.01/0.0/0.0/16_013/0001799; OP VVV [2.16/3.1.00/24503; Specific university research MSMT, grant number No 21-SVV/2019; by Martina Roeselová foundation and by L'Oréal -UNESCO for Women in Science 2019; MSMT LTAUSA19065.

**Acknowledgments:** We truly thank Craig Alfred Riddell for his tremendous effort and time spent on the English correction.

**Conflicts of Interest:** The authors declare no conflict of interest. The funders had no role in the design of the study; in the collection, analyses, or interpretation of data; in the writing of the manuscript, or in the decision to publish the results.

## Abbreviations

AB, 9,10-anthracenediyl-bis(methylene)dimalonic acid; ALA, δ-aminolevulinic acid; cis-PT, cisplatin; Colo-205, human cells from colorectal carcinoma; DIC, *N,N*-diisopropylcarbodiimide; DCM, dichloromethane; DMEM, Dulbecco's Modified Eagle medium; EDIPA, Hünig's base; FBS, fetal bovine serum; GFP, green fluorescent protein; HaCaT, human keratinocytes; HSA, human serum albumin; Hep G2, human cells from liver carcinoma; HeLa, human cells from cervical carcinoma; HOBt, *N*-hydroxybenzotriazole; HSA, human serum albumin; $IC_{50}$, half maximal inhibitory concentration of a compound; LNCaP, human cells from prostate carcinoma; MCF-7, human cells from breast carcinoma; MIA PaCa-2, human cells from pancreatic carcinoma; NIH 3T3, mouse fibroblasts; PBS, phosphate buffered saline; PC-3, human cells from prostate carcinoma; PDT, photodynamic therapy; PEG, polyethylene glycol; PEGDA, poly(ethylene glycol) diamine; PI, propidium iodide; PS, photosensitizers; RB, Rose Bengal; TFA, trifluoroacetic acid; U-2 OS, human cells from osteosarcoma; VH10, human foreskin fibroblasts.

# References

1. Lim, S.H.; Yam, M.L.; Lam, M.L.; Kamarulzaman, F.A.; Samat, N.; Kiew, L.V.; Chung, L.Y.; Lee, H.B. Photodynamic characterization of amino acid conjugated 15(1)-hydroxypurpurin-7-lactone for cancer treatment. *Mol. Pharm.* **2014**, *11*, 3164–3173. [CrossRef] [PubMed]

2. Bible, K.C.; Buytendorp, M.; Zierath, P.D.; Rinehart, K.L. Tunichlorin: A nickel chlorin isolated from the caribbean tunicate *Trididemnum solidum. Proc. Natl. Acad. Sci. USA* **1988**, *85*, 4582–4586. [CrossRef] [PubMed]

3. Tang, P.M.; Chan, J.Y.; Au, S.W.; Kong, S.K.; Tsui, S.K.; Waye, M.M.; Mak, T.C.; Fong, W.P.; Fung, K.P. Pheophorbide a, an active compound isolated from *Scutellaria barbata*, possesses photodynamic activities by inducing apoptosis in human hepatocellular carcinoma. *Cancer Biol. Ther.* **2006**, *5*, 1111–1116. [CrossRef] [PubMed]

4. Juzeniene, A. Chlorin e6-based photosensitizers for photodynamic therapy and photodiagnosis. *Photodiagnosis Photodyn. Ther.* **2009**, *6*, 94–96. [CrossRef] [PubMed]

5. Pavlíčková, V.; Jurášek, M.; Rimpelová, S.; Záruba, K.; Sedlák, D.; Šimková, M.; Kodr, D.; Staňková, E.; Fähnrich, J.; Rottnerová, Z.; et al. Oxime-based 19-nortestosterone–pheophorbide a conjugate: Bimodal controlled release concept for PDT. *J. Mater. Chem. B* **2019**, *7*, 5465–5477. [CrossRef] [PubMed]

6. Darmostuk, M.; Jurášek, M.; Lengyel, K.; Zelenka, J.; Rumlová, M.; Drašar, P.; Ruml, T. Conjugation of chlorins with spermine enhances phototoxicity to cancer cells in vitro. *J. Photochem. Photobiol. B* **2017**, *168*, 175–184. [CrossRef]

7. Kwon, J.-G.; Song, I.-S.; Kim, M.-S.; Lee, B.H.; Kim, J.H.; Yoon, I.; Shim, Y.K.; Kim, N.; Han, J.; Youm, J.B. Pu-18-N-butylimide-NMGA-GNP conjugate is effective against hepatocellular carcinoma. *Integr. Med. Res.* **2013**, *2*, 106–111. [CrossRef]

8. Stefano, A.D.; Ettorre, A.; Sbrana, S.; Giovani, C.; Neri, P. Purpurin-18 in combination with light leads to apoptosis or necrosis in HL60 leukemia cells. *Photochem. Photobiol.* **2001**, *73*, 290–296. [CrossRef]

9. Dougherty, T.J.; Gomer, C.J.; Henderson, B.W.; Jori, G.; Kessel, D.; Korbelik, M.; Moan, J.; Peng, Q. Photodynamic therapy. *J. Natl. Cancer Inst.* **1998**, *90*, 889–905. [CrossRef]

10. Banfi, S.; Caruso, E.; Caprioli, S.; Mazzagatti, L.; Canti, G.; Ravizza, R.; Gariboldi, M.; Monti, E. Photodynamic effects of porphyrin and chlorin photosensitizers in human colon adenocarcinoma cells. *Bioorganic Med. Chem.* **2004**, *12*, 4853–4860. [CrossRef]

11. Lee, S.-J.H.; Jagerovic, N.; Smith, K.M. Use of the chlorophyll derivative, purpurin-18, for syntheses of sensitizers for use in photodynamic therapy. *J. Chem. Soc.* **1993**, *19*, 2369–2377. [CrossRef]

12. Robertson, C.A.; Evans, D.H.; Abrahamse, H. Photodynamic therapy (PDT): A short review on cellular mechanisms and cancer research applications for PDT. *J. Photochem. Photobiol. B* **2009**, *96*, 1–8. [CrossRef] [PubMed]

13. Castano, A.P.; Demidova, T.N.; Hamblin, M.R. Mechanisms in photodynamic therapy: Part three-Photosensitizer pharmacokinetics, biodistribution, tumor localization and modes of tumor destruction. *Photodiagnosis Photodyn. Ther.* **2005**, *2*, 91–106. [CrossRef]

14. Krosl, G.; Korbelik, M.; Dougherty, G.J. Induction of immune cell infiltration into murine SCCVII tumour by photofrin-based photodynamic therapy. *Br. J. Cancer* **1995**, *71*, 549–555. [CrossRef]

15. St. Denis, T.G.; Aziz, K.; Waheed, A.A.; Huang, Y.-Y.; Sharma, S.K.; Mroz, P.; Hamblin, M.R. Combination approaches to potentiate immune response after photodynamic therapy for cancer. *Photochem. Photobiol. Sci.* **2011**, *10*, 792–801. [CrossRef]

16. Taniguchi, M.; Ptaszek, M.; McDowell, B.E.; Lindsey, J.S. Sparsely substituted chlorins as core constructs in chlorophyll analogue chemistry. Part 2: Derivatization. *Tetrahedron* **2007**, *63*, 3840–3849. [CrossRef]

17. Kimani, S.; Ghosh, G.; Ghogare, A.; Rudshteyn, B.; Bartusik, D.; Hasan, T.; Greer, A. Synthesis and characterization of mono-, di-, and tri-poly(ethylene glycol) chlorin e(6) conjugates for the photokilling of human ovarian cancer cells. *J. Org. Chem.* **2012**, *77*, 10638–10647. [CrossRef]

18. Rapozzi, V.; Zorzet, S.; Zacchigna, M.; Drioli, S.; Xodo, L.E. The PDT activity of free and pegylated pheophorbide a against an amelanotic melanoma transplanted in C57/BL6 mice. *Investig. New Drugs* **2013**, *31*, 192–199. [CrossRef]

19. Hamblin, M.R.; Miller, J.L.; Rizvi, I.; Ortel, B.; Maytin, E.V.; Hasan, T. Pegylation of a chlorin(e6) polymer conjugate increases tumor targeting of photosensitizer. *Cancer Res.* **2001**, *61*, 7155–7162.

20. Rapozzi, V.; Zacchigna, M.; Biffi, S.; Garrovo, C.; Cateni, F.; Stebel, M.; Zorzet, S.; Bonora, G.M.; Drioli, S.; Xodo, L.E. Conjugated PDT drug photosensitizing activity and tissue distribution of PEGylated pheophorbide a. *Cancer Biol. Ther.* **2010**, *10*, 471–482. [CrossRef]

21. Srivatsan, A.; Ethirajan, M.; Pandey, S.K.; Dubey, S.; Zheng, X.; Liu, T.-H.; Shibata, M.; Missert, J.; Morgan, J.; Pandey, R.K. Conjugation of cRGD peptide to chlorophyll a based photosensitizer (HPPH) alters its pharmacokinetics with enhanced tumor-imaging and photosensitizing (PDT) efficacy. *Mol. Pharm.* **2011**, *8*, 1186–1197. [CrossRef] [PubMed]

22. Thomas, N.; Bechet, D.; Becuwe, P.; Tirand, L.; Vanderesse, R.; Frochot, C.; Guillemin, F.; Barberi-Heyob, M. Peptide-conjugated chlorin-type photosensitizer binds neuropilin-1 in vitro and in vivo. *J. Photochem. Photobiol. B* **2009**, *96*, 101–108. [CrossRef] [PubMed]

23. Tirand, L.; Frochot, C.; Vanderesse, R.; Thomas, N.; Trinquet, E.; Pinel, S.; Viriot, M.-L.; Guillemin, F.; Barberi-Heyob, M. A peptide competing with VEGF165 binding on neuropilin-1 mediates targeting of a chlorin-type photosensitizer and potentiates its photodynamic activity in human endothelial cells. *J. Control. Release* **2006**, *111*, 153–164. [CrossRef] [PubMed]

24. Thomas, N.; Tirand, L.; Chatelut, E.; Plenat, F.; Frochot, C.; Dodeller, M.; Guillemin, F.; Barberi-Heyob, M. Tissue distribution and pharmacokinetics of an ATWLPPR-conjugated chlorin-type photosensitizer targeting neuropilin-1 in glioma-bearing nude mice. *Photochem. Photobiol. Sci.* **2008**, *7*, 433–441. [CrossRef] [PubMed]

25. Zhang, X.; Meng, Z.; Ma, Z.; Liu, J.; Han, G.; Ma, F.; Jia, N.; Miao, Y.; Zhang, W.; Sheng, C.; et al. Design and synthesis of novel water-soluble amino acid derivatives of chlorin p6 ethers as photosensitizer. *Chinese Chem. Lett.* **2019**, *30*, 247–249. [CrossRef]

26. Jinadasa, R.G.W.; Zhou, Z.H.; Vicente, M.G.H.; Smith, K.M. Syntheses and cellular investigations of di-aspartate and aspartate-lysine chlorin e(6) conjugates. *Org. Biomol. Chem.* **2016**, *14*, 1049–1064. [CrossRef]

27. Meng, Z.; Yu, B.; Han, G.Y.; Liu, M.H.; Shan, B.; Dong, G.Q.; Miao, Z.Y.; Jia, N.Y.; Tan, Z.; Li, B.H.; et al. Chlorin p(6)-based water-soluble amino acid derivatives as potent photosensitizers for photodynamic therapy. *J. Med. Chem.* **2016**, *59*, 4999–5010. [CrossRef]

28. Hirohara, S.; Oka, C.; Totani, M.; Obata, M.; Yuasa, J.; Ito, H.; Tamura, M.; Matsui, H.; Kakiuchi, K.; Kawai, T.; et al. Synthesis, photophysical properties, and biological evaluation of *trans*-bisthioglycosylated tetrakis(fluorophenyl)chlorin for photodynamic therapy. *J. Med. Chem.* **2015**, *58*, 8658–8670. [CrossRef]

29. Tanaka, M.; Kataoka, H.; Mabuchi, M.; Sakuma, S.; Takahashi, S.; Tujii, R.; Akashi, H.; Ohi, H.; Yano, S.; Morita, A.; et al. Anticancer effects of novel photodynamic therapy with glycoconjugated chlorin for gastric and colon cancer. *Anticancer Res.* **2011**, *31*, 763–769.

30. Kato, A.; Kataoka, H.; Yano, S.; Hayashi, K.; Hayashi, N.; Tanaka, M.; Naitoh, I.; Ban, T.; Miyabe, K.; Kondo, H.; et al. Maltotriose conjugation to a chlorin derivative enhances the antitumor effects of photodynamic therapy in peritoneal dissemination of pancreatic cancer. *Mol. Cancer Ther.* **2017**, *16*, 1124–1132. [CrossRef]

31. Murakami, G.; Nanashima, A.; Nonaka, T.; Tominaga, T.; Wakata, K.; Sumida, Y.; Akashi, H.; Okazaki, S.; Kataoka, H.; Nagayasu, T. Photodynamic therapy using novel glucose-conjugated chlorin increases apoptosis of cholangiocellular carcinoma in comparison with talaporfin sodium. *Anticancer Res.* **2016**, *36*, 4493–4501. [CrossRef] [PubMed]

32. Demberelnyamba, D.; Ariunaa, M.; Shim, Y.K. Newly synthesized water soluble cholinium-purpurin photosensitizers and their stabilized gold nanoparticles as promising anticancer agents. *Int. J. Mol. Sci.* **2008**, *9*, 864–871. [CrossRef] [PubMed]

33. Zenkevich, E.; Sagun, E.; Knyukshto, V.; Shulga, A.; Mironov, A.; Efremova, O.; Bonnett, R.; Songca, S.P.; Kassem, M. Photophysical and photochemical properties of potential porphyrin and chlorin photosensitizers for PDT. *J. Photochem. Photobiol. B* **1996**, *33*, 171–180. [CrossRef]

34. Hoober, J.K.; Sery, T.W.; Yamamoto, N. Photodynamic sensitizers from chlorophyll-purpurin-18 and chlorin-P6. *Photochem. Photobiol.* **1988**, *48*, 579–582. [CrossRef]

35. Pandey, S.K.; Sajjad, M.; Chen, Y.; Pandey, A.; Missert, J.R.; Batt, C.; Yao, R.; Nabi, H.A.; Oseroff, A.R.; Pandey, R.K. Compared to purpurinimides, the pyropheophorbide containing an iodobenzyl group showed enhanced PDT efficacy and tumor imaging (124I-PET) ability. *Bioconjugate Chem.* **2009**, *20*, 274–282. [CrossRef]

36. Sharma, S.; Dube, A.; Bose, B.; Gupta, P.K. Pharmacokinetics and phototoxicity of purpurin-18 in human colon carcinoma cells using liposomes as delivery vehicles. *Cancer Chemother. Pharmacol.* **2006**, *57*, 500–506. [CrossRef]

37. Olshevskaya, V.A.; Savchenko, A.N.; Golovina, G.V.; Lazarev, V.V.; Kononova, E.G.; Petrovskii, P.V.; Kalinin, V.N.; Shtil', A.A.; Kuz'min, V.A. New boronated derivatives of purpurin-18: Synthesis and intereaction with serum albumin. *Dokl. Chem.* **2010**, *435*, 328–333. [CrossRef]

38. Hoebeke, M.; Damoiseau, X. Determination of the singlet oxygen quantum yield of bacteriochlorin a: A comparative study in phosphate buffer and aqueous dispersion of dimiristoyl-l-α-phosphatidylcholine liposomes. *Photochem. Photobiol. Sci.* **2002**, *1*, 283–287. [CrossRef]

39. Jiang, G.-Y.; Lei, W.-H.; Zhou, Q.-X.; Hou, Y.-J.; Wang, X.-S.; Zhang, B.-W. A new phenol red-modified porphyrin as efficient protein photocleaving agent. *Phys. Chem. Chem. Phys.* **2010**, *12*, 12229–12236. [CrossRef]

40. Gottfried, V.; Peled, D.; Winkelman, J.W.; Kimel, S. Photosensitizers in organized media: Singlet oxygen production and spectral properties. *Photochem. Photobiol.* **1988**, *48*, 157–163. [CrossRef]

41. Redmond, R.W.; Gamlin, J.N. A compilation of singlet oxygen yields from biologically relevant molecules. *Photochem. Photobiol.* **1999**, *70*, 391–475. [CrossRef] [PubMed]

42. Zhang, Y.; Zhang, H.; Wang, Z.; Jin, Y. pH-Sensitive graphene oxide conjugate purpurin-18 methyl ester photosensitizer nanocomplex in photodynamic therapy. *New J. Chem.* **2018**, *42*, 13272–13284. [CrossRef]

43. Cheng, J.; Tan, G.; Li, W.; Li, J.; Wang, Z.; Jin, Y. Preparation, characterization and in vitro photodynamic therapy of a pyropheophorbide a conjugated $Fe_3O_4$ multifunctional magnetofluorescence photosensitizer. *RSC Adv.* **2016**, *6*, 37610–37620. [CrossRef]

44. Bohmer, R.M.; Morstyn, G. Uptake of hematoporphyrin derivative by normal and malignant cells: Effect of serum, pH, temperature, and cell size. *Cancer Res.* **1985**, *45*, 5328–5334.

45. Friberg, E.G.; Čunderlíková, B.; Pettersen, E.O.; Moan, J. pH effects on the cellular uptake of four photosensitizing drugs evaluated for use in photodynamic therapy of cancer. *Cancer Lett.* **2003**, *195*, 73–80. [CrossRef]

46. Sharma, M.; Dube, A.; Bansal, H.; Kumar Gupta, P. Effect of pH on uptake and photodynamic action of chlorin p6 on human colon and breast adenocarcinoma cell lines. *Photochem. Photobiol. Sci.* **2004**, *3*, 231–235. [CrossRef]

47. Bříza, T.; Králová, J.; Rimpelová, S.; Havlík, M.; Kaplánek, R.; Kejík, Z.; Reddy, B.; Záruba, K.; Ruml, T.; Mikula, I.; et al. Dimethinium heteroaromatic salts as building blocks for dual-fluorescence intracellular probes. *ChemPhotoChem* **2017**, *1*, 442–450. [CrossRef]

48. Luo, W.; Liu, R.S.; Zhu, J.G.; Li, Y.C.; Liu, H.C. Subcellular location and photodynamic therapeutic effect of chlorin e6 in the human tongue squamous cell cancer Tca8113 cell line. *Oncol. Lett.* **2015**, *9*, 551–556. [CrossRef]

49. Huang, Y.-Y.; Mroz, P.; Zhiyentayev, T.; Sharma, S.K.; Balasubramanian, T.; Ruzié, C.; Krayer, M.; Fan, D.; Borbas, K.E.; Yang, E.; et al. In vitro photodynamic therapy and quantitative structure–activity relationship studies with stable synthetic near-infrared-absorbing bacteriochlorin photosensitizers. *J. Med. Chem.* **2010**, *53*, 4018–4027. [CrossRef]

50. Li, Y.; Yu, Y.; Kang, L.; Lu, Y. Effects of chlorin e6-mediated photodynamic therapy on human colon cancer SW480 cells. *Int. J. Clin. Exp. Med.* **2014**, *7*, 4867–4876.

51. Mojzisova, H.; Bonneau, S.; Vever-Bizet, C.; Brault, D. Cellular uptake and subcellular distribution of chlorin e6 as functions of pH and interactions with membranes and lipoproteins. *Biochim. Biophys. Acta Biomembr.* **2007**, *1768*, 2748–2756. [CrossRef] [PubMed]

52. Lkhagvadulam, B.; Kim, J.H.; Yoon, I.; Shim, Y.K. Synthesis and photodynamic activities of novel water soluble purpurin-18-*N*-methyl-D-glucamine photosensitizer and its gold nanoparticles conjugate. *J. Porphyr. Phthalocyanines* **2012**, *16*, 331–340. [CrossRef]

53. Yoon, I.; Sung, H.; Cui, B.; Kim, J.; Shim, Y. Synthesis and photodynamic activities of pyrazolyl and cyclopropyl derivatives of purpurin-18 methyl ester and purpurin-18-*N*-butylimide: Synthesis and photodynamic activities of chlorins. *Bull. Korean Chem. Soc.* **2011**, *32*, 169–174. [CrossRef]

54. Cui, B.C.; Yoon, I.; Li, J.Z.; Lee, W.K.; Shim, Y.K. Synthesis and characterization of novel purpurinimides as photosensitizers for photodynamic therapy. *Int. J. Mol. Sci.* **2014**, *15*, 8091–8105. [CrossRef] [PubMed]

55. Lin, Y.-X.; Wang, Y.; Qiao, S.-L.; An, H.-W.; Wang, J.; Ma, Y.; Wang, L.; Wang, H. "In vivo self-assembled" nanoprobes for optimizing autophagy-mediated chemotherapy. *Biomaterials* **2017**, *141*, 199–209. [CrossRef] [PubMed]

56. Klein, O.J.; Yuan, H.; Nowell, N.H.; Kaittanis, C.; Josephson, L.; Evans, C.L. An integrin-targeted, highly diffusive construct for photodynamic therapy. *Sci. Rep.* **2017**, *7*, 13375. [CrossRef]

57. Sibrian-Vazquez, M.; Nesterova, I.V.; Jensen, T.J.; Vicente, M.G.H. Mitochondria targeting by guanidine–and biguanidine–porphyrin photosensitizers. *Bioconjugate Chem.* **2008**, *19*, 705–713. [CrossRef]

58. Gryshuk, A.; Chen, Y.; Goswami, L.N.; Pandey, S.; Missert, J.R.; Ohulchanskyy, T.; Potter, W.; Prasad, P.N.; Oseroff, A.; Pandey, R.K. Structure–activity relationship among purpurinimides and bacteriopurpurinimides: trifluoromethyl substituent enhanced the photosensitizing efficacy. *J. Med. Chem.* **2007**, *50*, 1754–1767. [CrossRef]

59. Agostinis, P.; Berg, K.; Cengel, K.A.; Foster, T.H.; Girotti, A.W.; Gollnick, S.O.; Hahn, S.M.; Hamblin, M.R.; Juzeniene, A.; Kessel, D.; et al. Photodynamic therapy of cancer: An update. *CA Cancer J. Clin.* **2011**, *61*, 250–281. [CrossRef]

60. Mroz, P.; Yaroslavsky, A.; Kharkwal, G.B.; Hamblin, M.R. Cell death pathways in photodynamic therapy of cancer. *Cancers* **2011**, *3*, 2516–2539. [CrossRef]

61. Foo, J.B.; Ng, L.S.; Lim, J.H.; Tan, P.X.; Lor, Y.Z.; Loo, J.S.E.; Low, M.L.; Chan, L.C.; Beh, C.Y.; Leong, S.W.; et al. Induction of cell cycle arrest and apoptosis by copper complex Cu(SBCM)$_2$ towards oestrogen-receptor positive MCF-7 breast cancer cells. *RSC Adv.* **2019**, *9*, 18359–18370. [CrossRef]

62. Yan, W.; Ma, X.; Zhao, X.; Zhang, S. Baicalein induces apoptosis and autophagy of breast cancer cells via inhibiting PI3K/AKT pathway in vivo and in vitro. *Drug Des. Dev. Ther.* **2018**, *12*, 3961–3972. [CrossRef] [PubMed]

63. Tsai, T.; Hong, R.-L.; Tsai, J.-C.; Lou, P.-J.; Ling, I.-F.; Chen, C.-T. Effect of 5-aminolevulinic acid-mediated photodynamic therapy on MCF-7 and MCF-7/ADR cells. *Laser. Surg. Med.* **2004**, *34*, 62–72. [CrossRef] [PubMed]

64. Chaves, O.A.; Amorim, A.P.D.O.; Castro, L.H.E.; Sant'Anna, C.M.R.; De Oliveira, M.C.C.; Cesarin-Sobrinho, D.; Netto-Ferreira, J.C.; Ferreira, A.B.B. Fluorescence and docking studies of the interaction between human serum albumin and pheophytin. *Molecules* **2015**, *20*, 19526–19539. [CrossRef] [PubMed]

65. Zunszain, P.A.; Ghuman, J.; Komatsu, T.; Tsuchida, E.; Curry, S. Crystal structural analysis of human serum albumin complexed with hemin and fatty acid. *BMC Struct. Biol.* **2003**, *3*, 6. [CrossRef] [PubMed]

66. Wardell, M.; Wang, Z.; Ho, J.X.; Robert, J.; Ruker, F.; Ruble, J.; Carter, D.C. The atomic structure of human methemalbumin at 1.9 Å. *Biochem. Biophys. Res. Commun.* **2002**, *291*, 813–819. [CrossRef]

67. Akimova, A.; Rychkov, G.N.; Grin, M.A.; Filippova, N.A.; Golovina, G.V.; Durandin, N.A.; Vinogradov, A.M.; Kokrashvili, T.A.; Mironov, A.F.; Shtil, A.A.; et al. Interaction with serum albumin as a factor of the photodynamic efficacy of novel bacteriopurpurinimide derivatives. *Acta Nat.* **2015**, *7*, 109–116. [CrossRef]

68. Kamal, J.K.; Behere, D.V. Binding of heme to human serum albumin: Steady-state fluorescence, circular dichroism and optical difference spectroscopic studies. *Indian J. Biochem. Biophys.* **2005**, *42*, 7–12.

69. Ol'shevskaya, V.A.; Nikitina, R.G.; Zaitsev, A.V.; Luzgina, V.N.; Kononova, E.G.; Morozova, T.G.; Drozhzhina, V.V.; Ivanov, O.G.; Kaplan, M.A.; Kalinin, V.N.; et al. Boronated protohaemins: Synthesis and *in vivo* antitumour efficacy. *Org. Biomol. Chem.* **2006**, *4*, 3815–3821. [CrossRef]

70. Ol'shevskaya, V.A.; Nikitina, R.G.; Savchenko, A.N.; Malshakova, M.V.; Vinogradov, A.M.; Golovina, G.V.; Belykh, D.V.; Kutchin, A.V.; Kaplan, M.A.; Kalinin, V.N.; et al. Novel boronated chlorin e6-based photosensitizers: Synthesis, binding to albumin and antitumour efficacy. *Bioorganic Med. Chem.* **2009**, *17*, 1297–1306. [CrossRef]

71. Ol'shevskaya, V.A.; Savchenko, A.N.; Zaitsev, A.V.; Kononova, E.G.; Petrovskii, P.V.; Ramonova, A.A.; Tatarskiy, V.V.; Uvarov, O.V.; Moisenovich, M.M.; Kalinin, V.N.; et al. Novel metal complexes of boronated chlorin e6 for photodynamic therapy. *J. Organomet. Chem.* **2009**, *694*, 1632–1637. [CrossRef]

72. Pshenkina, N.N. Structure of albumin and transport of drugs. *Med. Acad. J.* **2011**, *11*, 3–15.

73. Sharman, W.M.; van Lier, J.E.; Allen, C.M. Targeted photodynamic therapy via receptor mediated delivery systems. *Adv. Drug Deliv. Rev.* **2004**, *56*, 53–76. [CrossRef] [PubMed]

74. Tsuchida, T.; Zheng, G.; Pandey, R.K.; Potter, W.R.; Bellnier, D.A.; Henderson, B.W.; Kato, H.; Dougherty, T.J. Correlation between site II-specific human serum albumin (HSA) binding affinity and murine in vivo photosensitizing efficacy of some photofrin components. *Photochem. Photobiol.* **1997**, *66*, 224–228. [CrossRef] [PubMed]

75. Khodaei, A.; Bolandnazar, S.; Valizadeh, H.; Hasani, L.; Zakeri-Milani, P. Interactions between sirolimus and anti-inflammatory drugs: Competitive binding for human serum albumin. *Adv. Pharm. Bull.* **2016**, *6*, 227–233. [CrossRef] [PubMed]

76. Yang, Z.; Zhou, T.; Cheng, Y.; Li, M.; Tan, X.; Xu, F. Weakening impact of excessive human serum albumin (eHSA) on cisplatin and etoposide anticancer effect in C57BL/6 mice with tumor and in human NSCLC A549 cells. *Front. Pharmacol.* **2016**, *7*, 434. [CrossRef]

77. Liu, C.; Liu, Z.; Wang, J. Uncovering the molecular and physiological processes of anticancer leads binding human serum albumin: A physical insight into drug efficacy. *PLoS ONE* **2017**, *12*, e0176208. [CrossRef]

78. Plika, V.; Testa, B.; van de Waterbeemd, H. Lipophilicity: The empirical tool and the fundamental objective. An introduction. In *Lipophilicity in Drug Action and Toxicology*; Wiley-VCH Verlag GmbH: Weinheim, Germany, 1996; pp. 1–6.

79. Lázníček, M.; Lázníčková, A. The effect of lipophilicity on the protein binding and blood cell uptake of some acidic drugs. *J. Pharm. Biomed. Anal.* **1995**, *13*, 823–828. [CrossRef]

80. Henderson, B.W.; Bellnier, D.A.; Greco, W.R.; Sharma, A.; Pandey, R.K.; Vaughan, L.A.; Weishaupt, K.R.; Dougherty, T.J. An in vivo quantitative structure-activity relationship for a congeneric series of pyropheophorbide derivatives as photosensitizers for photodynamic therapy. *Cancer Res.* **1997**, *57*, 4000–4007.

81. Pucelik, B.; Paczyński, R.; Dubin, G.; Pereira, M.M.; Arnaut, L.G.; Dąbrowski, J.M. Properties of halogenated and sulfonated porphyrins relevant for the selection of photosensitizers in anticancer and antimicrobial therapies. *PLoS ONE* **2017**, *12*, e0185984. [CrossRef]

82. Pucelik, B.; Gürol, I.; Ahsen, V.; Dumoulin, F.; Dąbrowski, J.M. Fluorination of phthalocyanine substituents: Improved photoproperties and enhanced photodynamic efficacy after optimal micellar formulations. *Eur. J. Med. Chem.* **2016**, *124*, 284–298. [CrossRef]

83. Ezzeddine, R.; Al-Banaw, A.; Tovmasyan, A.; Craik, J.D.; Batinic-Haberle, I.; Benov, L.T. Effect of molecular characteristics on cellular uptake, subcellular localization, and phototoxicity of Zn(II) N-alkylpyridylporphyrins. *J. Biol. Chem.* **2013**, *288*, 36579–36588. [CrossRef]

84. Velapoldi, R.A.; Tønnesen, H.H. Corrected emission spectra and quantum yields for a series of fluorescent compounds in the visible spectral region. *J. Fluoresc.* **2004**, *14*, 465–472. [CrossRef] [PubMed]

85. Rimpelová, S.; Jurášek, M.; Peterková, L.; Bejček, J.; Spiwok, V.; Majdl, M.; Jirásko, M.; Buděšínský, M.; Harmatha, J.; Kmoníčková, E.; et al. Archangelolide: Sesquiterpene lactone with immunobiological potential from *Laserpitium archangelica*. *Beilstein J. Org. Chem.* **2019**, *15*, 1933–1944. [CrossRef] [PubMed]

86. Vermes, I.; Haanen, C.; Steffens-Nakken, H.; Reutellingsperger, C. A novel assay for apoptosis flow cytometric detection of phosphatidylserine expression on early apoptotic cells using fluorescein labelled annexin V. *J. Immunol. Methods* **1995**, *184*, 39–51. [CrossRef]

87. Kirakci, K.; Zelenka, J.; Rumlová, M.; Martinčík, J.; Nikl, M.; Ruml, T.; Lang, K. Octahedral molybdenum clusters as radiosensitizers for X-ray induced photodynamic therapy. *J. Mater. Chem. B* **2018**, *6*, 4301–4307. [CrossRef]

88. Rumlová, M.; Křížová, I.; Keprová, A.; Hadravová, R.; Doležal, M.; Strohalmová, K.; Pichová, I.; Hájek, M.; Ruml, T. HIV-1 protease-induced apoptosis. *Retrovirology* **2014**, *11*, 37. [CrossRef]

89. Wildman, S.A.; Crippen, G.M. Prediction of physicochemical parameters by atomic contributions. *J. Chem. Inf. Comp. Sci.* **1999**, *39*, 868–873. [CrossRef]

90. Daina, A.; Michielin, O.; Zoete, V. iLOGP: A simple, robust, and efficient description of n-octanol/water partition coefficient for drug design using the GB/SA approach. *J. Chem. Inf. Model.* **2014**, *54*, 3284–3301. [CrossRef]

*molecules*

**MDPI**

*Article*

# Design, Synthesis and Anti-Platelet Aggregation Activity Study of Ginkgolide-1,2,3-triazole Derivatives

Jian Cui [1,†], Le'An Hu [1,2,†], Wei Shi [3], Guozhen Cui [3,*], Xumu Zhang [2,*] and Qing-Wen Zhang [1,*]

[1]   State Key Laboratory of Quality Research in Chinese Medicine, Institute of Chinese Medical Sciences, University of Macau, Macau 999078, China; mb65822@umac.mo (J.C.); yb67526@umac.mo (L.H.)
[2]   Department of Chemistry and Shenzhen Grubbs Institute, Southern University of Science and Technology, Shenzhen 518000, China
[3]   Department of Bioengineering, Zhuhai Campus of Zunyi Medical University, Zhuhai 519041, Guangdong, China; diana1354726@outlook.com
*   Correspondence: cgzum@hotmail.com (G.C.); zhangxm@sustc.edu.cn (X.Z.); qwzhang@umac.mo (Q.-W.Z.)
†   These authors contributed equally to the work.

Received: 7 May 2019; Accepted: 3 June 2019; Published: 7 June 2019

check for updates

**Abstract:** Ginkgolides are the major active component of *Ginkgo biloba* for inhibition of platelet activating factor receptor. An azide-alkyne Huisgen cycloaddition reaction was used to introduce a triazole nucleus into the target ginkgolide molecules. A series of ginkgolide-1,2,3-triazole conjugates with varied functional groups including benzyl, phenyl and heterocycle moieties was thus synthesized. Many of the designed derivatives showed potent antiplatelet aggregation activities with $IC_{50}$ values of 5~21 nM.

**Keywords:** ginkgolide; platelet-activating factor receptor; inhibitor

---

## 1. Introduction

*Ginkgo biloba*, also named maidenhair tree, the only surviving species from the family Ginkgoaceae has existed for more than 180 million years, and for this reason it was called a "living fossil" by Darwin. *G. biloba* has been used as a traditional Chinese medicine for a long time for the treatment of lung weakness, asthma, coughing, cancer, etc [1,2]. Ginkgo has also been popular in the Western world since 1965, when a German company developed from ginkgo extracts a botanical medicine named EGB761, with various effects on central nervous system (CNS) diseases, including Alzheimer's disease, dementia, hypomnesia, etc [3,4]. The main components of ginkgo extracts are terpene trilactones (including ginkgolides and bilobalide) and flavonoids [5,6]. Since flavonoids are deemed to hardly penetrate the blood-brain barrier, it is assumed the terpene trilactones from ginkgo extracts should be the major active components for the CNS effects and cardiovascular activity [7,8]. As a natural phospholipid agonist of the platelet activating factor of platelet activating factor receptor (PAFR), platelet activating factor (PAF) regulates various physiological activities of the CNS and peripheral nervous system, including platelet aggregation, blood pressure regulation, inflammation, long-term enhancement of CNS, etc [9]. It's reported that ginkgolides could competitively inhibit the platelet-activating factor receptor (PAFR), resulting in the observed CNS protection and antithrombotic effects [10,11].

To date, 10 ginkgolides (ginkgolide A~Q) [5,6,12] and two bilobalides (bilobalide and bilobanol) [13,14] have been isolated and from *G. biloba* and their structures elucidated. Most natural ginkgolides displayed significant activity against PAFR, while the bilobalides didn't. In particular, ginkgolide B is the most potent PAFR antagonist discovered in nature [1]. Since the 1970s, many investigations on the structural modifications and structure-activity relationship of ginkgolides have been conducted. It is revealed that both ring C and ring D are essential for anti-PAFR activity [15]. Substituents at the C-7 position decrease the activity [16]. It's noteworthy that the introduction of bulky or aromatic substituents at 10-OH could help to increase activity against PAFR [17,18]. The 1,2,3-triazole moiety, also known simply as triazole, acts as an important structural fragment widely used to construct new drug molecules [19]. The triazole is an electron isostere of the amide group, which easily forms hydrogen bonds, coordination bonds, etc., helping to form a variety of non-covalent bond interactions with target proteins [19,20].

In this paper, a series of ginkgolide derivatives with 1,2,3-triazole moieties connected with various benzyl, phenyl and heterocycle moieties at the C-10 position were designed and synthetized. Their antiplatelet aggregation activities were also evaluated and several derivatives displayed more potent inhibitory effects against PAFR than the natural ginkgolide B, with $IC_{50}$ values of 5~21 nM, or about 10 to 20 times higher than the natural compound.

## 2. Results and Discussion

### 2.1. Chemistry

Drugability is improved when the triazole moiety is introduced into some leads [20]. The azide-alkyne Huisgen cycloaddition reaction, also known as the Huisgen 1,3-dipolar cycloaddition, has been proved to be a powerful tool in construction of triazoles [21]. In this reaction, the azide moiety reacts to a terminal alkyne group to form the triazole ring. The copper(I) catalyst improves both the reaction rate and selectivity. A series of ginkgolide-1,2,3-triazole conjugates with varied functional groups including benzyl, phenyl and heterocycle moieties was synthesized via this method.

10-*O*-Propargylated ginkgolide B (**3**) was synthesized in 65% yield by mixing ginkgolide B (**1**) and 3-bromoprop-1-yne (**2**) with potassium carbonate in anhydrous acetonitrile for several hours (Scheme 1). 10-*O*-Propargylated ginkgolide A (**3′**) and 10-*O*-propargylated ginkgolide C (**3″**) were obtained in 32% and 69% yield, respectively from the corresponding ginkgolide A (**1′**) and ginkgolide C (**1″**) following the same scheme.

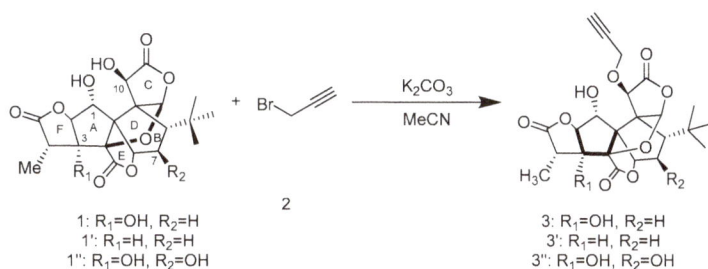

1: R<sub>1</sub>=OH, R<sub>2</sub>=H
1': R<sub>1</sub>=H, R<sub>2</sub>=H
1": R<sub>1</sub>=OH, R<sub>2</sub>=OH

3: R<sub>1</sub>=OH, R<sub>2</sub>=H
3': R<sub>1</sub>=H, R<sub>2</sub>=H
3": R<sub>1</sub>=OH, R<sub>2</sub>=OH

**Scheme 1.** Formation of 10-*O*-propargylated ginkgolides **3**, **3′** and **3″**.

The target molecules, 10-substituted 1,2,3-triazole-ginkgolide B derivatives **5**, were synthesized using the copper(I)-catalyzed Huisgen 1,3-dipolar cycloaddition reaction [21] by mixing 10-*O*-propargylated ginkgolide B (**3**) with a series of azides **4** in a three-phase solvent, using copper(I) as catalyst (Scheme 2). The 1,2,3-triazole-ginkgolide A derivatives (**5′**) and 1,2,3-triazole-ginkgolide ginkgolide C derivatives (**5″**) were obtained by the same route.

Azides **4a**–**4ee** were synthesized by firstly mixing the corresponding aniline, sodium nitrite and concentrated hydrochloride in ethyl acetate at 0 °C for 30 min, then adding sodium azide to the system at room temperature and stirring for another two hours (Scheme 3) [22–24]. The resulting products are listed in Table 1.

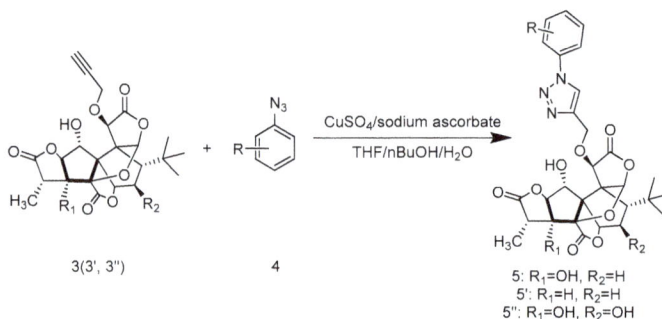

**Scheme 2.** Formation of 10-substituted 1,2,3-triazole-ginkgolide derivatives **5**, **5′** and **5″**.

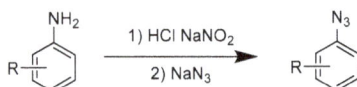

**Scheme 3.** Preparation of azides **4a–4ee**.

**Table 1.** Synthesized azides **4a–4ee**.

| Entry | Azide | Ar | Yield |
|---|---|---|---|
| 1 | 4a | Ph | 77 |
| 2 | 4b | 2-Cl-Ph | 88 |
| 3 | 4c | 3-Cl-Ph | 91 |
| 4 | 4d | 4-Cl-Ph | 75 |
| 5 | 4e | 2-Cl, 6Cl-Ph | 72 |
| 6 | 4f | 2-Cl, 4Cl-Ph | 71 |
| 7 | 4g | 3-Cl, 4Cl-Ph | 87 |
| 8 | 4h | 3-F-Ph | 88 |
| 9 | 4i | 3-CF$_3$, 4-Cl-Ph | 62 |
| 10 | 4j | 4-OCF$_3$-Ph | 86 |
| 11 | 4k | 3-CF$_3$, 5-CF$_3$Ph | 91 |
| 12 | 4l | 4-tertbutyl-Ph | 76 |
| 13 | 4m | 3-Br-Ph | 62 |
| 14 | 4n | 2-Me-Ph | 75 |
| 15 | 4o | 2-OH-Ph | 92 |
| 16 | 4p | 2-CN-Ph | 88 |
| 17 | 4q | 3-Me-Ph | 74 |
| 18 | 4r | 3-isopropyl-Ph | 87 |
| 19 | 4s | 3-COOEt-Ph | 78 |
| 20 | 4t | 3-COOH-Ph | 78 |
| 21 | 4u | 4-Me-Ph | 65 |
| 22 | 4v | 4-OMe-Ph | 87 |
| 23 | 4w | 4-F-Ph | 74 |
| 24 | 4x | 4-Br-Ph | 63 |
| 25 | 4y | 4-CF$_3$-Ph | 77 |
| 26 | 4z | 4-NO$_2$-Ph | 67 |
| 27 | 4aa | 4-COOMe-Ph | 87 |
| 28 | 4bb | 4-COOH-Ph | 76 |
| 29 | 4cc | 3-OMe,5-OMe-Ph | 79 |
| 30 | 4dd | 3-Cl, 4F-Ph | 87 |
| 31 | 4ee | 3-pyridine | 68 |

Azides **4ff–4gg** were synthesized by mixing the corresponding benzyl bromide and sodium azide in DMF for two hours (Scheme 4) [22,25]. The resulting products are listed in Table 2.

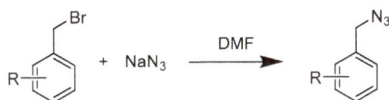

**Scheme 4.** Preparation of azides **4ff**, **4gg**.

**Table 2.** Synthesized azides **4ff–4gg**.

| Entry | Azide | Ar | Yield |
|-------|-------|---------|-------|
| 1 | **4ff** | Bn | 82 |
| 2 | **4gg** | 2-Me-Bn | 77 |

Following the reaction shown in Scheme 2, using 10-*O*-propargylated ginkgolide B **3** and azides **4a-4gg**, compounds **5a-5gg** were obtained (Figure 1).

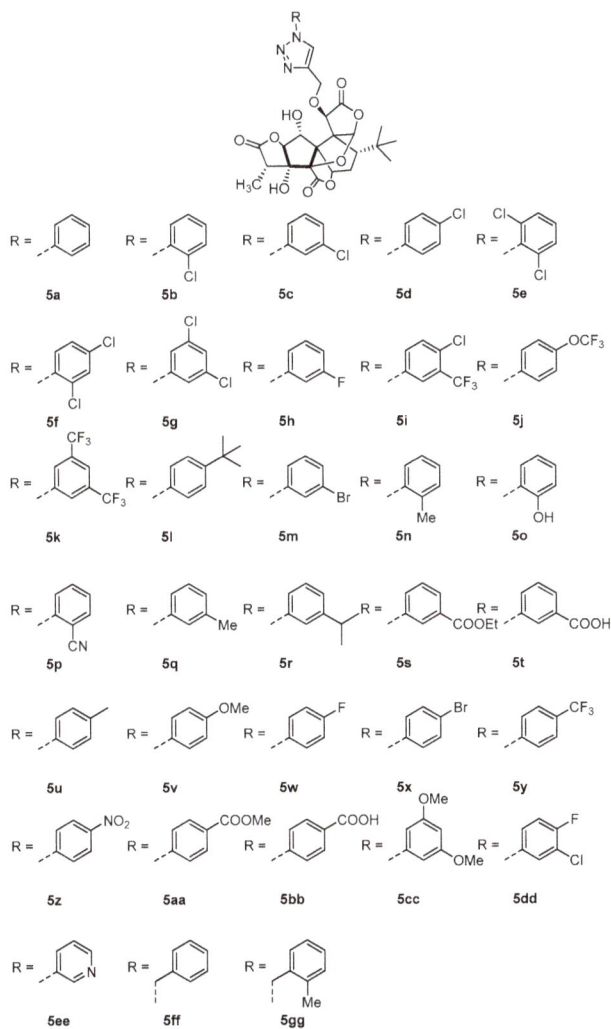

**Figure 1.** Structures of 10-substituted 1,2,3-triazole-ginkgolide B derivatives **5a–5gg**.

Following the reaction shown in Scheme 2, using 10-*O*-propargylated ginkgolide A **3′**, 10-*O*-propargylated ginkgolide C **3″** and azides **4a, 4n, 4p, 4ff, 4gg**, compounds **5′a-5′gg**, **5″a-5″gg** were obtained as illustrated in Figure 2.

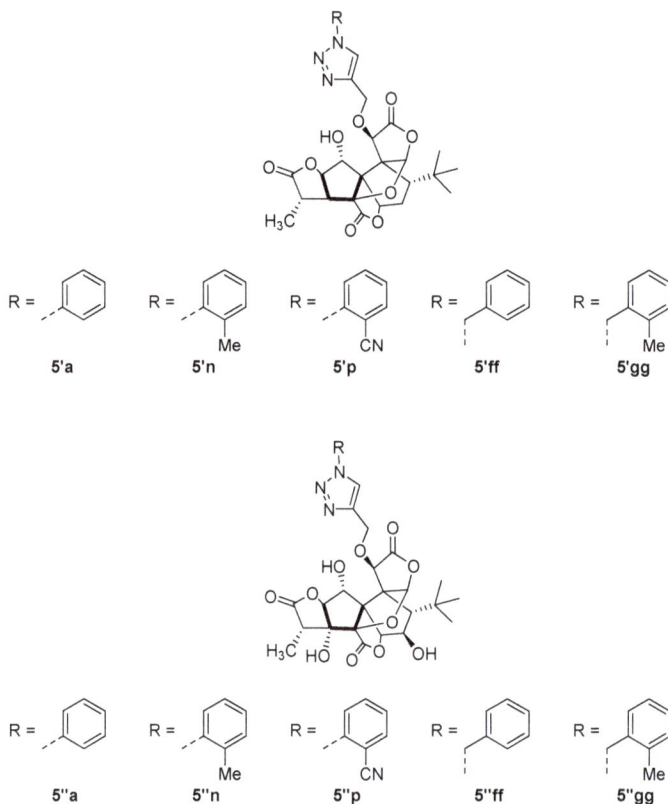

**Figure 2.** Structures of 10-substituted 1,2,3-triazole-ginkgolide A derivatives **5′a-5′gg** and 10-substituted 1,2,3-triazole-ginkgolide C derivatives **5″a-5″gg**.

## 2.2. Biology

Since ginkgolide derivatives have shown antiplatelet aggregation activities as reported before [15–18], the as-synthesized ginkgolide-1,2,3-triazole derivatives were expected to show improved biological potency.

Firstly, the newly synthesized 10-substituted 1,2,3-triazole-ginkgolide B derivatives (**5**) were tested by the method of Born [26,27]. The result was given in Figure 3. Which showed some of them exhibited considerable activity better than ginkgolide B. The best results were obtained with compounds **5a, 5n, 5p, 5ff** and **5gg**.

From the results shown above, we can preliminarily conclude that non-substituted benzyl (compound **5a**) and phenyl (compound **5ff**) 1,2,3-triazole conjugates have significantly enhanced antiplatelet aggregation activities compared to ginkgolide B (**1**). We can also see that the substitution at the *meta-* or/and *para-* positions of the benzyl group reduce the activity. Some small steric hindrance groups (such as methyl and cyano groups) substituted at the *ortho*-position of the phenyl (compounds **5n** and **5p**) and benzyl (compound **5gg**) groups maintain or slightly reduce the activity.

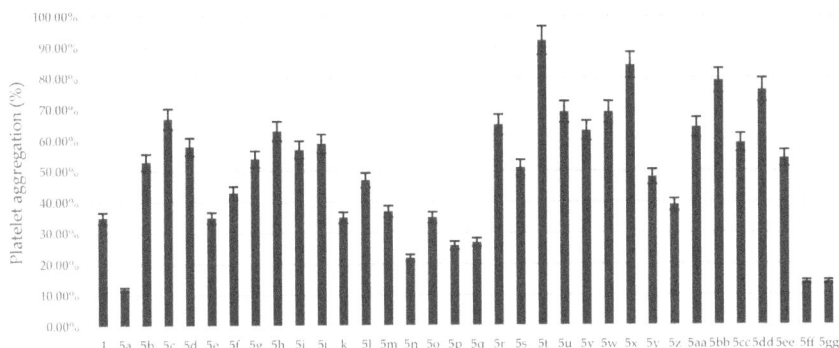

**Figure 3.** Antiplatelet aggregation activities of the 10-substituted 1,2,3-triazole-ginkgolide B derivatives. Ginkgolide B (**1**) was used as reference.

In further studies, 10-substituted 1,2,3-triazole-ginkgolide A **5′** and ginkgolide C **5″** derivatives, having the same moieties as the most active ginkgolide B derivatives **5a**, **5n**, **5p**, **5ff** and **5gg**, were synthesized and their antiplatelet aggregation activities tested and reported as inhibition ratios at 50 nM). The results are shown in Figure 4. Some of them exhibit better activity than not only their precursors **1′** or **1″**, but also ginkgolide B (**1**). The best results were obtained with compounds **5′a**, **5′ff**, **5′gg** and **5″gg**.

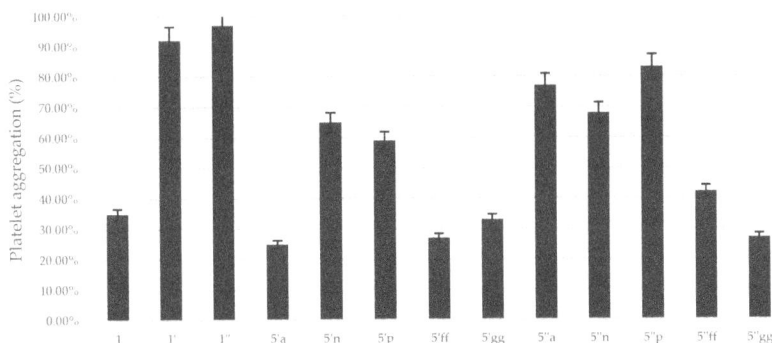

**Figure 4.** Antiplatelet aggregation activity activities of the 10-substituted 1,2,3-triazole-ginkgolide A and ginkgolide C derivatives. Ginkgolide B (**1**), ginkgolide A (**1′**) and ginkgolide C (**1″**) were used as reference.

The most active compounds obtained by method above, **5a**, **5n**, **5p**, **5q**, **5ff**, **5gg**, **5′a**, **5′ff**, **5′gg** and **5″gg**, were further examined in order to get their activity expressed as an $IC_{50}$ value. The results are listed in Table 3.

As we can see, compounds **5a**, **5p**, **5ff**, **5gg** and **5′a** display promising antiplatelet aggregation activity with $IC_{50}$ values ranging from 5–21 nM. Among them compounds **5ff** and **5gg** were the best among the series of compounds, showing about a 20-fold increase in comparison with the natural ginkgolide B (**1**).

**Table 3.** In vitro antiplatelet aggregation activity study of some of ginkgolide-1,2,3-triazole derivatives.

| Compound | PAF-Induced Platelet Aggregation IC$_{50}$ (nM) | Compound | PAF-Induced Platelet Aggregation IC$_{50}$ (nM) |
|---|---|---|---|
| 1 | 130.1 ± 20.0 | 5ff | 5.31 ± 1.84 |
| 1′ | 580.4 ± 50.3 | 5gg | 5.68 ± 0.71 |
| 1″ | 7540 ± 662 | 5′a | 18.55 ± 2.25 |
| 5a | 7.74 ± 1.89 | 5′ff | 25.37 ± 4.25 |
| 5n | 20.63 ± 2.84 | 5′gg | 71.47 ± 6.26 |
| 5p | 10.19 ± 2.03 | 5″gg | 37.68 ± 4.17 |
| 5q | 44.33 ± 4.42 | | |

In order to verify if the most active ginkgolide-1,2,3-triazole derivatives could also be considered as potential antiplatelet aggregation therapeutics, compounds **5a**, **5n**, **5p**, **5q**, **5ff**, **5gg**, **5′a**, **5′ff** and **5″gg** were examined to confirm their cytotoxicity using an LDH assay [28]. The results are shown in Figure 5.

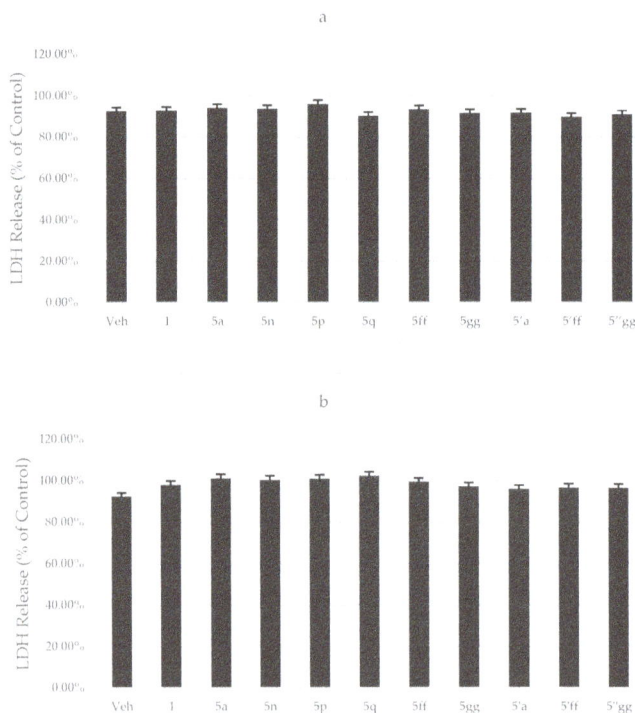

**Figure 5.** Cytotoxicity of some of ginkgolide-1,2,3-triazole derivatives: (**a**) Tested at 1 μM concentration. (**b**) Tested at 10 μM concentration. Ginkgolide B (**1**) was used as reference.

In addition, compounds **5a**, **5p**, **5ff**, **5gg** and **5′a** were examined to confirm their toxicity on platelets using the LDH assay [29]. The results are shown in Figure 6.

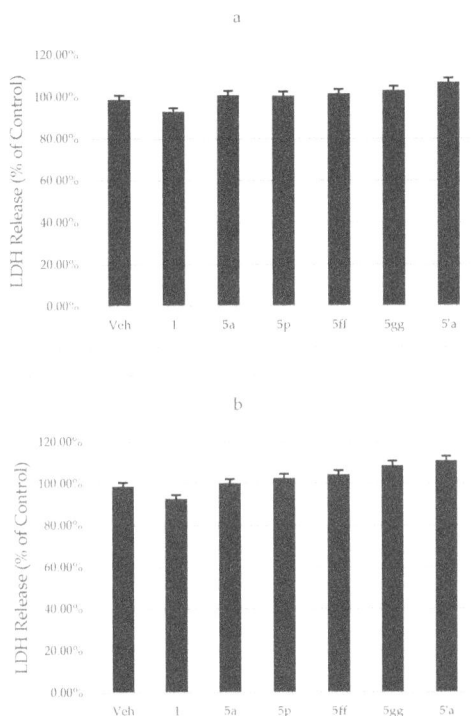

**Figure 6.** Toxicity on platelets of some of ginkgolide-1,2,3-triazole derivatives: (**a**) Tested at 1 μM concentration. (**b**) Tested at 10 μM concentration. Ginkgolide B (**1**) was used as reference.

As shown in results above, these most active compounds did not demonstrate toxicity towards cardiomyocytes and platelets ($p > 0.05$) up to 10 μM (almost two order of magnitude higher than $IC_{50}$ of ginkgolide B), which suggest that they have a broad therapeutic window/safety window.

## 3. Materials and Methods

### 3.1. Compound Synthesis

#### 3.1.1. General Experimental Procedures

All solvents and reagents of analytical grade were obtained from commercial sources. Flash chromatography was performed using silica gel (200–300 mesh, Qingdao Marine Chemical Group Co., Qingdao, China). All reactions were monitored by TLC on silica gel plates (Merck, Darmstadt, Germany). NMR spectra were recorded in $CDCl_3$ or DMSO at 400 or 600 MHz for $^1$H-NMR and 125 or 150 MHz for $^{13}$C-NMR on an Ascent 400 or 600 spectrometer (Bruker, Fallanden, Switzerland). The solvent signal was used as an internal standard. ESI-MS were recorded on an 1200/MSD mass spectrometer (Agilent, Santa Clara, CA, USA). HREIMS were recorded on a LTQ Orbitrap XL mass spectrometer (Thermo, Bremen, Germany).

#### 3.1.2. General Procedures for the Preparation 10-*O*-propargylated Ginkgolides

Propargyl bromide (2.4 mmol) were slowly added to a mixture of ginkgolide (**1**, **1′** or **1″**, 2.0 mmol) and $K_2CO_3$ (4.0 mmol) in acetonitrile (15 mL). The reaction mixture was refluxed for 24 h under an argon atmosphere and then was extracted with EtOAc three times. The combined organic phases were dried over anhydrous $MgSO_4$, filtered, and concentrated in vacuo. The residue was purified by chromatography ($SiO_2$, petroleum ether (PE)/EtOAc stepwise elution, 1:1 to EtOAc).

*10-O-Propargylated ginkgolide B* (**3**). Following the described procedure, 602 mg (65%) of compound **3** were obtained from 829 mg (2.0 mmol) of ginkgolide B (**1**). $^1$H-NMR (600 MHz, DMSO-$d_6$) δ 6.46 (s, 1H), 6.14 (s, 1H), 5.32 (d, *J* = 4.1 Hz, 1H), 5.19 (s, 1H), 5.15 (d, *J* = 4.9 Hz, 1H), 4.69 (ddd, *J* = 70.7, 15.8, 2.5 Hz, 2H), 4.61 (d, *J* = 6.8 Hz, 1H), 4.09 (dd, *J* = 6.9, 4.9 Hz, 1H), 3.62 (t, *J* = 2.4 Hz, 1H), 2.84 (q, *J* = 7.1 Hz, 1H), 2.13 (dd, *J* = 13.6, 4.6 Hz, 1H), 1.99 (s, 1H), 1.87 (td, *J* = 13.9, 4.2 Hz, 1H), 1.71 (dd, *J* = 14.4, 4.5 Hz, 1H), 1.11 (d, *J* = 7.2 Hz, 3H), 1.05 (s, 9H). $^{13}$C-NMR (150 MHz, DMSO-$d_6$) δ 176.77, 172.56, 170.63, 110.02, 99.51, 93.37, 83.06, 79.71, 79.29, 74.96, 74.12, 72.41, 67.84, 60.23, 57.91, 49.08, 41.96, 36.96, 32.29, 29.38 (3C), 8.44. HRMS (ESI): *m/z* calcd for $C_{23}H_{26}O_{10}$ [M + H]$^+$: 463.1599, found 463.1531.

*10-O-Propargylated ginkgolide A* (**3′**), Following the described procedure, 286 mg (32%) of compound **3′** were obtained from 797 mg (2.0 mmol) of ginkgolide A (**1′**). $^1$H-NMR (600 MHz, DMSO-$d_6$) δ 6.43 (s, 1H), 6.12 (s, 1H), 5.10 (s, 1H), 5.00 (d, *J* = 4.1 Hz, 1H), 4.86 (t, *J* = 7.7 Hz, 1H), 4.72 (dd, *J* = 15.6, 2.5 Hz, 1H), 4.50 (dd, *J* = 15.5, 2.5 Hz, 1H), 3.59 (s, 1H), 3.17 (d, *J* = 5.2 Hz, 1H), 2.91 (q, *J* = 7.2 Hz, 1H), 2.75 (dd, *J* = 15.1, 7.2 Hz, 1H), 2.08 (dd, *J* = 13.7, 4.8 Hz, 1H), 1.99 (s, 1H), 1.95 (dd, *J* = 13.8, 4.3 Hz, 1H), 1.82 (dd, *J* = 15.1, 8.2 Hz, 1H), 1.72 (dd, *J* = 14.1, 4.8 Hz, 1H), 1.23 (s, 2H), 1.12 (d, *J* = 7.2 Hz, 3H), 1.03 (s, 9H). $^{13}$C-NMR (150 MHz, DMSO-$d_6$) δ 176.91, 172.75, 170.96, 110.08, 100.94, 87.95, 86.48, 85.31, 79.49, 78.95, 75.32, 68.67, 66.83, 57.81, 49.07, 49.03, 36.78, 36.45, 32.25, 29.24 (3C), 8.64. HRMS (ESI): *m/z* calcd for $C_{23}H_{26}O_9$ [M + H]+: 447.1650, found 447.1582.

*10-O-Propargylated ginkgolide C* (**3″**). Following the described procedure, 669 mg (69%) of compound **3″** were obtained from 861 mg (2.0 mmol) of ginkgolide A (**1″**). $^1$H-NMR (600 MHz, DMSO-$d_6$) δ 6.47 (s, 1H), 6.17 (s, 1H), 5.68 (d, *J* = 6.1 Hz, 1H), 5.26 (d, *J* = 5.0 Hz, 1H), 5.18 (s, 1H), 4.71 (ddd, *J* = 67.5, 15.8, 2.5 Hz, 2H), 4.59 (d, *J* = 6.6 Hz, 1H), 4.06–4.01 (m, 2H), 3.99 (ddd, *J* = 12.6 Hz, 6.1 Hz, 4.3 Hz, 1H), 3.63 (s, 1H), 2.82 (q, *J* = 7.1 Hz, 1H), 2.50 (s, 2H), 1.99 (s, 2H), 1.54 (d, *J* = 12.5 Hz, 1H), 1.18 (t, *J* = 7.1 Hz, 2H), 1.12 (s, 9H). $^{13}$C-NMR (150 MHz, DMSO-$d_6$) δ 176.73, 172.50, 170.90, 109.86, 99.26, 93.45, 83.06, 79.82, 79.29, 74.79, 74.17, 73.95, 67.05, 64.01, 60.23, 58.01, 49.26, 41.94, 32.20, 21.23 (3C), 8.48. HRMS (ESI): *m/z* calcd for $C_{23}H_{26}O_{11}$ [M + H]$^+$: 479.1548, found 479.1472.

3.1.3. General Procedures for the Preparation of 10-Substituted 1,2,3-triazole-Ginkgolide Derivatives

Sodium ascorbate (0.03 mmol) and $CuSO_4$ (0.01 mmol) were added in single portions to a solution of 10-O-propargylated ginkgolide **3**, **3′** or **3″** (0.1 mmol) and corresponding azide **4** (0.11 mmol) in 1:1:1 t-BuOH/ $H_2O$/THF (3 mL). The reaction mixture was stirred at room temperature for 48 h under an argon atmosphere and then was extracted with EtOAc three times. The combined organic phases were dried over anhydrous $MgSO_4$, filtered, and concentrated in vacuo. The residue was purified by chromatography ($SiO_2$, PE/EtOAc 1:3 stepwise elution to EtOAc) to afford the appropriate compound **5**.

*10-O-(1-Phenyl-1H-1,2,3-triazole) ginkgolide B* (**5a**). Following the described procedure, 42.3 mg (73%) of compound **5a** were obtained from 46.2 mg (0.1 mmol) of 10-O-propargylated ginkgolide B (**3**). $^1$H-NMR (400 MHz, DMSO-$d_6$) δ 8.798 (1H, s), 7.88 (d, *J* = 7.6 Hz, 2H), 7.50–7.64 (m, 3H), 6.47 (s, 1H), 6.20 (s, 1H), 5.52 (d, *J* = 4.4 Hz, 1H), 5.45 (d, *J* = 12.0 Hz, 1H), 4.93 (d, *J* = 12.0 Hz, 1H), 4.63 (d, *J* = 6.8 Hz, 1H), 4.21 (dd, *J* = 6.8 Hz, 4.8 Hz, 1H), 2.88 (q, *J* = 6.8 Hz, 1H), 2.11 (dd, *J* = 13.6 Hz, 4.0 Hz, 1H), 1.84 (ddd, *J* = 14.0 Hz, 13.6 Hz, 4.0 Hz, 1H), 1.73 (dd, *J* =14.4 Hz, 4.0 Hz, 1H), 1.13 (d, *J* = 7.2 Hz, 3H), 1.02 (s, 9H). $^{13}$C-NMR (125 MHz, DMSO-$d_6$) δ 176.32, 172.51, 170.12, 143.76, 136.46, 129.96 (2C), 128.94, 121.59, 120.23 (2C), 109.68, 98.88, 92.62, 82.47, 78.49, 75.16, 73.77, 71.96, 67.34, 63.00, 48.73, 41.56, 36.54, 31.73, 28.64 (3C), 7.84. HRMS (ESI): *m/z* calcd for $C_{29}H_{31}N_3O_{10}$ [M + H]$^+$: 582.2082, found 582.2062.

*10-O-(1-(2-Chlorophenyl)-1H-1,2,3-triazole) ginkgolide B* (**5b**). Following the described procedure, 21.0 mg (34%) of compound **5b** were obtained from 46.2 mg (0.1 mmol) of 10-O-propargylated ginkgolide B (**3**). $^1$H-NMR (400 MHz, DMSO-$d_6$) δ 7.78 (d, *J* = 1.2 Hz, 1H), 7.58–7.71 (m,2H), 6.47 (s, 1H), 6.20 (s, 1H), 5.50 (d, *J* = 4.8 Hz, 1H), 5.46 (d, *J* = 12.0 Hz, 1H), 5.33–5.35 (m, 2H), 4.97 (d, *J* = 12.0 Hz, 1H), 4.64 (d, *J* = 7.2 Hz, 1H), 4.21 (dd, *J* = 7.2 Hz, 4.8 Hz, 1H), 2.88 (q, *J* = 7.0 Hz, 1H), 2.12 (dd, *J* = 13.6 Hz, 4.0 Hz, 1H), 1.85 (ddd, *J* = 14.4 Hz, 13.4 Hz, 4.0 Hz, 1H), 1.74 (dd, *J* = 14.4 Hz, 4.0 Hz,1H), 1.12 (d,

*J* = 6.8 Hz, 3H), 1.02 (s, 9H), 176.30, 172.47, 170.10, 142.91, 134.30, 131.77, 130.60, 128.52, 128.39, 128.31, 125.43, 109.67, 98.86, 92.58, 82.46, 78.47, 75.25, 73.74, 71.93, 67.32, 63.07, 48.72, 41.55, 36.51, 31.72, 28.65 (3C), 8.57 (s,1H). $^{13}$C-NMR (125 MHz, DMSO-d6) $\delta$7.82, 7.80 (d, *J* = 1.2 Hz, 1H). HRMS (ESI): *m/z* calcd for $C_{29}H_{30}ClN_3O_{10}$ [M + H]$^+$: 616.1692, found 616.1670.

*10-O-(1-(3-Chlorophenyl)-1H-1,2,3-triazole) ginkgolide B* (**5c**). Following the described procedure, 40.1 mg (65%) of compound **5c** were obtained as method 1, from 46.2 mg (0.1 mmol) of 10-*O*-propargylated ginkgolide B (**3**).$^1$H-NMR (400 MHz, DMSO-$d_6$) $\delta$ 8.87 (s, 1H), 8.02 (s, 1H), 7.59–7.91 (m, 3H), 6.47 (s, 1H), 6.20 (s, 1H), 5.43–5.46 (m, 2H), 5.33 (s, 2H), 4.93 (d, *J* = 12.0 Hz, 1H), 4.64 (d, *J* = 6.8 Hz, 1H), 4.20 (dd, *J* = 7.0 Hz, 4.6 Hz, 1H), 2.88 (q, *J* = 6.8 Hz, 1H), 2.11 (dd, *J* = 13.0 Hz, 3.8 Hz, 1H), 1.86 (ddd, *J* = 14.2 Hz, 13.4 Hz, 3.8 Hz, 1H), 1.73 (dd, *J* = 14.4 Hz, 4.0 Hz, 1H), 1.12(d, *J* = 6.8 Hz, 3H), 1.02 (s, 9H). $^{13}$C-NMR (125 MHz, DMSO-$d_6$) $\delta$ 176.28, 172.46, 170.08, 143.82, 137.48, 134.20, 131.69, 128.74, 121.91, 120.02, 118.84, 109.66, 98.88, 92.60, 82.47, 78.45, 75.14, 73.75, 71.93, 67.29, 62.90, 48.71, 41.53, 36.51, 31.70, 28.62 (3C), 7.82. HRMS (ESI): *m/z* calcd for $C_{29}H_{30}ClN_3O_{10}$ [M + H]+: 616.1692, found 616.1674.

*10-O-(1-(4-Chlorophenyl)-1H-1,2,3-triazole) ginkgolide B* (**5d**). Following the described procedure, 20.3 mg (33%) of compound **5d** were obtained from 46.2 mg (0.1 mmol) of 10-*O*-propargylated ginkgolide B (**3**). $^1$H-NMR (400 MHz, DMSO-$d_6$) $\delta$8.82 (s,1H), 7.68–7.94 (dd, 4H), 6.47 (s, 1H), 6.20 (s, 1H), 5.43–5.48 (m, 2H), 5.32–5.33 (m, 2H), 4.92 (d, *J* = 8.0 Hz, 1H), 4.63 (d, *J* = 7.2 Hz, 1H), 4.20 (dd, *J* = 7.0 Hz, 4.6 Hz, 1H), 2.88 (q, *J* = 6.8 Hz, 1H), 2.11 (dd, *J* = 13.0 Hz, 3.8 Hz, 1H), 1.83 (ddd, *J* = 14.4Hz, 13.2Hz, 4.0 Hz, 1H), 1.73 (dd, *J* = 14.4 Hz, 4.0 Hz, 1H), 1.13 (d, *J* = 7.2 Hz, 3H), 1.02 (s, 9H). $^{13}$C-NMR (125 MHz, DMSO-$d_6$) $\delta$ 176.28, 172.46, 170.08, 143.87, 135.24, 133.19, 129.91 (2C), 121.90 (2C), 121.73, 109.66, 98.87, 92.60, 82.46, 78.45, 75.16, 73.74, 71.92, 67.30, 62.93, 48.70, 41.53, 36.51, 31.70, 28.62 (3C), 7.81. HRMS (ESI): *m/z* calcd for $C_{29}H_{30}ClN_3O_{10}$ [M + H]+: 616.1692, found 616.1671.

*10-O-(1-(2,6-Dichlorophenyl)-1H-1,2,3-triazole) ginkgolide B* (**5e**). Following the described procedure, 34.5 mg (53%) of compound **5e** were obtained from 46.2 mg (0.1 mmol) of 10-*O*-propargylated ginkgolide B (**3**). $^1$H-NMR (400 MHz, DMSO-$d_6$) $\delta$8.54 (s,1H), 7.68–7.81(m, 3H), 6.48 (s, 1H), 6.20 (s, 1H), 5.47 (d, *J* = 12.0 Hz, 1H), 5.32–5.54 (m, 2H), 5.24 (d, *J* = 4.4 Hz, 1H), 4.99 (d, *J* = 12.4 Hz, 1H), 4.64 (d, *J* = 7.2 Hz, 1H), 4.19 (dd, *J* = 7.2 Hz, 4.6 Hz, 1H), 2.87 (q, *J* = 7.2 Hz, 1H), 2.13 (dd, *J* = 13.2 Hz, 4.0 Hz, 1H), 1.84 (ddd, *J* = 14.2 Hz, 13.4 Hz, 4.0 Hz, 1H), 1.74 (dd, *J* = 14.4 Hz, 4.0 Hz, 1H), 1.12 (d, *J* = 7.2 Hz, 3H), 1.03 (s, 9H). $^{13}$C-NMR (125 MHz, DMSO-$d_6$) $\delta$ 176.27, 172.38, 170.07, 143.11, 133.04, 132.58 (2C), 132.42, 129.26 (2C), 125.88, 109.68, 98.82, 92.40, 82.48, 78.45, 75.51, 73.74, 71.91, 67.28, 63.20, 48.68, 41.54, 36.50, 31.74, 28.69 (3C), 7.80. HRMS (ESI): *m/z* calcd for $C_{29}H_{29}Cl_2N_3O_{10}$ [M + H]$^+$: 650.1303, found 650.1287.

*10-O-(1-(2,4-Dichlorophenyl)-1H-1,2,3-triazole) ginkgolide B* (**5f**). Following the described procedure, 29.3 mg (45%) of compound **5f** were obtained from 46.2 mg (0.1 mmol) of 10-*O*-propargylated ginkgolide B (**3**). $^1$H-NMR (400 MHz, DMSO-$d_6$) $\delta$ 8.57 (s,1H), 8.02(d, *J* = 2.4 Hz, 1H), 7.68–7.76(m, 2H), 6.47 (s, 1H), 6.20 (s, 1H), 5.43–5.48 (m, 2H), 5.33–5.40 (m, 2H), 4.96 (d, *J* = 12.4Hz, 1H), 4.63 (d, *J* = 7.2 Hz, 1H), 4.20 (dd, *J* = 7.2 Hz, 4.8Hz, 1H), 2.88 (q, *J* = 6.8 Hz, 1H), 2.12 (dd, *J* = 12.8 Hz, 4.0Hz, 1H), 1.84 (ddd, *J* = 14.2 Hz, 13.2 Hz, 4.0 Hz, 1H), 1.74 (dd, *J* = 14.4 Hz, 4.0 Hz, 1H), 1.12(d, *J* = 6.8 Hz, 3H), 1.02 (s, 9H).$^{13}$C-NMR (125 MHz, DMSO-$d_6$) $\delta$ 176.27, 172.44, 170.07, 143.02, 135.51, 133.34, 130.16, 129.64, 129.49, 128.65, 125.51, 109.65, 98.85, 92.58, 82.45, 78.44, 75.27, 73.71, 71.91, 67.30, 63.03, 59.71, 48.69, 41.53, 36.49, 31.71, 28.64 (3C), 7.81. HRMS (ESI): *m/z* calcd for $C_{29}H_{29}Cl_2N_3O_{10}$ [M + H]$^+$: 650.1303, found 650.1289.

*10-O-(1-(3,5-Dichlorophenyl)-1H-1,2,3-triazole) ginkgolide B* (**5g**). Following the described procedure, 38.3 mg (34%) of compound **5g** were obtained from 46.2 mg (0.1 mmol) of 10-*O*-propargylated ginkgolide B (**3**). $^1$H-NMR (400 MHz, DMSO-$d_6$) $\delta$ 8.92 (s,1H), 8.05 (d, *J* = 1.6 Hz, 2H), 7.80 (t, *J* = 1.6 Hz, 1H), 6.48 (s, 1H), 6.20 (s, 1H), 5.44 (d, *J* = 12.0 Hz,1H), 5.36 (d, *J* = 4.8 Hz, 1H), 5.32–5.35 (m, 2H), 4.93 (d, *J* = 12.0 Hz, 1H), 4.63 (d, *J* = 7.2 Hz, 1H), 4.20 (dd, *J* = 7.2 Hz, 4.6Hz, 1H), 2.87 (q, *J* = 7.2 Hz, 1H), 2.11 (dd, *J* = 12.8 Hz, 4.0 Hz, 1H), 1.83 (ddd, *J* = 14.4 Hz, 13.2 Hz, 4.0 Hz, 1H), 1.73 (dd, *J* = 14.4 Hz, 4.0 Hz, 1H), 1.12 (d, *J* = 7.2 Hz, 3H), 1.02 (s, 9H). $^{13}$C-NMR (125 MHz, DMSO-$d_6$) $\delta$ 176.28, 172.43, 170.08, 143.90, 138.05, 135.26 (2C), 128.27, 122.22, 118.98, 118.88, 109.66, 98.91, 92.60, 82.49, 78.44,

75.17, 73.75, 71.93, 67.28, 62.84, 48.71, 41.52, 36.51, 31.71, 28.62 (3C), 7.84. HRMS (ESI): *m/z* calcd for $C_{29}H_{29}Cl_2N_3O_{10}$ $[M + H]^+$: 650.1303, found 650.1287.

*10-O-(1-(3-Fluorophenyl)-1H-1,2,3-triazole) ginkgolide B* (**5h**). Following the described procedure, 32.4 mg (54%) of compound **5h** were obtained from 46.2 mg (0.1 mmol) of 10-*O*-propargylated ginkgolide B (**3**).[1]H-NMR (400 MHz, DMSO-$d_6$) δ 7.99 (s, 1H), 7.16–7.54 (m, 4H), 6.01 (s, 1H), 5.67 (d, *J* = 11.6 Hz, 1H), 5.53 (d, *J* = 3.6 Hz, 1H), 5.05 (d, *J* = 4.4 Hz, 1H), 4.96 (s, 1H), 4.91 (d, *J* = 11.2 Hz, 1H), 4.64 (d, *J* = 7.6 Hz, 1H), 4.40 (dd, *J* = 7.6 Hz, 4.4 Hz, 1H), 3.07 (q, *J* = 6.8 Hz, 1H), 2.84 (s, 1H), 2.28 (dd, *J* = 12.8 Hz, 4.2 Hz, 1H), 2.05 (ddd, *J* = 16.8 Hz, 10.8 Hz, 4.0 Hz, 1H), 1.98 (dd, *J* = 14.2 Hz, 4.2 Hz, 1H), 1.30 (d, *J* = 6.8 Hz, 3H), 1.13 (s, 9H). [13]C-NMR (125 MHz, DMSO-$d_6$) δ 176.28, 172.46, 170.08, 143.86, 137.64, 131.93, 121.85, 116.14, 115.66, 109.66, 107.83, 107.62, 92.60, 82.47, 78.45, 75.17, 73.75, 71.93, 67.30, 62.92, 48.71, 41.53, 36.51, 31.71, 28.62 (3C), 7.82. HRMS (ESI): *m/z* calcd for $C_{29}H_{31}FN_3O_{10}$ $[M + H]^+$: 600.1993, found 600.2014.

*10-O-(1-(4-Chloro-3-trifluoromethylphenyl)-1H-1,2,3-triazole) ginkgolide B* (**5i**). Following the described procedure, 27.9 mg (41%) of compound **5i** were obtained from 46.2 mg (0.1 mmol) of 10-*O*-propargylated ginkgolide B (**3**). [1]H-NMR (400 MHz, DMSO-$d_6$) δ 8.07 (s,1H), 7.70–8.09 (m, 3H), 6.01 (s, 1H,), 5.65 (d, *J* = 11.6 Hz, 1H), 5.52 (d, *J* = 4.0 Hz, 1H), 5.06 (d, *J* = 4.0 Hz, 1H), 4.97 (s, 1H), 4.91 (d, *J* = 11.6 Hz, 1H), 4.62 (d, *J* = 8.0 Hz, 1H), 4.38 (dd, *J* = 7.6 Hz, 4.4 Hz, 1H), 3.07 (q, *J* = 7.0 Hz, 1H), 2.98 (s, 1H), 2.28 (dd, *J* = 12.8 Hz, 4.0 Hz, 1H), 2.04 (ddd, *J* = 14.0 Hz, 13.2 Hz, 4.0 Hz, 1H), 1.95 (dd, *J* = 14.4 Hz, 4.0 Hz, 1H), 1.30 (d, *J* = 6.8 Hz, 3H), 1.12 (s, 9H). [13]C-NMR (125 MHz, DMSO-$d_6$) δ 176.27, 172.44, 170.07, 143.99, 135.40, 133.39, 130.69, 128.15, 127.83, 125.51, 122.25, 119.60, 109.66, 98.88, 92.56, 82.47, 78.43, 75.19, 73.74, 71.93, 67.28, 62.84, 48.70, 41.53, 36.50, 31.70, 28.62 (3C), 7.82. HRMS (ESI): *m/z* calcd for $C_{30}H_{29}ClF_3N_3O_{10}$ $[M + H]^+$: 684.1566, found 684.1544.

*10-O-(1-(4-Trifluoromethoxyphenyl)-1H-1,2,3-triazole) ginkgolide B* (**5j**). Following the described procedure, 20.6 mg (31%) of compound **5j** were obtained from 46.2 mg (0.1 mmol) of 10-*O*-propargylated ginkgolide B (**3**). [1]H-NMR (400 MHz, DMSO-$d_6$) δ 8.83 (s,1H), 8.03 (d, *J* = 9.0 Hz, 2H), 7.65 (d, *J* = 8.4 Hz, 2H), 6.47 (s, 1H), 6.20 (s, 1H), 5.43–5.49 (m, 2H), 5.32–5.34 (m, 2H), 4.93 (d, *J* = 12.0 Hz, 1H), 4.62 (d, *J* = 7.2 Hz, 1H), 4.21 (dd, *J* = 7.2 Hz, 4.6 Hz, 1H), 2.88 (q, *J* = 7.0 Hz, 1H), 2.11 (dd, *J* = 13.2 Hz, 4.0 Hz, 1H), 1.84 (ddd, *J* = 14.2 Hz, 13.4 Hz, 4.0 Hz, 1H), 1.73 (dd, *J* = 14.4 Hz, 4.0 Hz, 1H), 1.13 (d, *J* = 7.2 Hz, 3H), 1.02 (s, 9H). [13]C-NMR (125 MHz, DMSO-$d_6$) δ 176.28, 172.46, 170.08, 148.02, 143.91, 135.30, 122.63 (2C), 122.30 (2C), 121.97, 121.00, 109.66, 98.87, 92.61, 82.46, 78.45, 75.17, 73.74, 71.93, 67.30, 62.94, 48.71, 41.53, 36.51, 31.70, 28.62 (3C), 7.81. HRMS (ESI): *m/z* calcd for $C_{30}H_{30}F_3N_3O_{11}$ $[M + H]^+$: 666.1905, found 666.1887.

*10-O-(1-(3,5-Ditrifluoromethylphenyl)-1H-1,2,3-triazole) ginkgolide B* (**5k**). Following the described procedure, 32.3 mg (45%) of compound **5k** were obtained from 46.2 mg (0.1 mmol) of 10-*O*-propargylated ginkgolide B (**3**). [1]H-NMR (400 MHz, DMSO-$d_6$) δ 9.14 (s, 1H), 8.62 (s, 2H), 8.31 (s, 1H), 6.48 (s, 1H), 6.21 (s, 1H), 5.47 (d, *J* = 12.0 Hz, 1H), 5.35 (s, 1H), 5.30–5.33 (m, 2H), 4.96 (d, *J* = 12.0 Hz, 1H), 4.64 (d, *J* = 7.2 Hz, 1H), 4.21 (dd, *J* = 7.2 Hz, 4.8 Hz, 1H), 2.88 (q, *J* = 6.8 Hz, 1H), 2.12 (dd, *J* = 12.8 Hz, 4.0 Hz,1H), 1.71–1.87 (m, 2H), 1.13 (d, *J* = 6.8 Hz, 3H), 1.03 (s, 9H). [13]C-NMR (125 MHz, DMSO-$d_6$) δ 176.28, 172.44, 170.07, 144.01, 137.73, 132.00 (2C), 131.73 (2C), 123.81, 122.64, 121.64, 121.07, 109.68, 98.90, 92.54, 82.49, 78.42, 75.22, 73.77, 71.94, 62.78, 48.71, 41.53, 36.50, 31.71, 28.63 (3C), 7.83. HRMS (ESI): *m/z* calcd for $C_{31}H_{29}F_6N_3O_{10}$ $[M + H]^+$: 718.1830, found 718.1808.

*10-O-(1-(4-tert-Butylphenyl)-1H-1,2,3-triazole) ginkgolide B* (**5l**). Following the described procedure, 35.0 mg (55%) of compound **5l** were obtained from 46.2 mg (0.1 mmol) of 10-*O*-propargylated ginkgolide B (**3**). [1]H-NMR (400 MHz, DMSO-$d_6$) δ 8.75 (s, 1H), 7.78 (d, *J* = 8.4 Hz, 2H), 7.62 (d, *J* = 8.8 Hz, 2H), 6.47 (s, 1H), 6.20 (s, 1H), 5.58 (d, *J* = 4.4 Hz, 1H), 5.44 (d, *J* = 12.0 Hz, 1H), 5.33 (s, 2H), 4.93 (d, *J* = 12.0 Hz, 1H), 4.63 (d, *J* = 7.2 Hz, 1H), 4.21 (dd, *J* = 7.2 Hz, 4.6 Hz, 1H), 2.88 (q, *J* = 7.2 Hz, 1H), 2.11 (dd, *J* = 13.2 Hz, 4.0 Hz, 1H), 1.83 (ddd, *J* = 14.4 Hz, 13.2 Hz, 4.0 Hz, 1H), 1.73 (dd, *J* = 14.4 Hz, 3.8 Hz, 1H), 1.33 (s, 9H), 1.13 (d, *J* = 7.2 Hz, 3H), 1.01 (s, 9H). [13]C-NMR (125 MHz, DMSO-$d_6$) δ 176.29, 172.50, 170.08, 150.60, 143.60, 134.13, 126.64 (2C), 121.49, 119.98 (2C), 109.65, 98.84, 92.60, 82.43, 78.46, 75.09, 73.73,

71.93, 67.32, 62.97, 48.71, 41.54, 36.51, 34.50, 31.70, 30.95 (3C), 28.62 (3C), 7.80. HRMS (ESI): *m/z* calcd for $C_{33}H_{39}N_3O_{10}$ [M + H]$^+$: 638.2708, found 638.2689.

*10-O-(1-(3-Bromophenyl)-1H-1,2,3-triazole) ginkgolide B* (**5m**). Following the described procedure, 38.3 mg (58%) of compound **5m** were obtained from 46.2 mg (0.1 mmol) of 10-*O*-propargylated ginkgolide B (**3**). $^1$H-NMR (400 MHz, DMSO-$d_6$) δ 8.87 (s, 1H), 8.14 (t, *J* = 2.0 Hz, 1H), 7.94 (dd, *J* = 8.2 Hz, 1.4 Hz, 1H), 7.73 (d, *J* = 8.4 Hz, 1H), 7.58 (t, *J* = 8.0 Hz, 1H), 6.47 (s, 1H), 6.20 (s, 1H), 5.43–5.46 (m, 2H), 5.33 (s, 2H), 4.93 (d, *J* = 12.0 Hz, 1H), 4.63 (d, *J* = 7.2 Hz, 1H), 4.20 (dd, *J* = 7.2 Hz, 4.6 Hz, 1H), 2.88 (q, *J* = 7.2 Hz, 1H), 2.11 (dd, *J* = 13.2 Hz, 4.0 Hz, 1H), 1.83 (ddd, *J* = 14.4 Hz, 13.2 Hz, 4.0 Hz, 1H), 1.73 (dd, *J* = 14.4 Hz, 4.2 Hz, 1H), 1.12 (d, *J* = 7.2 Hz, 3H), 1.01 (s, 9H). $^{13}$C-NMR (125 MHz, DMSO-$d_6$) δ 176.28, 172.46, 170.08, 143.79, 137.55, 131.90, 131.66, 122.77, 122.43, 121.92, 119.23, 109.66, 98.88, 92.60, 82.47, 78.45, 75.13, 73.75, 71.93, 67.29, 62.89, 48.71, 41.53, 36.51, 31.70, 28.62 (3C), 7.83. HRMS (ESI): *m/z* calcd for $C_{29}H_{31}BrN_3O_{10}$ [M + H]$^+$: 660.1193, found 660.1207.

*10-O-(1-(2-Methylphenyl)-1H-1,2,3-triazole) ginkgolide B* (**5n**). Following the described procedure, 39.3 mg (66%) of compound **5n** were obtained from 46.2 mg (0.1 mmol) of 10-*O*-propargylated ginkgolide B (**3**). $^1$H-NMR (600 MHz, DMSO-$d_6$) δ 8.46 (s, 1H), 7.52–7.47 (m, 2H), 7.46–7.41 (m, 2H), 6.47 (s, 1H), 6.20 (s, 1H), 5.57 (d, *J* = 4.6 Hz, 1H), 5.45 (d, *J* = 12.2 Hz, 1H), 5.35–5.33 (m, 2H), 4.97 (d, *J* = 12.1 Hz, 1H), 4.64 (d, *J* = 7.2 Hz, 1H), 4.21 (dd, *J* = 7.2, 4.7 Hz, 1H), 2.89 (q, *J* = 7.0 Hz, 1H), 2.15 (s, 3H), 2.13 (dd, *J* = 13.7, 4.5 Hz, 1H), 1.91–1.80 (m, 1H), 1.75 (dd, *J* = 14.4 Hz, 4.4 Hz, 1H), 1.13 (d, *J* = 7.1 Hz, 3H), 1.03 (s, 9H). $^{13}$C-NMR (150 MHz, DMSO-$d_6$) δ 176.83, 173.02, 170.63, 143.35, 136.56, 133.54, 131.90, 130.45, 127.53, 126.45, 125.28, 110.16, 99.33, 93.08, 82.94, 79.01, 75.62, 74.23, 72.43, 67.83, 63.57, 49.20, 41.90, 37.01, 32.22, 29.14 (3C), 17.84, 8.32. HRMS (ESI): *m/z* calcd for $C_{30}H_{34}N_3O_{10}$ [M + H]+: 596.2244, found 596.2249.

*10-O-(1-(2-Hydroxyphenyl)-1H-1,2,3-triazole) ginkgolide B* (**5o**). Following the described procedure, 22.1 mg (37%) of compound **5o** were obtained from 46.2 mg (0.1 mmol) of 10-*O*-propargylated ginkgolide B (**3**). $^1$H-NMR (600 MHz, DMSO-$d_6$) δ 10.59 (s, 1H), 8.51 (s, 1H), 7.60 (dd, *J* = 7.9, 1.6 Hz, 1H), 7.36 (ddd, *J* = 8.2, 7.5, 1.7 Hz, 1H), 7.12 (dd, *J* = 8.2, 1.2 Hz, 1H), 7.00 (td, *J* = 7.8, 1.3 Hz, 1H), 6.46 (s, 1H), 6.20 (s, 1H), 5.64 (d, *J* = 4.7 Hz, 1H), 5.45 (d, *J* = 12.2 Hz, 1H), 5.34–5.32 (m, 2H), 4.93 (d, *J* = 12.1 Hz, 1H), 4.63 (d, *J* = 7.2 Hz, 1H), 4.22 (dd, *J* = 7.2, 4.5 Hz, 1H), 2.92–2.84 (m, 1H), 2.12 (dd, *J* = 13.3 Hz, 4.4 Hz, 1H), 1.84 (td, *J* = 13.8 Hz, 4.2 Hz, 1H), 1.74 (dd, *J* = 14.3 Hz, 4.3 Hz, 1H), 1.13 (d, *J* = 7.1 Hz, 3H), 1.02 (s, 9H). $^{13}$C-NMR (150 MHz, DMSO-$d_6$) δ 175.74, 171.96, 169.55, 149.06, 141.97, 129.74, 124.60, 124.12, 123.73, 118.97, 116.45, 109.07, 98.23, 91.99, 81.83, 77.91, 74.66, 73.13, 71.35, 66.74, 62.61, 48.10, 40.98, 35.93, 31.15, 28.07 (3C), 7.23. HRMS (ESI): *m/z* calcd for $C_{29}H_{32}N_3O_{11}$ [M + H]$^+$: 598.2037, found 598.2051.

*10-O-(1-(2-Cyanophenyl)-1H-1,2,3-triazole) ginkgolide B* (**5p**). Following the described procedure, 25.5 mg (42%) of compound **5p** were obtained from 46.2 mg (0.1 mmol) of 10-*O*-propargylated ginkgolide B (**3**). $^1$H-NMR (600 MHz, DMSO-$d_6$) δ 8.74 (s, 1H), 8.16 (dd, *J* = 7.8, 1.3 Hz, 1H), 7.98 (td, *J* = 7.9, 1.4 Hz, 1H), 7.90–7.86 (m, 1H), 7.79 (td, *J* = 7.7, 1.1 Hz, 1H), 6.49 (s, 1H), 6.20 (s, 1H), 5.48 (d, *J* = 12.4 Hz, 2H), 5.37–5.30 (m, 2H), 4.99 (d, *J* = 12.3 Hz, 1H), 4.64 (d, *J* = 7.1 Hz, 1H), 4.22 (d, *J* = 7.1 Hz, 1H), 2.92–2.85 (m, 1H), 2.11 (dd, *J* = 13.5, 4.5 Hz, 1H), 1.86 (dt, *J* = 13.7, 7.0 Hz, 1H), 1.74 (dd, *J* = 14.4, 4.5 Hz, 1H), 1.13 (d, *J* = 7.1 Hz, 3H), 1.02 (s, 9H). $^{13}$C-NMR (150 MHz, DMSO-$d_6$) δ 176.81, 172.97, 170.63, 144.21, 138.10, 135.41, 135.30, 130.92, 126.18, 124.91, 116.22, 110.16, 107.43, 99.41, 93.18, 82.97, 79.00, 75.76, 74.22, 72.41, 67.82, 63.48, 49.20, 42.04, 37.01, 32.22, 29.13 (3C), 8.35. HRMS (ESI): *m/z* calcd for $C_{30}H_{31}N_4O_{10}$ [M + H]$^+$: 607.2040, found 607.2050.

*10-O-(1-(3-Methylphenyl)-1H-1,2,3-triazole) ginkgolide B* (**5q**). Following the described procedure, 26.8 mg (45%) of compound **5q** were obtained from 46.2 mg (0.1 mmol) of 10-*O*-propargylated ginkgolide B (**3**). $^1$H-NMR (600 MHz, DMSO-$d_6$) δ 8.78 (s, 1H), 7.72 (t, *J* = 2.6 Hz, 1H), 7.68–7.63 (m, 1H), 7.50 (t, *J* = 7.8 Hz, 1H), 7.34 (d, *J* = 7.6 Hz, 1H), 6.47 (s, 1H), 6.20 (s, 1H), 5.50 (d, *J* = 4.5 Hz, 1H), 5.44 (d, *J* = 12.1 Hz, 1H), 5.33 (d, *J* = 4.8 Hz, 2H), 4.93 (d, *J* = 12.1 Hz, 1H), 4.64 (d, *J* = 7.1 Hz, 1H), 4.22–4.18 (m, 1H), 2.88 (q, *J* = 7.1 Hz, 1H), 2.42 (s, 3H), 2.11 (dd, *J* = 13.4 Hz, 4.4 Hz, 1H), 1.83 (td, *J* = 13.8 Hz, 4.2 Hz, 1H), 1.74 (dd, *J* = 14.4 Hz, 4.3 Hz, 1H), 1.13 (d, *J* = 7.1 Hz, 3H), 1.02 (s, 9H). $^{13}$C-NMR (150 MHz, DMSO-$d_6$) δ 175.72, 171.93, 169.53, 143.06, 139.14, 135.83, 129.16, 128.93, 120.98, 120.04, 116.73,

109.08, 98.26, 92.00, 81.87, 77.89, 74.53, 73.16, 71.35, 66.72, 62.36, 48.11, 40.97, 35.93, 31.13, 28.04 (3C), 20.32, 7.25. HRMS (ESI): *m/z* calcd for $C_{30}H_{34}N_3O_{10}$ [M + H]$^+$: 596.2244, found 596.2255.

*10-O-(1-(3-Isopropylphenyl)-1H-1,2,3-triazole) ginkgolide B* (**5r**). Following the described procedure, 34.3 mg (55%) of compound **5r** were obtained from 46.2 mg (0.1 mmol) of 10-*O*-propargylated ginkgolide B (**3**). $^1$H-NMR (600 MHz, DMSO-$d_6$) δ 8.81 (s, 1H), 7.75 (t, *J* = 1.8 Hz, 1H), 7.68–7.67 (m, 1H), 7.53 (t, *J* = 7.9 Hz, 1H), 7.41 (d, *J* = 7.7 Hz, 1H), 6.50 (s, 1H), 6.20 (s, 1H), 5.54 (d, *J* = 4.6 Hz, 1H), 5.45 (d, *J* = 12.1 Hz, 1H), 5.34 (d, *J* = 3.4 Hz, 2H), 4.94 (d, *J* = 12.1 Hz, 1H), 4.64 (d, *J* = 7.1 Hz, 1H), 4.22 (dd, *J* = 7.1 Hz, 4.5 Hz, 1H), 3.06–2.99 (m, 1H), 2.89 (q, *J* = 7.0 Hz, 1H), 2.12 (dd, *J* = 13.4 Hz, 4.4 Hz, 1H), 1.83 (td, *J* = 13.8, 4.2 Hz, 1H), 1.74 (dd, *J* = 14.4 Hz, 4.3 Hz, 1H), 1.27 (d, *J* = 6.9 Hz, 6H), 1.13 (d, *J* = 7.1 Hz, 3H), 1.02 (s, 9H). $^{13}$C-NMR (150 MHz, DMSO-$d_6$) δ 176.83, 173.02, 170.63, 150.14, 144.04, 136.99, 130.39, 127.50, 122.12, 118.67, 118.27, 110.16, 99.35, 93.06, 82.94, 78.99, 75.60, 74.25, 72.44, 67.81, 63.41, 49.19, 42.04, 37.00, 33.87, 32.20, 29.11 (3C), 24.11 (2C), 8.32. HRMS (ESI): *m/z* calcd for $C_{32}H_{38}N_3O_{10}$ [M + H]$^+$: 624.2557, found 624.2565.

*10-O-(1-(3-Ethylcarboxyphenyl)-1H-1,2,3-triazole) ginkgolide B* (**5s**). Following the described procedure, 17.6 mg (27%) of compound **5s** were obtained from 46.2 mg (0.1 mmol) of 10-*O*-propargylated ginkgolide B (**3**). $^1$H-NMR (600 MHz, DMSO-$d_6$) δ 8.94 (s, 1H), 8.40 (s, 1H), 8.18 (dd, *J* = 8.1, 0.8 Hz, 1H), 8.08 (dd, *J* = 7.8, 0.9 Hz, 1H), 7.79 (t, *J* = 8.0 Hz, 1H), 6.49 (s, 1H), 6.21 (s, 1H), 5.48 (d, *J* = 4.6 Hz, 1H), 5.46 (d, *J* = 12.2 Hz, 1H), 5.36–5.31 (m, 2H), 4.95 (d, *J* = 12.1 Hz, 1H), 4.64 (d, *J* = 7.1 Hz, 1H), 4.38 (q, *J* = 7.1 Hz, 2H), 4.22 (dd, *J* = 7.1, 4.6 Hz, 1H), 2.89 (q, *J* = 7.0 Hz, 1H), 2.11 (dd, *J* = 13.4 Hz, 4.4 Hz, 1H), 1.84 (td, *J* = 13.8 Hz, 4.2 Hz, 1H), 1.74 (dd, *J* = 14.4 Hz, 4.4 Hz, 1H), 1.36 (t, *J* = 7.1 Hz, 3H), 1.13 (d, *J* = 7.1 Hz, 3H), 1.02 (s, 9H). $^{13}$C-NMR (150 MHz, DMSO-$d_6$) δ 176.81, 173.00, 170.61, 165.21, 144.41, 137.17, 132.08, 131.16, 129.77, 125.17, 122.46, 120.88, 110.16, 99.37, 93.11, 82.96, 78.98, 75.64, 74.23, 72.43, 67.80, 63.39, 61.85, 49.20, 42.04, 37.00, 32.21, 29.12 (3C), 14.59, 8.34. HRMS (ESI): *m/z* calcd for $C_{32}H_{36}N_3O_{12}$ [M + H]$^+$: 654.2299, found 654.2309.

*10-O-(1-(3-Carboxyphenyl)-1H-1,2,3-triazole) ginkgolide B* (**5t**). Following the described procedure, 15.6 mg (25%) of compound **5t** were obtained from 46.2 mg (0.1 mmol) of 10-*O*-propargylated ginkgolide B (**3**). $^1$H-NMR (600 MHz, DMSO-$d_6$) δ 13.45 (s, 1H), 8.94 (s, 1H), 8.38 (s, 1H), 8.16–8.14 (m, 1H), 8.11–7.99 (m, 1H), 7.76 (t, *J* = 7.9 Hz, 1H), 6.47 (s, 1H), 6.21 (s, 1H), 5.47–5.44 (m, 2H), 5.35–5.28 (m, 2H), 4.94 (d, *J* = 12.1 Hz, 1H), 4.64 (d, *J* = 7.1 Hz, 1H), 4.21 (dd, *J* = 7.1, 4.6 Hz, 1H), 2.89 (q, *J* = 7.1 Hz, 1H), 2.11 (dd, *J* = 13.4 Hz, 4.4 Hz, 1H), 1.84 (td, *J* = 13.8 Hz, 4.2 Hz, 1H), 1.74 (dd, *J* = 14.4 Hz, 4.4 Hz, 1H), 1.13 (d, *J* = 7.1 Hz, 3H), 1.02 (s, 9H). $^{13}$C-NMR (150 MHz, DMSO-$d_6$) δ 176.81, 173.01, 170.62, 166.75, 144.21, 136.99, 133.01, 130.95, 129.88, 124.72, 122.30, 120.85, 110.07, 99.24, 92.99, 82.94, 78.93, 75.54, 74.21, 72.38, 67.74, 63.34, 49.20, 42.04, 36.74, 32.20, 29.12 (3C), 8.26. HRMS (ESI): *m/z* calcd for $C_{30}H_{32}N_3O_{12}$ [M + H]$^+$: 626.1986, found 626.1998.

*10-O-(1-(4-Methylphenyl)-1H-1,2,3-triazole) ginkgolide B* (**5u**). Following the described procedure, 31.5 mg (53%) of compound **5u** were obtained from 46.2 mg (0.1 mmol) of 10-*O*-propargylated ginkgolide B (**3**). $^1$H-NMR (600 MHz, DMSO-$d_6$) δ 8.75 (s, 1H), 7.76 (d, *J* = 8.4 Hz, 2H), 7.42 (d, *J* = 8.1 Hz, 2H), 6.46 (s, 1H), 6.20 (s, 1H), 5.53 (d, *J* = 4.6 Hz, 1H), 5.44 (d, *J* = 12.2 Hz, 1H), 5.33 (d, *J* = 3.2 Hz, 2H), 4.92 (d, *J* = 12.1 Hz, 1H), 4.63 (d, *J* = 7.1 Hz, 1H), 4.21 (dd, *J* = 7.2, 4.6 Hz, 1H), 2.88 (q, *J* = 7.0 Hz, 1H), 2.39 (s, 3H), 2.11 (dd, *J* = 13.4 Hz, 4.4 Hz, 1H), 1.83 (td, *J* = 13.8 Hz, 4.2 Hz, 1H), 1.73 (dd, *J* = 14.4 Hz, 4.3 Hz, 1H), 1.13 (d, *J* = 7.1 Hz, 3H), 1.02 (s, 9H). $^{13}$C-NMR (150 MHz, DMSO-$d_6$) δ 175.72, 171.94, 169.52, 142.94, 137.81, 133.34, 129.57 (2C), 120.81, 119.40 (2C), 109.01, 98.21, 91.97, 81.84, 77.84, 74.37, 73.03, 71.26, 66.65, 62.28, 47.99, 40.81, 35.83, 30.92, 28.04 (3C), 19.99, 7.01. HRMS (ESI): *m/z* calcd for $C_{30}H_{34}N_3O_{10}$ [M + H]$^+$: 596.2244, found 596.2251.

*10-O-(1-(4-Methoxyphenyl)-1H-1,2,3-triazole) ginkgolide B* (**5v**). Following the described procedure, 20.8 mg (34%) of compound **5v** were obtained from 46.2 mg (0.1 mmol) of 10-*O*-propargylated ginkgolide B (**3**). $^1$H-NMR (600 MHz, DMSO-$d_6$) δ 8.69 (s, 1H), 7.78 (d, *J* = 9.0 Hz, 2H), 7.15 (d, *J* = 9.1 Hz, 2H), 6.47 (s, 1H), 6.20 (s, 1H), 5.57 (d, *J* = 4.6 Hz, 1H), 5.44 (d, *J* = 12.1 Hz, 1H), 5.33 (s, 2H), 4.92 (d, *J* = 12.1 Hz, 1H), 4.63 (d, *J* = 7.2 Hz, 1H), 4.21 (dd, *J* = 7.2, 4.6 Hz, 1H), 3.84 (s, 3H), 2.88 (q, *J* = 7.1 Hz, 1H), 2.11 (dd, *J* = 13.4 Hz, 4.4 Hz, 1H), 1.83 (td, *J* = 13.8 Hz, 4.2 Hz, 1H), 1.74 (dd, *J* = 14.4

Hz, 4.3 Hz, 1H), 1.13 (d, $J$ = 7.1 Hz, 3H), 1.02 (s, 9H). $^{13}$C-NMR (150 MHz, DMSO-$d_6$) $\delta$ 176.83, 173.02, 170.63, 159.92, 144.00, 130.31, 122.40 (2C), 121.94, 115.42 (2C), 110.16, 99.34, 93.08, 82.93, 79.00, 75.43, 74.13, 72.34, 67.75, 63.27, 55.84, 49.09, 41.96, 36.94, 32.01, 28.95 (3C), 8.19. HRMS (ESI): *m/z* calcd for $C_{30}H_{34}N_3O_{11}$ [M + H]$^+$: 612.2193, found 612.2205.

*10-O-(1-(4-Fluorophenyl)-1H-1,2,3-triazole) ginkgolide B* (**5w**). Following the described procedure, 13.5 mg (22%) of compound **5w** were obtained from 46.2 mg (0.1 mmol) of 10-*O*-propargylated ginkgolide B (3). $^1$H-NMR (600 MHz, DMSO-$d_6$) $\delta$ 8.69 (s, 1H), 7.93–7.89 (m, 2H), 7.49–7.44 (m, 2H), 6.64 (s, 1H), 6.19 (s, 1H), 5.63 (d, $J$ = 4.5 Hz, 1H), 5.46 (d, $J$ = 12.2 Hz, 1H), 5.37 (d, $J$ = 4.1 Hz, 1H), 5.31 (s, 1H), 4.97 (d, $J$ = 12.2 Hz, 1H), 4.65 (d, $J$ = 7.1 Hz, 1H), 4.24 (dd, $J$ = 7.1 Hz, 4.4 Hz, 1H), 2.91 (q, $J$ = 7.1 Hz, 1H), 2.14 (dd, $J$ = 13.5 Hz, 4.5 Hz, 1H), 1.86 (td, $J$ = 13.9 Hz, 4.2 Hz, 1H), 1.75 (dd, $J$ = 14.4 Hz, 4.5 Hz, 1H), 1.15 (d, $J$ = 7.1 Hz, 3H), 1.03 (s, 9H). $^{13}$C-NMR (150 MHz, DMSO-$d_6$) $\delta$ 177.01, 172.98, 170.81, 162.28 (d, $J_{C-F}$ = 246.4 Hz), 144.27, 133.33 (d, $J_{C-F}$ = 2.5 Hz), 123.18 (d, $J_{C-F}$ = 8.9 Hz, 2C), 122.14, 117.30 (d, $J_{C-F}$ = 23.3 Hz, 2C), 110.18, 99.41, 93.08, 82.95, 79.16, 75.63, 74.28, 72.44, 67.85, 63.41, 49.21, 42.02, 36.95, 32.11, 29.03 (3C), 7.94. HRMS (ESI): *m/z* calcd for $C_{29}H_{31}FN_3O_{10}$ [M + H]$^+$: 600.1993, found 600.2010.

*10-O-(1-(4-Bromophenyl)-1H-1,2,3-triazole) ginkgolide B* (**5x**). Following the described procedure, 18.5 mg (28%) of compound **5x** were obtained from 46.2 mg (0.1 mmol) of 10-*O*-propargylated ginkgolide B (3). $^1$H-NMR (600 MHz, DMSO-$d_6$) $\delta$ 8.83 (s, 1H), 7.88–7.85 (m, 2H), 7.85–7.81 (m, 2H), 6.47 (s, 1H), 6.20 (s, 1H), 5.47 (d, $J$ = 4.6 Hz, 1H), 5.45 (d, $J$ = 12.2 Hz, 1H), 5.33 (d, $J$ = 4.1 Hz, 1H), 4.93 (d, $J$ = 12.1 Hz, 1H), 4.63 (d, $J$ = 7.1 Hz, 1H), 4.21 (dd, $J$ = 7.1 Hz, 4.6 Hz, 1H), 2.88 (q, $J$ = 7.1 Hz, 1H), 2.11 (dd, $J$ = 13.4 Hz, 4.4 Hz, 1H), 1.83 (td, $J$ = 13.8 Hz, 4.2 Hz, 1H), 1.74 (dd, $J$ = 14.4 Hz, 4.3 Hz, 1H), 1.13 (d, $J$ = 7.1 Hz, 3H), 1.02 (s, 9H). $^{13}$C-NMR (150 MHz, DMSO-$d_6$) $\delta$ 176.80, 172.99, 170.60, 144.39, 136.14, 133.35 (2C), 122.63 (2C), 122.20, 122.09, 110.16, 99.36, 93.10, 82.96, 78.97, 75.65, 74.23, 72.42, 67.79, 63.33, 49.02, 41.91, 36.81, 32.10, 28.98 (3C), 8.26. HRMS (ESI): *m/z* calcd for $C_{29}H_{31}BrN_3O_{10}$ [M + H]$^+$: 660.1193, found 660.1208.

*10-O-(1-(4-Trifluoromethylphenyl)-1H-1,2,3-triazole)ginkgolide B* (**5y**). Following the described procedure, 20.1 mg (31%) of compound **5y** were obtained from 46.2 mg (0.1 mmol) of 10-*O*-propargylated ginkgolide B (3). $^1$H-NMR (600 MHz, DMSO-$d_6$) $\delta$ 8.95 (s, 1H), 8.15 (d, $J$ = 8.4 Hz, 2H), 8.02 (d, $J$ = 8.6 Hz, 2H), 6.47 (s, 1H), 6.21 (s, 1H), 5.47 (d, $J$ = 12.2 Hz, 1H), 5.45 (d, $J$ = 4.6 Hz, 1H), 5.35–5.32 (m, 2H), 4.95 (d, $J$ = 12.1 Hz, 1H), 4.64 (d, $J$ = 7.1 Hz, 1H), 4.21 (dd, $J$ = 7.1 Hz, 4.6 Hz, 1H), 2.89 (q, $J$ = 7.1 Hz, 1H), 2.11 (dd, $J$ = 13.4 Hz, 4.4 Hz, 1H), 1.84 (td, $J$ = 13.8 Hz, 4.2 Hz, 1H), 1.74 (dd, $J$ = 14.4 Hz, 4.4 Hz, 1H), 1.13 (d, $J$ = 7.1 Hz, 3H), 1.02 (s, 9H). $^{13}$C-NMR (150 MHz, DMSO-$d_6$) $\delta$ 176.81, 172.98, 170.61, 144.59, 139.72, 129.34 (q, $J_{C-F}$ = 32.1 Hz), 127.78 (q, $J_{C-F}$ = 3.6 Hz, 2C), 124.27 (q, $J$ = 270.5 Hz), 122.50, 121.21 (2C), 110.16, 99.38, 93.12, 82.97, 78.97, 75.68, 74.24, 72.42, 67.80, 63.39, 49.19, 42.04, 37.01, 32.21, 29.12 (3C), 8.34. HRMS (ESI): *m/z* calcd for $C_{30}H_{31}F_3N_3O_{10}$ [M + H]$^+$: 650.1962, found 650.1967.

*10-O-(1-(4-Nitrophenyl)-1H-1,2,3-triazole) ginkgolide B* (**5z**). Following the described procedure, 21.9 mg (35%) of compound **5z** were obtained from 46.2 mg (0.1 mmol) of 10-*O*-propargylated ginkgolide B (3). $^1$H-NMR (600 MHz, DMSO-$d_6$) $\delta$ 9.00 (s, 1H), 8.49–8.47 (m, 2H), 8.23–8.20 (m, 2H), 6.47 (s, 1H), 6.20 (d, $J$ = 6.2 Hz, 1H), 5.46 (t, $J$ = 12.2 Hz, 1H), 5.40 (d, $J$ = 4.6 Hz, 1H), 5.36–5.32 (m, 2H), 4.95 (d, $J$ = 12.1 Hz, 1H), 4.64 (d, $J$ = 7.1 Hz, 1H), 4.21 (dd, $J$ = 7.1 Hz, 4.6 Hz, 1H), 2.93–2.84 (m, 1H), 2.11 (dd, $J$ = 13.4 Hz, 4.4 Hz, 1H), 1.84 (td, $J$ = 13.8 Hz, 4.2 Hz, 1H), 1.74 (dd, $J$ = 14.4 Hz, 4.4 Hz, 1H), 1.13 (d, $J$ = 7.1 Hz, 3H), 1.02 (s, 9H). $^{13}$C-NMR (150 MHz, DMSO-$d_6$) $\delta$ 176.79, 172.97, 170.60, 147.38, 144.82, 141.16, 126.13 (2C), 122.76, 121.30 (2C), 110.16, 99.39, 93.14, 82.98, 78.96, 75.74, 74.23, 72.41, 67.79, 63.37, 49.19, 42.04, 37.01, 32.22, 29.13 (3C), 8.35. HRMS (ESI): *m/z* calcd for $C_{29}H_{31}N_4O_{12}$ [M + H]$^+$: 627.1938, found 627.1948.

*10-O-(1-(4-Methylcarboxyphenyl)-1H-1,2,3-triazole) ginkgolide B* (**5aa**). Following the described procedure, 18.5 mg (29%) of compound **5aa** were obtained from 46.2 mg (0.1 mmol) of 10-*O*-propargylated ginkgolide B (3), $^1$H-NMR (600 MHz, DMSO-$d_6$) $\delta$ 8.94 (s, 1H), 8.20–8.16 (m, 2H), 8.09–8.06 (m, 2H), 6.47 (s, 1H), 6.21 (s, 1H), 5.46 (d, $J$ = 12.2 Hz, 1H), 5.42 (d, $J$ = 4.6 Hz, 1H), 5.36–5.28 (m, 2H), 4.94 (d, $J$ = 12.1 Hz, 1H), 4.64 (d, $J$ = 7.1 Hz, 1H), 4.21 (dd, $J$ = 7.1 Hz, 4.6 Hz, 1H), 3.90 (s, 3H), 2.88 (q, $J$ = 7.1 Hz, 1H), 2.11 (dd, $J$ = 13.4 Hz, 4.4 Hz, 1H), 1.84 (td, $J$ = 13.8 Hz, 4.2 Hz, 1H), 1.74 (dd, $J$ = 14.4 Hz, 4.4 Hz, 1H), 1.13 (d, $J$ = 7.1 Hz, 3H), 1.02 (s, 9H). $^{13}$C-NMR (150 MHz, DMSO-$d_6$) $\delta$ 176.80, 172.98, 170.60, 165.79, 144.58, 140.11, 131.53 (2C), 130.08, 122.40, 120.57 (2C), 110.16, 99.38, 93.12, 82.97,

78.97, 75.71, 74.23, 72.42, 67.79, 63.41, 52.64, 49.03, 42.04, 36.70, 32.12, 28.90 (3C), 8.29. HRMS (ESI): *m/z* calcd for $C_{31}H_{34}N_3O_{12}$ [M + H]$^+$: 640.2142, found 640.2154.

*10-O-(1-(4-Carboxyphenyl)-1H-1,2,3-triazole) ginkgolide B* (**5bb**). Following the described procedure, 16.3 mg (26%) of compound **5bb** were obtained from 46.2 mg (0.1 mmol) of 10-*O*-propargylated ginkgolide B (3). $^1$H-NMR (600 MHz, DMSO-$d_6$) δ 8.88 (s, 1H), 8.16 (d, *J* = 8.4 Hz, 2H), 8.16 (d, *J* = 8.4 Hz, 2H), 6.50 (s, 1H), 6.19 (s, 1H), 5.49–5.44 (m, 2H), 5.35–5.32 (m, 2H), 4.94 (d, *J* = 12.0 Hz, 1H), 4.63 (d, *J* = 7.2 Hz, 1H), 4.21 (dd, *J* = 7.2 Hz, 4.2 Hz, 1H), 2.88 (q, *J* = 7.2 Hz, 1H), 2.11 (dd, *J* = 13.4 Hz, 4.4 Hz, 1H), 1.83 (td, *J* = 13.8 Hz, 4.2 Hz, 1H), 1.73 (dd, *J* = 14.4 Hz, 4.4 Hz, 1H), 1.13 (d, *J* = 7.1 Hz, 3H), 1.02 (s, 9H). $^{13}$C-NMR (150 MHz, DMSO-$d_6$) δ 176.84, 172.98, 170.50, 166.86, 144.40, 139.78, 131.60 (2C), 131.29, 122.20, 120.28 (2C), 109.95, 99.28, 93.09, 82.95, 79.01, 75.55, 74.25, 72.32, 67.76, 63.29, 49.16, 41.92, 36.87, 32.10, 29.11 (3C), 8.25. HRMS (ESI): *m/z* calcd for $C_{30}H_{32}N_3O_{12}$ [M + H]$^+$: 626.1986, found 626.1998.

*10-O-(1-(3,5-Dimethoxyphenyl)-1H-1,2,3-triazole) ginkgolide B* (**5cc**). Following the described procedure, 22.5 mg (35%) of compound **5cc** were obtained from 46.2 mg (0.1 mmol) of 10-*O*-propargylated ginkgolide B (3). $^1$H-NMR (600 MHz, DMSO-$d_6$) δ 8.79 (s, 1H), 7.05 (d, *J* = 2.4 Hz, 2H), 6.63 (t, *J* = 2.4 Hz, 1H), 6.52 (s, 1H), 6.18 (s, 1H), 5.52 (s, *J* = 4.2 Hz, 1H), 5.43 (d, *J* = 12.0 Hz, 1H), 5.34 (d, *J* = 4.2 Hz, 1H), 5.32 (s, 1H), 4.92 (d, *J* = 12.0 Hz, 1H), 4.63 (d, *J* = 7.2 Hz, 1H), 4.22 (dd, *J* = 7.2 Hz, 4.2 Hz, 1H), 3.84 (s, 6H), 2.88 (q, *J* = 7.2 Hz, 1H), 2.11 (dd, *J* = 13.4 Hz, 4.4 Hz, 1H), 1.83 (td, *J* = 13.8, 4.2 Hz, 1H), 1.73 (dd, *J* = 14.4 Hz, 4.4 Hz, 1H), 1.13 (d, *J* = 7.1 Hz, 3H), 1.02 (s, 9H). $^{13}$C-NMR (150 MHz, DMSO-$d_6$) δ 176.87, 173.00, 170.66, 161.70 (2C), 144.08, 138.40, 122.15, 110.17, 100.86, 99.37, 98.98 (2C), 93.05, 82.94, 79.03, 75.60, 74.26, 72.43, 67.81, 63.37, 56.21 (2C), 49.20, 42.03, 36.98, 32.17, 29.09 (3C), 8.31. HRMS (ESI): *m/z* calcd for $C_{31}H_{36}N_3O_{12}$ [M + H]$^+$: 642.2299, found 642.2308.

*10-O-(1-(3-Chloro-4-fluorophenyl)-1H-1,2,3-triazole) ginkgolide B* (**5dd**). Following the described procedure, 29.2 mg (46%) of compound **5dd** were obtained from 46.2 mg (0.1 mmol) of 10-*O*-propargylated ginkgolide B (3). $^1$H-NMR (600 MHz, DMSO-$d_6$) δ 8.83 (s, 1H), 8.20–8.18 (m, 1H), 8.00–7.88 (m, 1H), 7.75–7.60 (m, 1H), 6.48 (s, 1H), 6.20 (s, 1H), 5.46–5.41 (m, 2H), 5.33 (s, 2H), 4.93 (d, *J* = 12.0 Hz, 1H), 4.64 (d, *J* = 7.1 Hz, 1H), 4.23–4.17 (m, 1H), 2.88 (q, *J* = 7.2 Hz, 1H), 2.11 (dd, *J* = 13.4 Hz, 4.4 Hz, 1H), 1.83 (td, *J* = 13.8 Hz, 4.2 Hz, 1H), 1.73 (dd, *J* = 14.4 Hz, 4.4 Hz, 1H), 1.13 (d, *J* = 7.1 Hz, 3H), 1.02 (s, 9H). HRMS (ESI): *m/z* calcd for $C_{29}H_{30}ClFN_3O_{10}$ [M + H]$^+$: 634.1604, found 634.1619.

*10-O-(1-(3-Pyridinyl)-1H-1,2,3-triazole) ginkgolide B* (**5ee**). Following the described procedure, 45.7 mg (67%) of compound **5ee** were obtained from 46.2 mg (0.1 mmol) of 10-*O*-propargylated ginkgolide B (3). $^1$H-NMR (600 MHz, DMSO-$d_6$) δ 9.13 (d, *J* = 2.5 Hz, 1H), 8.88 (s, 1H), 8.71 (d, *J* = 4.8 Hz, 1H), 8.35–8.30 (m, 1H), 7.68 (dd, *J* = 8.4, 4.8 Hz, 1H), 6.49 (s, 1H), 6.21 (s, 1H), 5.49 (d, *J* = 4.8 Hz, 1H), 5.47 (d, *J* = 12.0 Hz, 1H), 5.34 (d, *J* = 3.2 Hz, 2H), 4.95 (d, *J* = 12.1 Hz, 1H), 4.64 (d, *J* = 7.1 Hz, 1H), 4.21 (dd, *J* = 7.1 Hz, 4.5 Hz, 1H), 2.88 (q, *J* = 7.0 Hz, 1H), 2.12 (dd, *J* = 13.5 Hz, 4.5 Hz, 1H), 1.85 (td, *J* = 13.8 Hz, 4.2 Hz, 1H), 1.74 (dd, *J* = 14.4 Hz, 4.4 Hz, 1H), 1.13 (d, *J* = 7.1 Hz, 3H), 1.03 (s, 9H). $^{13}$C-NMR (150 MHz, DMSO-$d_6$) δ 176.80, 172.99, 170.59, 150.41, 144.45, 141.84, 133.59, 128.59, 125.10, 122.47, 110.07, 99.32, 92.92, 82.84, 78.99, 75.45, 73.96, 72.07, 67.61, 63.29, 48.94, 42.04, 36.71, 31.75, 29.12 (3C), 8.16. HRMS (ESI): *m/z* calcd for $C_{28}H_{31}N_4O_{10}$ [M + H]$^+$: 583.2040, found 583.2056.

*10-O-(1-Benzyl-1H-1,2,3-triazole) ginkgolide B* (**5ff**). Following the described procedure, 42.8 mg (72%) of compound **5ff** were obtained from 46.2 mg (0.1 mmol) of 10-*O*-propargylated ginkgolide B (3). $^1$H-NMR (600 MHz, DMSO-$d_6$) δ 8.17 (s, 1H), 7.41–7.30 (m, 5H), 6.47 (s, 1H), 6.17 (s, 1H), 5.64 (d, *J* = 4.4 Hz, 1H), 5.61 (s, 2H), 5.35–5.31 (m, 2H), 5.27 (s, 1H), 4.84 (d, *J* = 12.1 Hz, 1H), 4.62 (d, *J* = 7.2 Hz, 1H), 4.17 (dd, *J* = 7.1 Hz, 4.5 Hz, 1H), 2.86 (q, *J* = 7.0 Hz, 1H), 2.09 (dd, *J* = 12.9 Hz, 3.9 Hz, 1H), 1.77 (td, *J* = 13.8 Hz, 4.2 Hz, 1H), 1.70 (dd, *J* = 14.4 Hz, 4.4 Hz, 1H), 1.12 (d, *J* = 7.1 Hz, 3H), 0.98 (s, 9H). $^{13}$C-NMR (150 MHz, DMSO-$d_6$) δ 176.83, 172.99, 170.62, 143.50, 136.17, 129.26 (2C), 128.69, 128.48 (2C), 123.01, 109.77, 98.98, 92.78, 82.56, 78.85, 75.51, 74.03, 72.28, 67.54, 63.16, 53.19, 48.92, 41.92, 36.77, 31.99, 28.81 (3C), 8.15. HRMS (ESI): *m/z* calcd for $C_{30}H_{34}N_3O_{10}$ [M + H]$^+$: 596.2244, found 596.2266.

*10-O-(((Anthracen-2-yloxy)methyl)-1H-1,2,3-triazole) ginkgolide B* (**5gg**). Following the described procedure, 43.5 mg (70%) of compound **5gg** were obtained from 46.2 mg (0.1 mmol) of 10-*O*-propargylated ginkgolide B (3). $^1$H-NMR (600 MHz, DMSO-$d_6$) δ 8.22 (s, 1H), 7.93 (dd, *J*

= 7.8 Hz, 1.3 Hz, 1H), 7.73 (td, *J* = 7.7 Hz, 1.4 Hz, 1H), 7.58 (td, *J* = 7.7 Hz, 1.2 Hz, 1H), 7.38 (dd, *J* = 7.9 Hz, 1.1 Hz, 1H), 6.45 (s, 1H), 6.17 (s, 1H), 5.83 (s, 2H), 5.58 (d, *J* = 4.6 Hz, 1H), 5.36 (d, *J* = 12.2 Hz, 1H), 5.31 (d, *J* = 3.9 Hz, 1H), 5.28 (s, 1H), 4.86 (d, *J* = 12.1 Hz, 1H), 4.61 (d, *J* = 7.2 Hz, 1H), 4.17 (dd, *J* = 7.2 Hz, 4.2 Hz, 1H), 4.03 (q, *J* = 7.1 Hz, 1H), 2.87 (q, *J* = 7.1 Hz, 1H), 2.50 (s, 2H), 2.09 (dd, *J* = 13.0 Hz, 4.1 Hz, 1H), 1.99 (s, 2H), 1.83–1.68 (m, 2H), 1.18 (t, *J* = 7.1 Hz, 2H), 1.12 (d, *J* = 7.1 Hz, 3H), 0.98 (s, 9H). $^{13}$C NMR (150 MHz, DMSO-$d_6$) δ 176.91, 172.97, 170.70, 143.52, 138.80, 134.36, 133.90, 130.22, 129.89, 124.13, 117.45, 111.79, 110.15, 99.32, 82.87, 79.06, 75.62, 74.19, 72.42, 67.81, 63.54, 60.34, 51.71, 49.15, 42.02, 36.92, 32.12, 29.04(3C), 8.25. HRMS (ESI): *m/z* calcd for $C_{31}H_{32}N_4O_{10}$ [M + H]$^+$: 621.2198, found 621.2170.

*10-O-(1-Phenyl-1H-1,2,3-triazole) ginkgolide A* (**5'a**). Following the described procedure, 36.2 mg (64%) of compound **5'a** were obtained from 44.6 mg (0.1 mmol) of 10-*O*-propargylated ginkgolide A (**3'**). $^1$H-NMR (600 MHz, DMSO-$d_6$) δ 8.51 (s, 1H), 7.48 (s, 2H), 7.43 (d, *J* = 16.7 Hz, 2H), 6.45–6.32 (m, 1H), 6.15 (s, 1H), 5.36 (d, *J* = 12.0 Hz, 1H), 5.23 (s, 1H), 4.95 (d, *J* = 4.0 Hz, 1H), 4.89 (d, *J* = 11.9 Hz, 1H), 4.83 (dd, *J* = 8.4 Hz, 7.3 Hz, 1H), 3.85–3.71 (m, 2H), 3.62–3.56 (m, 1H), 3.17 (d, *J* = 5.3 Hz, 1H), 2.95 (q, *J* = 7.1 Hz, 1H), 2.75 (dd, *J* = 15.1 Hz, 7.2 Hz, 1H), 2.04 (dd, *J* = 13.6 Hz, 5.0 Hz, 1H), 2.01–1.91 (m, 2H), 1.91–1.80 (m, 2H), 1.79–1.70 (m, 3H), 1.70–1.63 (m, 1H), 1.13 (d, *J* = 7.1 Hz, 3H), 1.02 (s, 9H). $^{13}$C-NMR (150 MHz, DMSO-$d_6$) δ 176.98, 173.22, 171.05, 144.45, 130.41(2C), 122.91, 120.67(2C), 110.14, 100.81, 87.96, 86.50, 85.31, 75.74, 68.63, 66.81, 66.19, 63.35, 49.16, 36.78, 36.45, 33.62, 32.19, 29.17(3C), 23.75, 8.59. HRMS (ESI): *m/z* calcd for $C_{29}H_{31}N_3O_9$ [M + H]$^+$: 566.2133, found 566.2111.

*10-O-(1-(2-Methylphenyl)-1H-1,2,3-triazole) ginkgolide A* (**5'n**). Following the described procedure, 40.0 mg (69%) of compound **5'n** were obtained from 44.6 mg (0.1 mmol) of 10-*O*-propargylated ginkgolide A (**3'**). $^1$H-NMR (600 MHz, DMSO-$d_6$) δ 8.51 (s, 1H), 7.52–7.48 (m, 2H), 7.44 (dd, *J* = 2.2, 1.0 Hz, 2H), 6.40 (s, 1H), 6.15 (s, 1H), 5.40–5.33 (m, 2H), 5.23 (s, 1H), 4.95 (d, *J* = 4.1 Hz, 1H), 4.89 (d, *J* = 11.9 Hz, 1H), 4.83 (d, *J* = 1.1 Hz, 1H), 3.82–3.75 (m, 2H), 3.60 (s, 1H), 3.17 (d, *J* = 5.3 Hz, 1H), 2.95 (d, *J* = 7.2 Hz, 1H), 2.75 (dd, *J* = 15.1 Hz, 7.2 Hz, 1H), 2.15 (s, 3H), 1.97 (s, 1H), 1.78–1.69 (m, 3H), 1.13 (d, *J* = 7.1 Hz, 3H), 1.02 (s, 9H). $^{13}$C-NMR (150 MHz, DMSO-$d_6$) δ 176.97, 173.19, 171.05, 143.65, 136.71, 133.55, 131.87, 130.34, 127.50, 126.51, 126.03, 110.13, 100.79, 97.64, 87.94, 86.50, 85.35, 75.74, 67.49, 63.53, 49.07, 41.66, 40.47, 36.78, 32.20, 29.16(3C), 17.87, 8.59. HRMS (ESI): *m/z* calcd for $C_{30}H_{33}N_3O_9$ [M + H]$^+$: 580.2197, found 580.2266.

*10-O-(1-(2-Cyanophenyl)-1H-1,2,3-triazole) ginkgolide* A (**5'p**)., Following the described procedure, 34.3 mg (58%) of compound **5'p** were obtained from 446 mg (0.1 mmol) of 10-*O*-propargylated ginkgolide A (**3'**). $^1$H-NMR (600 MHz, DMSO-$d_6$) δ 8.82 (s, 1H), 8.16 (dd, *J* = 7.9, 1.5 Hz, 1H), 7.98 (td, *J* = 7.9, 1.5 Hz, 1H), 7.91 (dd, *J* = 8.2, 1.2 Hz, 1H), 7.79 (td, *J* = 7.7, 1.2 Hz, 1H), 6.41 (s, 1H), 6.16 (s, 1H), 5.41 (d, *J* = 11.9 Hz, 1H), 5.39–5.35 (m, 2H), 5.26 (s, 1H), 4.97–4.87 (m, 2H), 4.82 (t, *J* = 7.8 Hz, 1H), 2.95 (q, *J* = 7.1 Hz, 1H), 2.79–2.72 (m, 1H), 1.40 (s, 3H), 1.13 (d, *J* = 7.2 Hz, 3H), 1.04 (s, 9H). $^{13}$C-NMR (150 MHz, DMSO-$d_6$) δ 174.85, 171.03, 168.75, 142.36, 133.25, 133.10, 128.73, 128.69, 124.10, 123.44, 110.75, 108.04, 108.01, 98.77, 95.53, 85.83, 84.40, 83.30, 73.77, 66.57, 64.74, 61.33, 47.05, 39.68, 34.68, 34.37, 31.51(3C), 8.07. HRMS (ESI): *m/z* calcd for $C_{30}H_{30}N_4O_9$ [M + H] $^+$: 591.2093, found 591.2062.

*10-O-(1-Benzyl -1H-1,2,3-triazole) ginkgolide A* (**5'ff**). Following the described procedure, 39.4 mg (68%) of compound **5'ff** were obtained from 44.6 mg (0.1 mmol) of 10-*O*-propargylated ginkgolide A (**3'**). $^1$H-NMR (600 MHz, DMSO-$d_6$) δ 8.25 (s, 1H), 7.38 (dd, *J* = 8.0, 6.4 Hz, 2H), 7.34–7.22 (m, 3H), 6.40 (s, 1H), 6.12 (s, 1H), 5.61 (d, *J* = 6.9 Hz, 2H), 5.24 (d, *J* = 11.6 Hz, 1H), 5.16 (s, 1H), 4.92 (d, *J* = 4.0 Hz, 1H), 4.82 (t, *J* = 7.8 Hz, 1H), 4.76 (d, *J* = 11.6 Hz, 1H), 4.03 (q, *J* = 7.1 Hz, 1H), 2.94 (d, *J* = 7.2 Hz, 1H), 2.65 (dd, *J* = 15.1 Hz, 7.2 Hz, 1H), 2.50 (p, *J* = 1.8 Hz, 3H), 2.05–1.99 (m, 1H), 1.99 (s, 1H), 1.90 (d, *J* = 4.4 Hz, 1H), 1.84 (dd, *J* = 15.1, 8.4 Hz, 1H), 1.70 (dd, *J* = 14.1, 4.7 Hz, 1H), 1.17 (t, *J* = 7.1 Hz, 1H), 1.12 (dd, *J* = 7.1, 4.0 Hz, 3H), 0.98 (d, *J* = 3.6 Hz, 9H). $^{13}$C-NMR (150 MHz, DMSO-$d_6$) δ 177.20, 173.13, 171.32, 136.27, 129.28(2C), 128.74, 128.39, 124.82, 110.16, 110.03, 101.11, 88.10, 86.53, 85.77, 75.61, 68.76, 66.93, 63.50, 63.40, 53.55, 53.35, 49.15, 40.86, 36.82, 32.04, 29.02(3C), 8.71. HRMS (ESI): *m/z* calcd for $C_{30}H_{33}N_3O_9$ [M + H] $^+$: 580.2297, found 580.2267.

*10-O-(((Anthracen-2-yloxy)methyl)-1H-1,2,3-triazole) ginkgolide A* (**5'gg**). Following the described procedure, 36.9 mg (61%) of compound **5'gg** were obtained from 44.6 mg (0.1 mmol) of

10-*O*-propargylated ginkgolide A (**3'**). $^1$H-NMR (600 MHz, DMSO-$d_6$) δ 8.29 (s, 1H), 7.92 (dd, $J$ = 7.7, 1.4 Hz, 1H), 7.82–7.69 (m, 1H), 7.57 (td, $J$ = 7.7, 1.2 Hz, 1H), 7.44–7.26 (m, 1H), 6.40 (s, 1H), 6.13 (s, 1H), 5.83 (s, 2H), 5.27 (d, $J$ = 11.8 Hz, 1H), 5.18 (s, 1H), 4.90 (d, $J$ = 4.0 Hz, 1H), 4.83 (dd, $J$ = 8.2 Hz, 7.3 Hz, 1H), 4.78 (d, $J$ = 11.7 Hz, 1H), 4.03 (q, $J$ = 7.1 Hz, 19H), 2.95 (q, $J$ = 7.2 Hz, 1H), 2.66 (dd, $J$ = 15.2 Hz, 7.2 Hz, 1H), 2.51 (p, $J$ = 1.8 Hz, 4H), 1.90–1.82 (m, 2H), 1.72 (dd, $J$ = 14.1 Hz, 4.7 Hz, 2H), 1.40 (s, 3H), 1.13 (d, $J$ = 7.2 Hz, 3H), 0.99 (d, $J$ = 5.3 Hz, 9H). $^{13}$C-NMR (150 MHz, DMSO-$d_6$) δ 176.93, 173.14, 170.81, 143.66, 139.26, 134.29, 133.85, 129.91, 129.68, 125.44, 117.45, 111.68, 110.09, 100.82, 87.90, 86.50, 85.33, 75.74, 68.63, 66.81, 63.47, 60.22, 51.47, 49.03, 40.89, 36.40, 32.14, 29.10(3C), 8.56. HRMS (ESI): *m/z* calcd for $C_{31}H_{32}N_4O_9$ [M + H]$^+$: 605.2249, found 605.2219.

*10-O-(1-Phenyl-1H-1,2,3-triazole) ginkgolide C* (**5″a**). Following the described procedure, 31.1 mg (52%) of compound **5″a** were obtained from 47.8 mg (0.1 mmol) of 10-*O*-propargylated ginkgolide C (**3″**). $^1$H-NMR (600 MHz, DMSO-$d_6$) δ 8.83 (s, 1H), 7.88 (dt, $J$ = 7.9, 1.1 Hz, 2H), 7.63 (dd, $J$ = 8.8, 7.2 Hz, 3H), 7.53 (t, $J$ = 7.4 Hz, 1H), 6.48 (s, 1H), 6.23 (s, 1H), 5.64 (t, $J$ = 5.3 Hz, 2H), 5.46 (d, $J$ = 12.2 Hz, 1H), 5.32 (s, 1H), 4.99 (d, $J$ = 4.2 Hz, 2H), 4.63 (d, $J$ = 7.0 Hz, 1H), 4.32 (s, 1H), 4.17 (dd, $J$ = 7.0 Hz, 4.6 Hz, 1H), 2.98–2.78 (m, 1H), 1.99 (s, 1H), 1.57 (d, $J$ = 12.5 Hz, 1H), 1.43–1.35 (m, 2H), 1.32–1.27 (m, 1H), 1.24 (s, 2H), 1.17 (t, $J$ = 7.1 Hz, 1H), 1.13 (d, $J$ = 7.1 Hz, 4H), 1.10 (s, 3H), 1.07 (s, 9H). $^{13}$C-NMR (150 MHz, DMSO-$d_6$) δ 176.79, 172.94, 170.86, 162.28, 144.14, 136.89, 130.50, 122.14, 120.73, 110.04, 99.08, 93.14, 82.93, 75.47, 74.09, 67.05, 64.03, 63.52, 60.84, 49.35, 42.03, 35.12, 32.08, 19.09(3C), 14.32, 8.34. HRMS (ESI): *m/z* calcd for $C_{29}H_{31}N_3O_{11}$ [M + H]$^+$: 598.2039, found 598.2003.

*10-O-(1-(2-Methylphenyl)-1H-1,2,3-triazole) ginkgolide C* (**5″n**). Following the described procedure, 28.7 mg (47%) of compound **5″n** were obtained from 47.8 mg (0.1 mmol) of 10-*O*-propargylated ginkgolide C (**3″**). $^1$H-NMR (600 MHz, DMSO-$d_6$) δ 8.47 (s, 1H), 7.50 (d, $J$ = 2.2 Hz, 2H), 7.44 (d, $J$ = 1.4 Hz, 2H), 6.48 (s, 1H), 6.23 (s, 1H), 5.68 (d, $J$ = 4.7 Hz, 1H), 5.66 (dd, $J$ = 6.2 Hz, 2.5 Hz, 1H), 5.49–5.43 (m, 1H), 5.32 (s, 1H), 5.02–5.00 (m, 2H), 5.00–4.96 (m, 1H), 4.65–4.61 (m, 1H), 4.17 (dd, $J$ = 7.0 Hz, 4.7 Hz, 1H), 2.86 (d, $J$ = 7.1 Hz, 1H), 2.16 (s, 3H), 1.13 (d, $J$ = 7.1 Hz, 4H), 1.10 (d, $J$ = 3.7 Hz, 6H), 1.08 (s, 9H). $^{13}$C-NMR (150 MHz, DMSO-$d_6$) δ 176.81, 172.95, 170.89, 143.24, 136.51, 133.52, 131.93, 130.49, 127.57, 126.41, 125.27, 110.03, 99.05, 93.12, 82.92, 79.36, 75.41, 74.07, 69.35, 67.06, 64.05, 60.27, 49.38, 42.03, 32.10, 21.23(3C), 17.86, 8.34. HRMS (ESI): *m/z* calcd for $C_{30}H_{33}N_3O_{11}$ [M + H]$^+$: 612.2195, found 612.2157.

*10-O-(1-(2-Cyanophenyl) -1H-1,2,3-triazole) ginkgolide C* (**5″p**). Following the described procedure, 24.3 mg (39%) of compound **5″p** were obtained from 47.8 mg (0.1 mmol) of 10-*O*-propargylated ginkgolide C (**3″**). $^1$H-NMR (600 MHz, DMSO-$d_6$) δ 8.78 (s, 1H), 8.17 (dd, $J$ = 7.7, 1.4 Hz, 1H), 7.99 (td, $J$ = 7.8 Hz, 1.5 Hz, 1H), 7.88 (dd, $J$ = 8.2 Hz, 1.1 Hz, 1H), 7.79 (td, $J$ = 7.7 Hz, 1.1 Hz, 1H), 6.48 (s, 1H), 6.23 (s, 1H), 5.64 (d, $J$ = 6.1 Hz, 1H), 5.60 (d, $J$ = 4.7 Hz, 1H), 5.49 (d, $J$ = 12.4 Hz, 1H), 5.32 (s, 1H), 5.03 (d, $J$ = 12.4 Hz, 1H), 5.01 (d, $J$ = 4.2 Hz, 1H), 4.63 (d, $J$ = 7.0 Hz, 1H), 4.17 (dd, $J$ = 7.0 Hz, 4.7 Hz, 1H), 3.97 (ddd, $J$ = 12.5 Hz, 6.1 Hz, 4.2 Hz, 1H), 2.86 (q, $J$ = 7.1 Hz, 1H), 1.57 (d, $J$ = 12.5 Hz, 1H), 1.13 (d, $J$ = 7.1 Hz, 3H), 1.08 (s, 9H). $^{13}$C-NMR (150 MHz, DMSO-$d_6$) δ 176.78, 172.90, 170.87, 143.98, 135.47, 126.10, 124.92, 116.26, 110.05, 107.36, 99.11, 93.18, 82.95, 79.34, 75.46, 74.18, 74.07, 67.04, 64.02, 63.40, 60.25, 49.37, 42.03, 32.10, 27.01, 22.55, 21.24(3C), 8.35. HRMS (ESI): *m/z* calcd for $C_{30}H_{30}N_4O_{11}$ [M + H]$^+$: 623.1991, found 623.1956.

*10-O-(1-Benzyl-1H-1,2,3-triazole) ginkgolide C* (**5″ff**). Following the described procedure, 26.9 mg (44%) of compound **5″ff** were obtained from 47.8 mg (0.1 mmol) of 10-*O*-propargylated ginkgolide C (**3″**). $^1$H-NMR (600 MHz, DMSO-$d_6$) δ 8.19 (s, 1H), 7.40–7.37 (m, 2H), 7.35–7.32 (m, 3H), 6.46 (s, 1H), 6.20 (s, 1H), 5.73 (s, 1H), 5.62 (s, 3H), 5.35 (d, $J$ = 12.2 Hz, 1H), 5.25 (s, 1H), 4.97 (d, $J$ = 4.2 Hz, 1H), 4.87 (d, $J$ = 12.2 Hz, 1H), 4.61 (d, $J$ = 7.2 Hz, 1H), 4.13 (d, $J$ = 5.7 Hz, 1H), 3.90 (d, $J$ = 11.4 Hz, 1H), 2.84 (q, $J$ = 7.1 Hz, 1H), 1.55 (dd, $J$ = 12.5 Hz, 5.2 Hz, 1H), 1.17 (t, $J$ = 7.1 Hz, 3H), 1.12 (d, $J$ = 7.1 Hz, 4H), 1.10–1.09 (m, 2H), 1.03 (s, 9H). $^{13}$C-NMR (150 MHz, DMSO-$d_6$) δ 176.81, 172.93, 170.87, 143.32, 136.15, 129.29(2C), 128.75, 128.56, 123.60, 110.01, 98.96, 93.02, 82.83, 79.33, 75.34, 74.01, 67.04, 64.00, 63.56, 60.27, 53.51, 49.30, 42.03, 32.04, 29.07, 21.23(3C), 8.27. HRMS (ESI): *m/z* calcd for $C_{30}H_{33}N_4O_{11}$ [M + H]$^+$: 612.2195, found 612.2156.

*10-O-(((Anthracen-2-yloxy)methyl)-1H-1,2,3-triazole) ginkgolide C* (**5″gg**). Following the described procedure, 32.5 mg (51%) of compound **5″gg** were obtained from 47.8 mg (0.1 mmol) of 10-O-propargylated ginkgolide C (**3″**). $^1$H-NMR (600 MHz, DMSO-$d_6$) δ 8.25 (s, 1H), 7.93 (dd, $J$ = 7.7, 1.4 Hz, 1H), 7.74 (td, $J$ = 7.7, 1.4 Hz, 1H), 7.58 (td, $J$ = 7.7 Hz, 1.2 Hz, 1H), 7.40 (dd, $J$ = 7.8 Hz, 1.1 Hz, 1H), 6.46 (s, 1H), 6.20 (s, 1H), 5.84 (s, 2H), 5.70 (d, $J$ = 4.8 Hz, 1H), 5.62 (d, $J$ = 6.1 Hz, 1H), 5.37 (d, $J$ = 12.2 Hz, 1H), 5.26 (s, 1H), 4.98 (d, $J$ = 4.2 Hz, 1H), 4.90 (d, $J$ = 12.1 Hz, 1H), 4.61 (d, $J$ = 7.0 Hz, 1H), 4.13 (dd, $J$ = 7.1 Hz, 4.4 Hz, 1H), 3.90 (dt, $J$ = 12.4 Hz, 4.7 Hz, 1H), 2.84 (q, $J$ = 7.0 Hz, 1H), 2.50 (s, 3H), 1.55 (d, $J$ = 12.5 Hz, 1H), 1.18 (t, $J$ = 7.1 Hz, 3H), 1.12 (d, $J$ = 7.2 Hz, 3H), 1.04 (s, 9H). $^{13}$C NMR (150 MHz, DMSO-$d_6$) δ 176.83, 172.94, 170.92, 143.37, 138.90, 134.36, 133.88, 130.11, 129.85, 124.19, 117.47, 111.82, 110.02, 98.99, 82.84, 79.33, 75.38, 74.15, 74.01, 67.05, 64.01, 63.53, 60.27, 51.69, 49.33, 42.03, 32.04, 21.22 (3C), 8.28. HRMS (ESI): $m/z$ calcd for $C_{31}H_{32}N_4O_{11}$ [M + H]$^+$: 637.2148, found 637.2117

### 3.2. Antiplatelet Aggregation Activity Assay

The in vitro antiplatelet aggregation activity of the newly synthesized 10-substituted 1,2,3-triazole-ginkgolide derivatives was tested by the method of Born [26]. Blood samples (2 mL) from male New Zealand rabbits (2–2.5 kg body weight) were drawn into vacutainer tubes containing 200 μL of 3.2% sodium citrate. Platelet-rich plasma (PRP) was prepared by centrifuging the blood at 250× $g$ for 10 min at 4 °C. The PRP was diluted with platelet-poor plasma obtained by further centrifuging at 3000× $g$ for 10 min. The remaining blood was further centrifuged at 1600× $g$ for 5 min to obtain platelet-poor plasma (PPP) as control group. Platelet aggregation was induced by PAF (10 nM) after incubating platelets with different concentrations of samples, and the maximum rate of platelet aggregation (RPA%) within 5 min was measured with a Helena Platelet aggregometer instrument [27]. The inhibition ratio was calculated according to the following formula:

$$\text{Inhibition ratio (\%)} = (1-(\text{RPA\% of test group}) / (\text{RPA\% of control group})) \times 100\%$$

In primary screening, the activities were expressed directly as inhibition ratio at 50 nM concentration. The activity of the most active ginkgolide-1,2,3-triazole derivatives was further expressed as the IC$_{50}$ value (the concentration required to inhibit platelet aggregatory response by 50%). The values shown in the tables were calculated by linear regression from a single experimental curve with no less than four data points, each point being the mean of the percentage inhibition at a given concentration obtained from three independent experiments.

### 3.3. LDH Assay

#### 3.3.1. Preparation of H9c2 Cells

H9c2 cardio myoblast cells were grown in DMEM supplemented with 10% FBS, 100 U/mL penicillin-streptomycin. Cells were cultured at 37 °C with 5% CO$_2$. The cells were subcultured when they reached 70–80% confluence [28]. Then the cells were seeded at $1 \times 10^4$ cells per well in 96-well plate and incubated overnight. Then the cells were exposed to ginkgolide-1,2,3-triazole derivatives (1 μM and 10 μM).

#### 3.3.2. Preparation of Washed Platelets

Rat blood was collected in 3.8% sodium citrate vacuum anticoagulant tubes and centrifuged at 100× $g$ for 15 min to obtain platelet-rich plasma (PRP). The PRP was centrifuged at 1000× $g$ for 10 min at 37 °C. Then the platelet pellets were suspended in Tyrode's solution (pH 7.4). The washed platelets were adjusted to $3.6 \times 10^8$ platelets/mL. Washed platelets ($3.6 \times 10^8$ cells/mL) were pre-incubated with ginkgolide-1,2,3-triazole derivatives (1 μM and 10 μM) or 0.1% DMSO for 20 min at 37 °C, then centrifuged at 1700× $g$, 10 min and the supernatant collected [29].

3.3.3. Measurement of Lactate Dehydrogenase (LDH)

The assays to measure of LDH release were conducted in 96-well plates according to the manufacturer's protocol. The LDH levels were measured at 490 nm using a microplate reader (Thermo Fisher Scientific, Waltham, MA, USA). Cell cytotoxicity was also detected by the LDH activity assay kit. Cells incubated with 0.1% dimethyl sulfoxide (DMSO) served as the control group [28,29].

## 4. Conclusions

In summary, a series of ginkgolide-1,2,3-triazole conjugates were synthesized through a copper(I)-catalyzed Huisgen 1,3-dipolar cycloaddition reaction of the corresponding 10-*O*-propargylated ginkgolides with benzyl, phenyl and heterocyclic azides. Five of them (compounds **5a**, **5p**, **5ff**, **5gg** and **5'a**) displayed promising antiplatelet aggregation activities with IC$_{50}$ values ranging from 5–21 nM. Compounds **5ff** and **5gg**, having a benzyl group attached at the triazole nucleus were the best among the series of compounds. The most active compounds may be regarded as safe towards normal cells and platelets at therapeutic concentrations.

**Author Contributions:** Conceptualization, Q.-W.Z.; Methodology, J.C., L.H. and G.C.; Investigation, J.C., L.H. and W.S.; Writing-Review & Editing, J.C. and Q.-W.Z.; Project Administration, Q.-W.Z. and X.Z.; Funding Acquisition, Q.-W.Z.

**Funding:** This research was funded by Macao Science and Technology Development [FDCT/113/2013/A3] and Research Committee of the University of Macau (MYRG2016-00046-ICMS-QRCM).

**Acknowledgments:** We gratefully acknowledge the academic and technical assistance from Yu Chang.

**Conflicts of Interest:** The authors declare no conflict of interest.

## References

1. Maruyama, M.; Terahara, A.; Itagaki, Y.; Nakanishi, K. The ginkgolides. I. Isolation and characterization of the various groups. *Tetrahedron Lett.* **1967**, *8*, 299–302. [CrossRef]
2. Nakanishi, K.O.J.I. The ginkgolides. *Pure Appl. Chem.* **1967**, *14*, 89–114. [CrossRef] [PubMed]
3. Koch, E. Inhibition of platelet activating factor (PAF)-induced aggregation of human thrombocytes by ginkgolides: Considerations on possible bleeding complications after oral intake of *Ginkgo biloba* extracts. *Phytomedicine* **2005**, *12*, 10–16. [CrossRef] [PubMed]
4. Li, C.L.; Wong, Y.Y. The bioavailability of ginkgolides in *Ginkgo biloba* extracts. *Planta Med.* **1997**, *63*, 563–565. [CrossRef] [PubMed]
5. Zhang, X.; Li, Y.; Zhang, L.; Qin, M. A new ginkgolide from *Ginkgo biloba*. *J. China Pharm. Univ.* **2009**, *40*, 306–309.
6. Liao, H.J.; Zheng, Y.F.; Li, H.Y.; Peng, G.P. Two new ginkgolides from the leaves of *Ginkgo biloba*. *Planta Med.* **2011**, *77*, 1818–1821. [CrossRef] [PubMed]
7. Drago, F.; Floriddia, M.L.; Cro, M.; Giuffrida, S. Pharmacokinetics and bioavailability of a *Ginkgo biloba* extract. *J. Ocul. Pharmacol. Ther.* **2002**, *18*, 197–202. [CrossRef] [PubMed]
8. Marcheselli, V.L.; Rossowska, M.J.; Domingo, M.T.; Braquet, P.; Bazan, N.G. Distinct platelet-activating factor binding sites in synaptic endings and in intracellular membranes of rat cerebral cortex. *J. Biol. Chem.* **1990**, *265*, 9140–9145.
9. Ishii, S.; Shimizu, T. Platelet-activating factor (PAF) receptor and genetically engineered PAF receptor mutant mice. *Prog. Lipid Res.* **2000**, *39*, 41–82. [CrossRef]
10. Braquet, P.; Spinnewyn, B.; Braquet, M. BN 52021 and related compounds: A new series of highly specific PAF-acether receptor antagonists isolated from *Ginkgo biloba* L. *Blood Vessel* **1985**, *16*, 558–572. [CrossRef]
11. Braquet, P.; Drieu, K.; Etienne, A. Le Ginkgolide B (BN 52021): Un puissant inhibiteur du PAF-acether isolé du *Ginkgo biloba* L. *Actual. Chim. Ther.* **1986**, *13*, 237–254.
12. Nakanishi, K.; Habaguchi, K.; Nakadaira, Y.; Woods, M.C.; Maruyama, M.; Major, R.T.; Bähr, W. Structure of bilobalide, a rare tert-butyl containing sesquiterpenoid related to the C20-ginkgolides. *J. Am. Chem. Soc.* **1971**, *93*, 3544–3546. [CrossRef]

13. Yu, P.; Liang, J.Y. A new sesquiterpene trilactone from the roots of *Ginkgo biloba*. *Chin. Chem. Lett.* **2009**, *20*, 1224–1226. [CrossRef]

14. McKenna, D.J.; Jones, K.; Hughes, K. Efficacy, safety, and use of *Ginkgo biloba* in clinical and preclinical applications. *Altern. Ther. Health Med.* **2001**, *7*, 70–86.

15. Villhauer, E.B.; Anderson, R.C. Synthesis of the CDE ring system of the ginkgolides. *J. Org. Chem.* **1987**, *52*, 1186–1189. [CrossRef]

16. Vogensen, S.B.; Strømgaard, K.; Shindou, H.; Jaracz, S.; Suehiro, M.; Ishii, S.; Nakanishi, K. Preparation of 7-substituted ginkgolide derivatives: Potent platelet activating factor (PAF) receptor antagonists. *J. Med. Chem.* **2003**, *46*, 601–608. [CrossRef]

17. Hu, L.; Chen, Z.; Xie, Y.; Jiang, Y.; Zhen, H. Alkyl and alkoxycarbonyl derivatives of ginkgolide B: Synthesis and biological evaluation of PAF inhibitory activity. *Bioorgan. Med. Chem.* **2000**, *8*, 1505–1521. [CrossRef]

18. Park, H.K.; Lee, S.K.; Park, P.U.; Kwan, W.J. Ginkgolide Derivatives and a Process for Preparing Them. U.S. Patent Application No. 5,466,829, 14 November 1995.

19. Vatmurge, N.S.; Hazra, B.G.; Pore, V.S.; Shirazi, F.; Chavan, P.S.; Deshpande, M.V. Synthesis and antimicrobialactivity of beta-lactam-bile acid conjugates linked via triazole. *Bioorg. Med. Chem. Lett.* **2008**, *18*, 2043–2047. [CrossRef] [PubMed]

20. Agalave, S.G.; Maujan, S.R.; Pore, V.S. Click chemistry: 1,2,3-triazoles as pharmacophores. *Chem. Asian J.* **2011**, *6*, 2696–2718. [CrossRef]

21. Rostovtsev, V.V.; Green, L.G.; Fokin, V.V.; Sharpless, K.B. A stepwise huisgen cycloaddition process:copper(I)-catalyzed regioselective "ligation" of azides and terminal alkynes. *Angew. Chem. Int. Ed. Engl.* **2002**, *41*, 2596–2599. [CrossRef]

22. Xu, S.; Zhuang, X.; Pan, X.; Zhang, Z.; Duan, L.; Liu, Y.; Ding, K. 1-Phenyl-4-benzoyl-1 H-1, 2, 3-triazoles as orally bioavailable transcriptional function suppressors of estrogen-related receptor α. *J. Med. Chem.* **2013**, *56*, 4631–4640. [CrossRef] [PubMed]

23. Yang, H.; Li, Y.; Jiang, M.; Wang, J.; Fu, H. General copper-catalyzed transformations of functional groupsfrom arylboronic acids in water. *Chem. Eur. J.* **2011**, *17*, 5652–5660. [CrossRef] [PubMed]

24. Yuhong, J.; Dalip, K.; Rajender, S.V. Revisiting nucleophilic substitution reactions: Microwave-assistedsynthesis of azides, thiocyanates, and sulfones in an aqueous medium. *J. Org. Chem.* **2006**, *71*, 6697–6700.

25. Díaz, L.; Bujons, J.; Casas, J.; Llebaria, A.; Delgado, A. Click chemistry approach to new N-substitutedaminocyclitols as potential pharmacological chaperones for Gaucher disease. *J. Med. Chem.* **2010**, *53*, 5248–5255. [CrossRef] [PubMed]

26. Born, G.V.R. Aggregation of blood platelets by adenosine diphosphate and its reversal. *Nature* **1962**, *194*, 927–929. [CrossRef] [PubMed]

27. Jun, T.; Ke-yan, F. Experimental and clinical studies on inhibitory effect of *Ganoderma lucidum* on platelet aggregation. *J. Tongji Med. Univ.* **1990**, *10*, 240–243. [CrossRef]

28. Zhang, Y.; Deng, H.; Zhou, H.; Lu, Y.; Shan, L.; Lee, S.M.; Cui, G. A novel agent attenuates cardiotoxicity and improves antitumor activity of doxorubicin in breast cancer cells. *J. Cell. Biochem.* **2019**, *120*, 5913–5922. [CrossRef]

29. Zou, J.; Chen, Y.; Hoi, M.P.M.; Li, J.; Wang, T.; Zhang, Y.; Feng, Y.; Gao, J.; Lee, S.M.Y.; Cui, G. Discovery of a novel ERp57 inhibitor as antiplatelet agent from danshen (*Salvia miltiorrhiza*). *Evid. Based Complement. Altern. Med.* **2018**, *2018*, 9387568. [CrossRef]

**Sample Availability:** Samples of the compounds **5a–5gg**, **5′a-5′gg**, **5″a–5″gg** are available from the authors.

*molecules*

MDPI

Article

# Overexpression of the Melatonin Synthesis-Related Gene *SlCOMT1* Improves the Resistance of Tomato to Salt Stress

Dan-Dan Liu [1,†], Xiao-Shuai Sun [1,†], Lin Liu [1], Hong-Di Shi [1], Sui-Yun Chen [2,3,4,*] and Da-Ke Zhao [2,3,4,*]

[1] School of Agriculture, Yunnan University, Kunming 650091, China; liudandan@ynu.edu.cn (D.-D.L.); sunxiaoshuai2019@163.com (X.-S.S.); qiuqiu12-09@163.com (L.L.); dong19850412@163.com (H.-D.S.)
[2] Biocontrol Engineering Research Center of Plant Disease & Pest, Yunnan University, Kunming 650504, China
[3] Biocontrol Engineering Research Center of Crop Disease & Pest, Yunnan University, Kunming 650504, China
[4] School of Life Science, Yunnan University, Kunming 650504, China
* Correspondence: chensuiyun@ynu.edu.cn (S.-Y.C.); zhaodk2012@ynu.edu.cn (D.-K.Z.); Tel.: +86-0871-6522-7730 (D.-K.Z.)
† These authors contributed equally to this work.

Academic Editors: Pavel B. Drasar and Vladimir A. Khripach
Received: 25 March 2019; Accepted: 15 April 2019; Published: 17 April 2019

check for updates

**Abstract:** Melatonin can increase plant resistance to stress, and exogenous melatonin has been reported to promote stress resistance in plants. In this study, a melatonin biosynthesis-related *SlCOMT1* gene was cloned from tomato (*Solanum lycopersicum* Mill. cv. *Ailsa Craig*), which is highly expressed in fruits compared with other organs. The protein was found to locate in the cytoplasm. Melatonin content in *SlCOMT1* overexpression transgenic tomato plants was significantly higher than that in wild-type plants. Under 800 mM NaCl stress, the transcript level of *SlCOMT1* in tomato leaf was positively related to the melatonin contents. Furthermore, compared with that in wild-type plants, levels of superoxide and hydrogen peroxide were lower while the content of proline was higher in *SlCOMT1* transgenic tomatoes. Therefore, *SlCOMT1* was closely associated with melatonin biosynthesis confers the significant salt tolerance, providing a clue to cope with the growing global problem of salination in agricultural production.

**Keywords:** tomato; *SlCOMT1*; melatonin; genetical transformation; salt stress

## 1. Introduction

Melatonin (*N*-acetyl-5-methoxytryptamine) is also known as the pineal hormone, because it was first detected in the pineal gland of cattle by Lerner and colleagues in 1958 [1]. Since then, melatonin has been reported to regulate important physiological processes in mammals [2–5], such as circadian rhythm, mood, sleep, body temperature, activity, food intake, sexual behavior, and seasonal reproduction [6]. In addition to its physiochemical functions in mammals, melatonin also plays important roles in plant physiology [7]. Melatonin from plants, i.e., 'phytomelatonin', was originally regarded as an endogenous antioxidant molecule [8]. Based upon its powerful ROS scavenging activity, its roles in plant development as well as the resistance of plants to biotic and abiotic stress are also recognized. In detail, various physiological functions of melatonin in plants have been discovered, including promotion of explants growth, formation of the rhizome, induction of leaf senescence, and regulation of flowering, photosynthesis, circadian rhythms, and seed germination [9–17]. As mentioned above, phytomelatonin itself shows an effective antioxidative property, and it could clear the ROS generated under different kinds of abiotic stresses such as chemical pollution, ultraviolet radiation, herbicides, drought, heat, cold, and salinity, thus enhancing the abiotic resistances for plants [18–21].

The synthesis of melatonin is accomplished through four main reactions involving at least six enzymes, namely, tryptophan hydroxylase (TPH, EC 1.14.16.4), tryptophan decarboxylase (TDC, EC 4.1.1.28), tryptamine 5-hydroxylase (T5H, EC 1.1.13), serotonin *N*-acetyltransferase (SNAT, EC 2.3.1.87), *N*-acetylserotonin-*O*-methyltransferase (ASMT, EC 2.1.1.4), and caffeic acid *O*-methyl-transferase (COMT, EC 2.1.1.68) [22]. Tryptamine is produced by TDC in the cytoplasm, followed by serotonin generation in the endoplasmic reticulum [23]. Afterwards, serotonin is converted into *N*-acetylserotonin in the chloroplast and 5-methoxytryptamine in the cytoplasm by SNAT and ASMT, through acetylation and methylation, respectively. Then, these two intermediates are converted to melatonin by ASMT in the cytoplasm or SNAT in the chloroplast [24,25]. Similar to ASMT, COMT has been reported to play a pivotal role in the synthesis of phytomelatonin, specifically existing in the plant cytoplasm [26]. In *Arabidopsis*, serotonin can be converted to 5-methoxytryptamine by COMT methylation in the third step of melatonin synthesis, and then catalyzed by SNAT to melatonin [27]. On the basis of the type of enzymatic catalysis, COMT also belongs to the O-methyltransferase (OMT) family. The OMT family converts methylation sites on *S*-adenosyl-ʟ-methionine into various secondary metabolites, including flavonoids, phenyl- propanoids, and alkaloids [28]. Some plants, including dicots, lack ASMT homologs, suggesting that COMT plays an important role in the last step of melatonin synthesis [26,27]. In *Arabidopsis*, the catalytic activity of COMT is much higher than that of ASMT during the synthesis of melatonin, and the melatonin content is significantly reduced in mutant *Arabidopsis* when the *COMT* gene is silenced [26]. In plants, *COMT* gene improves melatonin production and positively contributes to strengthen both biotic and abiotic stress resistance in plants [29,30].

Tomato, one of the most highly consumed and extremely important horticultural plants, has been studied as a model plant for some aspects of plant growth and development [31]. Thus, it represents an ideal model organism to study the melatonin synthesis pathway. In recent years, several investigators have studied the effects of exogenous melatonin on tomato plants exposed to abiotic stress. For example, supplying additional melatonin improved the resistance to cadmium and water deficit in tomatoes [32,33]. Furthermore, exogenous melatonin promoted root development by regulating auxin and nitric oxide signaling in tomato [34]. In our study, *SlCOMT1*, a gene related to the biosynthesis of melatonin, was isolated from tomato. Phylogenetic relationships, subcellular localization and temporal-spatial expression were analyzed. To further characterize its potential stress-tolerant function in tomato, transgenic tomato plant with exogenous *SlCOMT1* was generated and analyzed. Furthermore, the relationship between melatonin production and salt resistance of transgenic tomato plants was investigated.

## 2. Results

### 2.1. Molecular Cloning and Sequence Analysis of SlCOMT1

To investigate the function of the melatonin biosynthesis-related gene *COMT* in tomato, *AtCOMT* gene sequence from *Arabidopsis* was used to search the tomato genome. Five similar sequences were identified, which were then aligned using an online tool (http//:www.ncbi.nlm.nih.gov/). The results show that the SlCOMT1 protein contains a dimerization domain (11–62aa) at the 5′ end and an adomet-MTase domain (105aa–321aa) at the 3′ end (Figure 1A). Phylogenetic tree analysis revealed protein homology between SlCOMT1 and AtCOMT (Figure 1B), showing that SlCOMT1 is the closest homolog of AtCOMT in tomato, with amino acid sequence similarity between them at 69.29%. Therefore, the gene was identified and designated as *SlCOMT1* (LOC101251452) (Figure 1C).

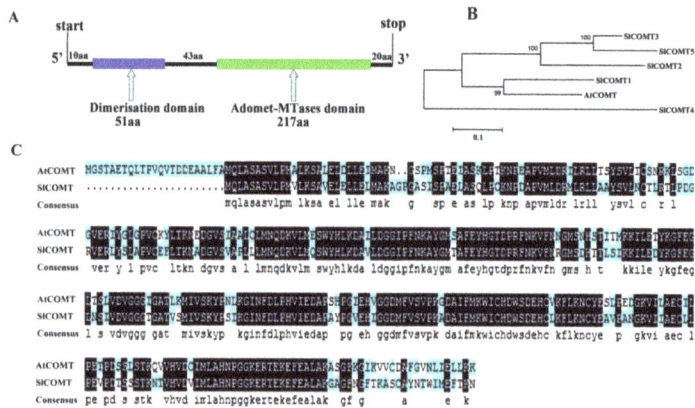

**Figure 1.** Bioinformatic analysis of the SlCOMT1 protein. (**A**) Two domains of the SlCOMT1 protein. (**B**) Phylogenetic tree was constructed using the five tomato COMT proteins and *Arabidopsis* AtCOMT protein. (**C**) Comparison of predicted SlCOMT1 protein sequence with AtCOMT. SlCOMT1 (XP_004235028.1), AtCOMT (NP_200227.1). Sl, *Solanum lycopersicum*; At, *Arabidopsis thaliana*.

To obtain the *SlCOMT1* gene from tomato, specific *SlCOMT1*-F/R primers were used for PCR (Supplementary Table S1), and a band of expected size was detected on a 1.2% agarose gel (Supplementary Figure S1), which was sequenced and characterized. The full-length *SlCOMT1* cDNA is 1023 bp, encoding a protein of 341 amino acids with a molecular weight of 37 kD and with an isoelectric point of 5.74. The *SlCOMT1*-PET32a recombinant vector was constructed to induce expression of the SlCOMT1 protein. The result shows that the SlCOMT1 protein is approximately 54 kD with a His-tag which is in line with our prediction (Figure 2).

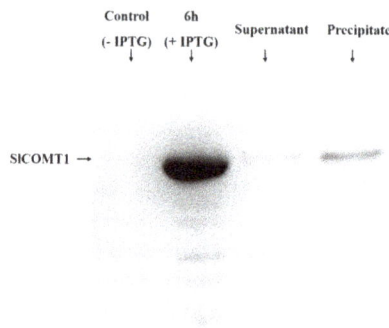

**Figure 2.** Induction of the SlCOMT1 protein in vitro. Lane 1, bacterial solution with no IPTG (control); lane 2, bacterial solution with IPTG cultured for 6 h at 37 °C; lane 3, supernatant derived from pET32a-SlCOMT cell lysate; and lane 4, precipitated SlCOMT1 protein.

Compared with COMT proteins in other plants, the SlCOMT1 protein sequence shows various conserved functional regions (Figure 3A). These proteins share five structurally conservative domains, including VVDVGGGTG, EHVGGDMF, GINFDLPHV, GGKERT, and NGKVI (Figure 3B), indicating that the sequence and function of SlCOMT1 are similar to these of other plant COMT proteins.

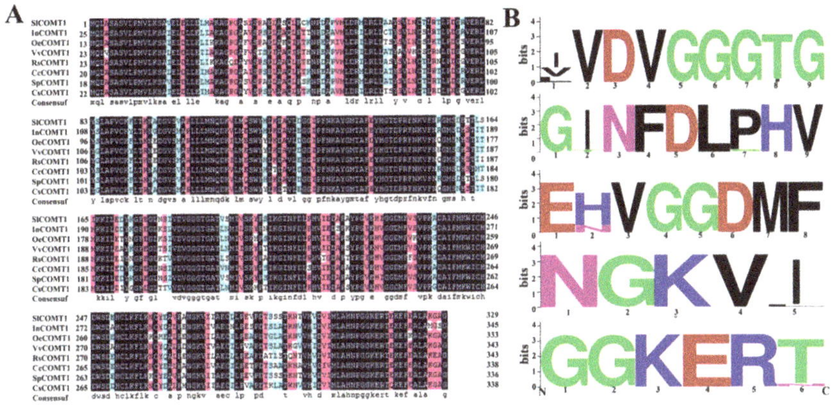

**Figure 3.** Evolution relationship of SlCOMT1 with other COMT proteins. (**A**) Comparison of amino acid sequences between tomato SlCOMT1 and COMT from other species, including *In, Ipomoea nil* (BAE94400.1); *Oe, Olea europaea* (XP_022844536.1); *Vv, Vitis vinifera* (XP_003634161.1); *Rs, Rauvolfia serpentine* (AOZ21153.1); *Cc, Capsicum chinense* (BAR88175.1); *Sp, Solaum pennellii* (XP_015070697.1); and *Cs, Camellia sinensis* (ADN27527.1); (**B**) Five conserved domains of the COMT proteins.

## 2.2. Phylogenetic Tree Analysis of SlCOMT1

To reveal the relationship between the SlCOMT1 protein and other plant COMT proteins, the phylogenetic tree of SlCOMT1 and other COMT proteins was constructed using MEGA 5.0 (Figure 4). The results show that SlCOMT1 and *S. tuberosum* COMT protein are classified into one category, suggesting that SlCOMT1 and StCOMT are derived from a recent ancestor, since reported in two congeneric species.

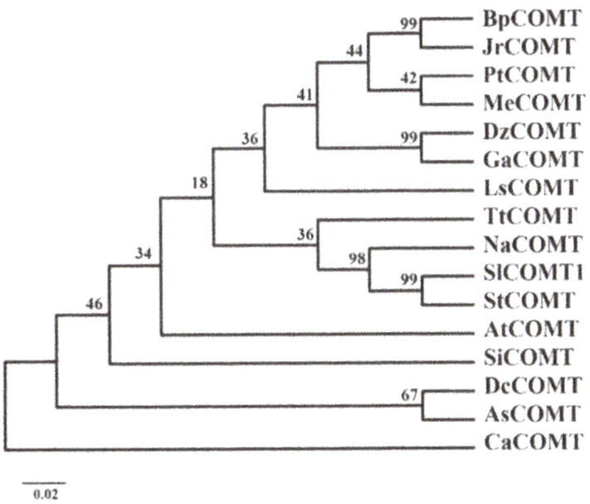

**Figure 4.** Phylogenetic analysis of the SlCOMT1 protein and its homologs. The proteins in the phylogenetic tree include *Bp, Betula pendula* (FJ667539.2); *Jr, Juglans regia* (XP_018828596.1); *Pt, Populus* Table 002321948. *Me, Manihot esculenta* (XP_021627291.1); *Dz, Durio zibethinus* (XP_022736469.1); *Ga, Gossypium arboretum* (XP_017611038.1); *Ls, Liquidambar styraciflua* (AF139533.1); *Tt, Thalictrum tuberosum* (AF064694.1); *Na, Nicotiana attenuata* (OIT03318.1); *St, Solanum tuberosum* (XP_015164331.1); *Si, Sesamum indicum* (XP_011075886.2); *Dc, Daucus carota* (XM_017381671.1); *As, Anthriscus sylvestris* (AB820126.1); and *Ca, Capsicum annuum* (NP_001311774.1).

## 2.3. Structure Prediction of SlCOMT1 Protein

The SOPMA online tool (https://npsa-prabi.ibcp.fr/cgi-bin/secpred_sopma.pl) was used to predict the secondary structure of the SlCOMT1 protein. The results suggest that the secondary structure of the protein is mainly composed of four parts, of which α-helices account for 46.2%, followed by random coils (29.24%), extended strands (16.08%), and β-turns (8.48%) (Figure 5A). Additionally, a three-dimensional structure of the SlCOMT1 protein was constructed to verify the above results using the Phyre 2 online tool (http://sbg.bio.ic.ac.phyre/) (Figure 5B).

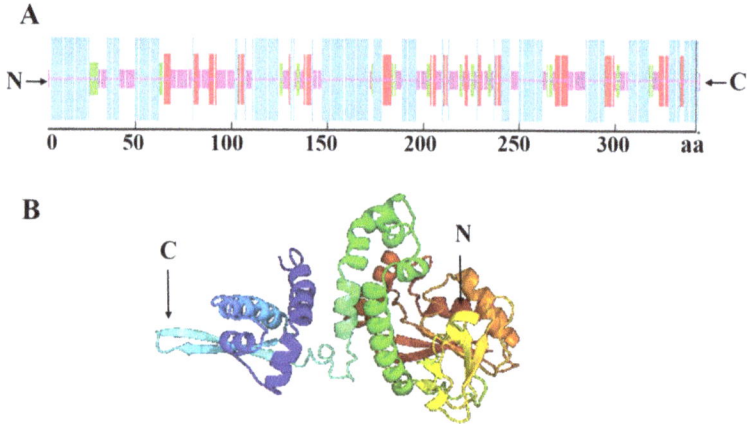

**Figure 5.** Structure prediction of the SlCOMT1 protein. (**A**) The secondary structure prediction of the SlCOMT1 protein. The blue indicates α-helices; purple indicates random coils; red indicates extended strands; green indicates β-turns; and the horizontal numbers indicate the positions of the amino acids. (**B**): The predicted three-dimensional structure of the SlCOMT1 protein.

## 2.4. SlCOMT1 Protein Subcellularly Localized in the Cytoplasm

The subcellular localization of SlCOMT1 in tobacco leaves was determined using a chimeric SlCOMT1-GFP fusion protein and a transient transfection assay. Green fluorescence was observed in the cytoplasm of epidermal cells transfected with the *35S:SlCOMT1-GFP* plasmid (Figure 6), revealing that SlCOMT1 is localized in the cytoplasm.

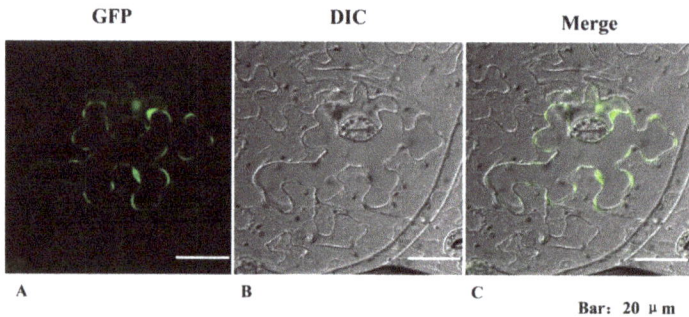

**Figure 6.** Localization of SlCOMT1. (**A**) Green fluorescence of SlCOMT1-PRI. (**B**) Bright-field image of *Agrobacterium*-infiltrated tobacco leaf. (**C**) The merged fluorescent images. The tobacco leaves were injected with the transgenic *Agrobacterium* liquid, and then cultivated in the culture chamber for 2 days and observed by confocal microscopy. GFP, Green Fluorescence Protein; DIC, Diascoptic Lighting Channel.

## 2.5. Temporal and Spatial Expression of SlCOMT1

The temporal and spatial expression of *SlCOMT1* was investigated by real-time PCR. The results show that *SlCOMT1* is constitutively expressed in tomato tissues, including roots, shoots, leaves, flowers, and fruits, with variable expression levels in these tissues. The lowest expression is in roots and the highest expression is in fruits (Figure 7), suggesting that *SlCOMT1* might be involved in the regulation of fruit development.

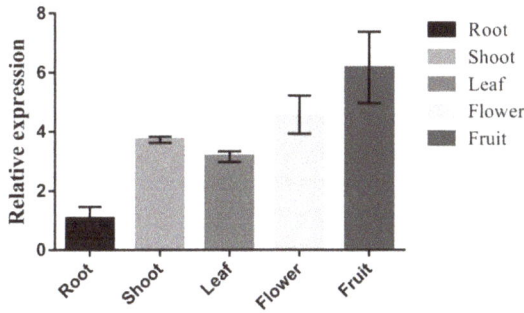

**Figure 7.** Expression levels of *SlCOMT1* in different tomato tissues.

## 2.6. SlCOMT1 Overexpression Increased the Melatonin Content and Salt Resistance in Tomato

To characterize the function of the *SlCOMT1* gene in plants, *SlCOMT1* driven by the 35S promoter was genetically transformed into tomato to generate transgenic lines, and two with different transcript levels were used for further study, namely, OE-1 and OE-2 (Supplementary Figure S2). In light of the role of COMT in the synthesis of melatonin, the melatonin content was measured in wild-type and transgenic tomato plants. The content of melatonin was higher (30–35 pg/mL) in overexpression transgenic plants, compared to 27.365 pg/mL in wild-type plants (Figure 8A), indicating that *SlCOMT1* functions in melatonin production.

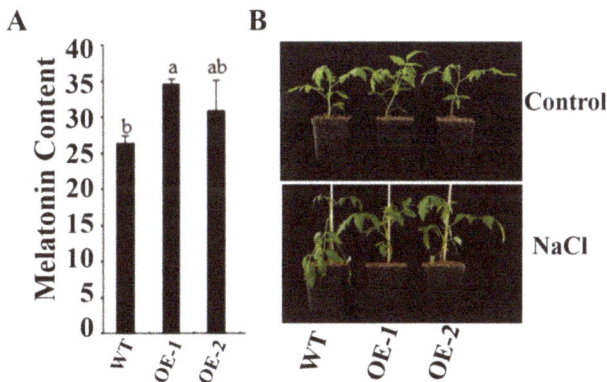

**Figure 8.** (**A**) Melatonin content in WT, OE-1, and OE-2 tomato plants. The same letter in the same growing season means no significant differences among three biological replicates ($p < 0.05$). Error bars represent standard error. (**B**) Growth status of WT, OE-1, and OE-2 after 800 mM NaCl treatment. WT, wild-type; OE, overexpression transgenic tomato.

It was reported that exogenous melatonin could increase the resistance of plants to salt stress in apple [35]. In the current study, wild-type and transgenic tomato plants were treated using 800 mM NaCl. One week later, the leaves appeared droop in wild-type tomato and displayed wilting.

In contrast, the leaves looked normal and healthy in transgenic plants (Figure 8B). Additionally, the levels of superoxide, hydrogen peroxide and proline were measured in both wild-type and transgenic plants. Under normal development, the levels of superoxide, hydrogen peroxide and proline varied slightly both in wild-type and transgenic tomato plants (Figure 9A–C). Compared with control plants, the levels of superoxide and hydrogen peroxide increased both in wild-type and transgenic tomato plants under treatment using 800 mM NaCl, but they were lower in transgenic plants (Figure 9A,B). On the contrary, the level of proline was significantly higher in transgenic plants than in control plants (Figure 9C). Thus, overexpression of *SlCOMT1* improves melatonin production and enhances salt tolerance in tomato.

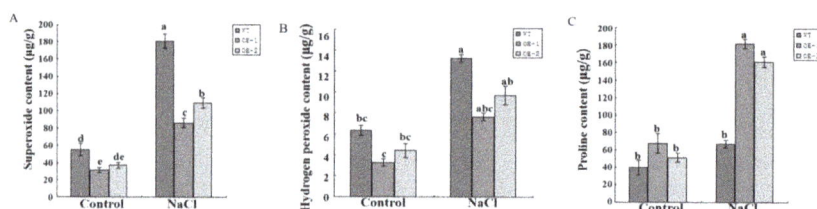

**Figure 9.** (**A–C**) were respectively the contents of superoxide, hydrogen peroxide, and proline in WT, OE-1, and OE-2 tomato plants, respectively. The same letter in the same growing season means no significant differences among three biological replicates ($p < 0.05$). Error bars represent SE.

## 3. Discussion

A high concentration of salt in soil is one of the most serious abiotic stresses for plants [36]. Melatonin plays important roles in various mechanisms that protect plants from external environmental stresses [17,37]. In this study, five *COMT* homologous genes were identified in the tomato genome, and the melatonin synthesis-related gene *SlCOMT1* was isolated based on homology comparison using *Arabidopsis* AtCOMT protein and the five tomato COMT proteins. Furthermore, SlCOMT1 protein was localized in the cytoplasm, suggesting that it might catalyze serotonin into 5-methoxytryptamine in the cytoplasm.

During normal cellular metabolism, ROS are generated by oxidative reaction process of mitochondrial respiration and photosynthesis process, and they act as signaling molecules during cellular repair processes at low amounts [38]. Once the plant is under environment stresses, its cells simultaneously initiate a series of response mechanisms and stress signals, such as the activation of cellular ROS scavenging mechanisms, which can trigger the production of reactive oxygen scavenging enzymes and antioxidants, including POD and SOD which work on scavenging excessive ROS, thereby alleviating or eliminating oxidative stress [39]. Under salt stress, the dynamic equilibrium of the production and elimination of reactive oxygen species in plant cells is disrupted, thereby causing the production of superoxide. Therefore, onset of cellular oxidative damage is the hallmark of salt stress, which is indicated by levels of superoxide and hydrogen peroxide [40]. Furthermore, proline in small amount plays multiple roles, such as stabilization of membrane and proteins, redox homeostasis and regulation of salt stress-responsive genes expression [41,42]. Superoxide, hydrogen peroxide and proline contents can respond to many environmental stresses in plants, including salt stress, and the accumulation of hydrogen peroxide and superoxide can disrupt the dynamic balance of cells under environmental stress [43–45]. Exogenous melatonin could have helped the tomato plants to bear the environmental stress by regulating the antioxidant system, proline and carbohydrates metabolism [46]. In this study, under treatment using NaCl, *SlCOMT1* overexpression transgenic plants displayed the increased proline and the decreased hydrogen peroxide and superoxide levels, which were resulted from the reduced oxidative damage by extra melatonin that can scavenge ROS in plant cells. As a result, melatonin produced by the *SlCOMT1* overexpression improved the growth characteristics of tomato compared to wild-type plants.

There are many cases suggesting that exogenous melatonin plays important roles in plant development and abiotic stress tolerances. For instance, exogenous melatonin has been reported to promote seed germination and seedling growth, and regulate the expression of growth-related genes involved in cell wall growth and expansion [46]. The molecule could improve plant tolerance to alkaline stress, drought stress, Cd stress and salinity stress by improving photosynthesis activity [47]. In addition, exogenous melatonin could confer cold tolerance in cucumber seedling by upregulating the expression of *ZAT12* gene accompanied by higher endogenous polyamine accumulation and higher ROS clearance system activity [48]. In this study, the melatonin content was elevated with the increased expression level of *SlCOMT1* in tomato, indicating that the *SlCOMT1* gene was involved in the synthesis of melatonin. Additionally, *SlCOMT1* overexpression transgenic tomato plants enhanced the resistance to salt stress. Therefore, the results indicate that *SlCOMT1* may be a key factor in regulating the response of plants against abiotic stresses by elevating melatonin production, therefore enhancing resistance to abiotic stress in tomato and other plants.

In conclusion, the present study shows that melatonin biosynthesis-related gene *SlCOMT1* isolated from tomato is localized in the cytoplasm, and is highly expressed in fruits. Melatonin content in *SlCOMT1* overexpression transgenic tomato plants is significantly higher than that in wild-type plants. The transgenic plants display increased proline levels and decreased hydrogen peroxide and superoxide levels, and the transgenic tomatoes tolerated salt stress better than the wild-type tomatoes. The results indicate that *SlCOMT1* is closely relate to melatonin production and functions in the improvement of plant resistance to abiotic stress.

## 4. Materials and Methods

### 4.1. Plant Materials

Tomato (*S. lycopersicum* Mill. cv. *Ailsa Craig*) was used for generating transgenic plant. After the surface of the tomato seeds was sterilized and soaked, the seeds were placed on a wet filter paper in a petri dish in a dark environment at 28 °C for germination. The seeds were transferred to a seedling tray containing sand and peat (1:1). After leaf growth, the seedlings were transplanted to pots containing matrix culture.

### 4.2. Cloning and Homology Analysis of the SlCOMT1 Gene

RNA Plant Plus reagent (Tiangen, Beijing, China) was used for total RNA extraction from tomato leaves. The total RNA served as the template for the synthesis of cDNA using the PrimeScript first-strand cDNA synthesis kit (Takara, Dalian, China). Primers (Supplementary Table S1) were designed based on sequences downloaded from the tomato genome website, and PCR was carried out using cDNA as a template. The reaction volume was 50 μL (Supplementary Table S2). The reaction consisted of 35 cycles (Supplementary Table S3). The PCR products were separated by 1.2% agarose gel electrophoresis, and the desired band was recovered and ligated to the pMD18-T cloning vector for sequencing.

### 4.3. Bioinformatics Analysis of the SlCOMT1 Gene

The sequence of the amplified gene was used to identify its ORF by DNAStar Lasergene EditSeq (7.1.0, DNAStar, Madison, WI, USA), and then the nucleotide sequence was translated into an amino acid sequence by DNAMAN 6.0.3.99 software (Lynnon Biosoft, San Ramon, CA, USA). Nucleotide and amino acid sequence similarity alignments were performed by Blast (http://blast.ncbi.nlm.nih.gov/). The ProtParam protein analysis tool (http://web.expasy.org/protparam/) was used to analyze the molecular weight, theoretical isoelectric point, and other protein properties. The 3D model of the encoded protein was generated using the Phyre 2 online tool (http://sbg.bio.ic.ac.phyre2/, London, England). The neighbor-joining phylogenetic tree of *SlCOMT1* was constructed using MEGA 5.0 software (Arizona State University, Tempe, AZ, USA).

*4.4. Expression Analysis of the SlCOMT1 Gene*

The BIO-RAD IQ5 (Bio-Rad, Hercules, CA, USA) was used for real-time PCR. The internal reference gene was 18S. All PCR reactions were performed three times. The reaction volume was 20 μL ( Supplementary Table S4). The real-time PCR reaction conditions were as follows: pre-denaturation at 95 °C for 10 min, followed by 40 cycles of denaturation at 95 °C for 15 s, annealing at 60 °C for 15 s, and extension at 60 °C for 45 s. The $2^{-\Delta\Delta CT}$ method was used for quantitative analysis.

*4.5. Construction of the Prokaryotic Expression Vector*

We designed specific restriction site primers based on the *SlCOMT1* gene sequence: *SlCOMT1*-Sma I–F (5′–CCCGGGAATGCAACTGGCGAGTGCC–3′, the underlined nucleotides correspond to the Sma I site) and *SlCOMT1*-Pst I-R (5′–CTGCAGAGAGATTCTTGGTGAATTCCA–3′, the underlined nucleotides correspond to the Pst I site) (Supplementary Table S1). The *SlCOMT1* pMD18-T plasmid served as the template for PCR, followed by product purification. In addition, the pET-32a expression vector and purified PCR product were digested with Sma I and Pst I, and the digested products were recovered. The recovered, digested products were ligated using *T4 ligase* at 16 °C overnight, and *E. coli* BL21 were transformed with the pET32a-*SlCOMT1* recombinant plasmid. Bacterial colonies were screened for ampicillin resistance. The positive clones were sent to Qingdao Qingke for sequencing and identification.

*4.6. Induction of SlCOMT1 Protein Expression*

To verify the substrate specificity of the SlCOMT1 protein, we introduced the *SlCOMT1* cDNA into the pET32a expression vector containing a His tag and expressed it in *E. coli*. We collected 1 mL of the bacterial solution, which served as the control. To the remaining bacterial solution, we added 1 mM IPTG, and after 6 h of induction at 37 °C, 1 mL of the bacterial solution was withdrawn, and the remaining bacterial solution was centrifuged to collect the bacterial cells. The bacterial cells were resuspended in 3 mL of 8 M urea and sonicated. The supernatant and precipitate were collected. After denaturing the precipitate, 1 mL of the solution was collected once again. All samples were subjected to SDS-PAGE.

*4.7. Establishment of Genetically Modified Tomatoes*

Firstly, construction of the SlCOMT1 expression vector was performed as follows. Specific restriction site primers were designed as shown in Supplementary Table S1. *SlCOMT1* was digested from the pMD18-T cloning vector and purified, followed by the digestion of the pCXSN-Myc vector with the same enzymes. Both fragments were ligated with *T4 ligase* at 16 °C for subsequent transformation into *E. coli*, and positive clones were identified. The *SlCOMT1-Myc* overexpression vector was transformed into *Agrobacterium* LBA4404 to obtain overexpressing (*OE-SlCOMT1*) transgenic tomato plants. Afterwards, *SlCOMT1* overexpression transgenic tomatoes were obtained as follows. Wild-type tomato seeds were sterilized with 70% ethanol, treated with 26% sodium hypochlorite, and then washed with sterile ddH$_2$O 4–5 times. The seeds were placed in seed germination medium for cultivation. After one week, the seeds germinated and reached the cotyledon stage. The cotyledons were cut into leaf discs and stem segments (Supplementary Figure S3A), respectively, transferred to pre-culture medium with the incision side down (Supplementary Table S5), and cultured in a dark environment for two days. A single *agrobacterium* colony carrying the expression plasmid was cultured in LB medium supplemented with antibiotics (kanamycin and rifampicin) at 28 °C under constant agitation. When the OD$_{600}$ reached 0.6, the bacterial cells were collected and suspended in MS medium. The pre-cultured explants were infected for 10 min, and the excess bacterial liquid on the surface of the explants was absorbed by a filter paper. The inoculated explants were placed on pre-culture medium and cultured in a dark environment at 28 °C for 1 day. Thereafter, the infected explants were placed on tomato differentiation medium (Supplementary Table S6) and cultured under normal conditions. When the adventitious buds grew into 2–3 cm seedlings (Supplementary Figure S3B,C), they were cut and transferred to rooting medium, with 3–4 strains per culture flask. The seedling roots (Supplementary Figure S3D) were washed, transplanted to pots containing vermiculite and perlite (1:1 ratio), and covered with a moisturizing film for 3–5 days. The moisturizing film was gradually removed to yield healthy tomato seedlings. Thereafter, the seedlings were transplanted for cultivation.

### 4.8. Subcellular Localization of the SlCOMT1

Confocal laser-scanning microscope (Zeiss LSM 510 META, Jena, Germany) was used to investigate the subcellular localization of SlCOMT1. Primers containing Sal I and Bam HI restriction sites (Supplementary Table S1) were used for PCR. The PCR product was gel purified and ligated into the 35S:PRI101-GFP vector. The expression plasmid was transformed into *Agrobacterium* 4401 by the freeze-thaw method, and then injected into two-week old tobacco leaves. After transient expression, images were acquired with a confocal microscope.

### 4.9. Molecular Identification of Transgenic Tomato

Transgenic tomatoes were performed by PCR and quantitative RT-PCR. To identify the transgenic tomatoes, the cDNA of both wild-type and transgenic plants were used as a template to detect the expression level of *SlCOMT1*, the plasmid DNA of *SlCOMT1* was used as positive control, and ddH$_2$O as negative control. The transgenic samples with increased *SlCOMT1* expression level were used for further study. Transcript level of *SlCOMT1* was detected using quantitative RT-PCR. Primers for quantitative RT-PCR (Supplementary Table S1) were used to detect the expression level of *SlCOMT1*. Quantitative RT-PCR reaction conditions were listed as Supplementary Table S4, with 30 cycles for fluorescence collection from denaturation to extension, and finally making a quantitative analysis by $2^{-\Delta\Delta CT}$ method.

### 4.10. Measurement of Melatonin Content

Wild-type and transgenic tomato plants with uniform growth potential were weighed and 0.1 g of fresh leaves were used to determine the melatonin content. Melatonin content in the leaves was measured using an enzyme-linked immunosorbent assay (Shanghai Enzyme Biotechnology, Shanghai, China). The standard, blank, and sample wells were assayed individually, and the absorbance at 450 nm was measured. The standard curve was generated after measuring the standard product, and the wild-type and transgenic tomato plants were assayed individually.

### 4.11. NaCl Treatment of Transgenic Tomato Plants and Detection of Hydrogen Peroxide, Superoxide and Proline

Wild-type and transgenic tomato seedlings with uniform growth potential were domesticated and then transplanted to medium containing matrix culture. After a period of healthy plant growth, we harvested 0.1 g of different plant leaves, and the contents of proline, hydrogen peroxide, and superoxide in the plants were determined. Two groups each made of both wild-type and transgenic tomato plants with uniform growth potential were selected. The first group served as the control; wild-type and transgenic tomato plants were treated using water. The second group served as the treatment group. Tomato plants were planted in 500 mL pots (matrix culture). Wild-type and *SlCOMT1* overexpression transgenic plants both growing five leaves were treated using 200 mL 800 mM NaCl twice a week at the same time. One week later, 0.1 g of different plant parts were harvested. Groups of three samples were pooled and used to measure the content of hydrogen peroxide, superoxide and proline. This experiment was repeated three times, and the results of three parallel experiments were averaged. The DPS data combing system and Tukey's multiple comparison method ($p < 0.05$; $p < 0.01$) were used for statistical analysis.

**Supplementary Materials:** The supplementary materials are available online.

**Author Contributions:** Data curation, X.-S.S.; Formal analysis, H.-D.S.; Funding acquisition, D.-D.L. and D.-K.Z.; Methodology, X.-S.S. and L.L.; Resources, S.-Y.C.; Software, L.L. and H.-D.S.; Supervision, S.-Y.C. and D.-K.Z.; Writing-original draft, X.-S.S.; Writing-review & editing, D.-D.L. and D.-K.Z.

**Funding:** This study was supported by Natural Science Foundation of China (Grant No. 31660566; No. 81560622).

**Acknowledgments:** We thank Lawrence Ji from University of Pittsburgh for help with editing the English in this paper.

**Conflicts of Interest:** The authors declare no conflicts of interest.

## References

1.  Lerner, A.B.; Case, J.D.; Takahashi, Y.; Lee, T.H.; Mori, W. Isolation of melatonin, the pineal gland factor that lightens melanocytes. *J. Am. Chem. Soc.* **1985**, *80*, 2587–2587. [CrossRef]
2.  Maronde, E.; Stehle, J. The mammalian pineal gland: Known facts, unknown facets. *Trends Endocrinol. Met.* **2007**, *18*, 142–149. [CrossRef]
3.  Pandi-Perumal, S.; Trakht, I.; Srinivasan, V.; Spence, D.W.; Maestroni, G.L.; Zisapel, N.; Cardinali, D.P. Physiological effects of melatonin: Role of melatonin receptors and signal transduction pathways. *Prog. Neurobiol.* **2008**, *85*, 335–353. [CrossRef]
4.  Jan, J.E.; Reiter, R.J.; Wasdell, M.B.; Bax, M. The role of the thalamus in sleep, pineal melatonin production, and circadian rhythm sleep disorders. *J. Pineal Res.* **2009**, *46*, 1–7. [CrossRef]
5.  Hardeland, R.; Madrid, J.A.; Tan, D.X.; Reiter, R.J. Melatonin, the circadian multioscillator system and health: The need for detailed analysis of peripheral melatonin signal. *J. Pineal Res.* **2012**, *52*, 139–166. [CrossRef]
6.  Carrillo, V.; Patricia, J.L.; Nuria, Á.S.; Ana, R.R.; Juan, M.G. Melatonin: Buffering the immune system. *Int. J. Mol. Sci.* **2013**, *14*, 8638–8683. [CrossRef]
7.  Kolar, J.; Machackova, I. Melatonin in higher plants: Occurrence and possible functions. *J. Pineal Res.* **2005**, *39*, 333–341. [CrossRef]
8.  Tan, D.X.; Manchester, L.C.; Esteban-Zubero, E.; Zhou, Z.; Reiter, R.J. Melatonin as a potent and inducible endogenous antioxidant: Synthesis and metabolism. *Molecules* **2015**, *20*, 18886–18906. [CrossRef]
9.  Murch, S.; Saxena, P. Melatonin: A potential regulator of plant growth and development. *In Vitro Cell. Dev. Biol. Plant* **2002**, *38*, 531–536. [CrossRef]
10. Hernández-Ruiz, J.; Cano, A.; Arnao, M.B. Melatonin: Growth-stimulating compound present in lupin tissues. *Planta* **2004**, *220*, 140–144. [CrossRef]
11. Hernández-Ruiz, J.; Cano, A.; Arnao, M.B. Melatonin acts as a growth stimulating compound in some monocot species. *J. Pineal Res.* **2005**, *39*, 137–142. [CrossRef]
12. Arnao, M.B.; Hernández-Ruiz, J. The physiological function of melatonin in plants. *Plant Signal. Behav.* **2006**, *1*, 89–95. [CrossRef]
13. Arnao, M.B.; Hernández-Ruiz, J. Melatonin promotes adventitious- and lateral root regeneration in etiolated hypocotyls of *Lupinus albus* L. *J. Pineal Res.* **2007**, *42*, 147–152. [CrossRef]
14. Arnao, M.B.; Hernández-Ruiz, J. Protective effect of melatonin against chlorophyll degradation during the senescence of barley leaves. *J. Pineal Res.* **2009**, *46*, 58–63. [CrossRef]
15. Tan, D.X.; Hardeland, R.; Manchester, L.C.; Korkmaz, A.; Ma, S.; Rosales-Corral, S.; Reiter, R.J. Functional roles of melatonin in plants, and perspectives in nutritional and agricultural science. *J. Exp. Bot.* **2012**, *63*, 577–597. [CrossRef]
16. Reiter, R.J.; Tan, D.X.; Zhou, Z.; Coelho Cruz, H.Z.; Fuentes-Broto, L.; Galano, A. Phytomelatonin: Assisting plants to survive and thrive. *Molecules* **2015**, *20*, 7396–7437. [CrossRef]
17. Arnao, M.B.; Hernández-Ruiz, J. Melatonin and its relationship to plant hormones. *Ann. Bot.* **2018**, *121*, 195–207. [CrossRef]
18. Janas, K.; Posmyk, M. Melatonin, an underestimated natural substance with great potential for agricultural application. *Acta Physiol. Plant* **2013**, *35*, 3285–3292. [CrossRef]
19. Shi, H.; Jiang, C.; Ye, T.; Tan, D.X.; Reiter, R.J.; Zhang, H.; Liu, R.; Chan, Z. Comparative physiological, metabolomic, and transcriptomic analyses reveal mechanisms of improved abiotic stress resistance in bermudagrass by exogenous melatonin. *J. Exp. Bot.* **2015**, *66*, 681–694. [CrossRef]
20. Zhang, N.; Sun, Q.; Zhang, H.; Cao, Y.; Weeda, S.; Ren, S.; Guo, Y.D. Roles of melatonin in abiotic stress resistance in plants. *J. Exp. Bot.* **2015**, *66*, 647–656. [CrossRef]

21. Wang, Y.P.; Reiter, R.J.; Chan, Z.L. Phytomelatonin: A universal abiotic stress regulator. *J. Exp. Bot.* **2018**, *69*, 963–974. [CrossRef]
22. Back, K.; Tan, D.X.; Reiter, R.J. Melatonin biosynthesis in plants: Multiple pathways catalyze tryptophan to melatonin in the cytoplasm or chloroplasts. *J. Pineal Res.* **2016**, *61*, 426–437. [CrossRef]
23. Kang, S.; Kang, K.; Lee, K.; Back, K. Characterization of rice tryptophan decarboxylases and their direct involvement in serotonin biosynthesis in transgenic rice. *Planta* **2007**, *227*, 263–272. [CrossRef]
24. Park, S.; Byeon, Y.; Back, K. Functional analyses of three ASMT gene family members in rice plants. *J. Pineal Res.* **2013**, *55*, 409–415. [CrossRef]
25. Kang, K.; Lee, K.; Park, S.; Byeon, Y.; Back, K. Molecular cloning of rice serotonin N-acetyltransferase, the penultimate gene in plant melatonin biosynthesis. *J. Pineal Res.* **2013**, *55*, 7–13. [CrossRef]
26. Byeon, Y.; Lee, H.Y.; Lee, K.; Back, K. Caffeic acid O-methyltransferase is involved in the synthesis of melatonin by methylating N-acetylserotonin in *Arabidopsis*. *J. Pineal Res.* **2014**, *57*, 219–227. [CrossRef]
27. Lee, H.Y.; Byeon, Y.; Lee, K.; Lee, H.J.; Back, K. Cloning of Arabidopsis serotonin N-acetyltransferase and its role with caffeic acid O-methyltransferase in the biosynthesis of melatonin in vitro despite their different subcellular localizations. *J. Pineal Res.* **2007**, *57*, 418–426. [CrossRef]
28. Lam, K.C.; Ibrahim, R.K.; Behdad, B.; Dayanandan, S. Structure, function, and evolution of plant O-methyltransferases. *Genome* **2007**, *50*, 1001–1013. [CrossRef]
29. Wang, M.; Zhu, X.; Wang, K.; Lu, C.; Luo, M.; Shan, T.; Zhang, Z. A wheat caffeic acid 3-O-methyltransferase TaCOMT-3D positively contributes to both resistance to sharp eyespot disease and stem mechanical strength. *Sci. Rep.* **2018**, *8*, 6543. [CrossRef]
30. Yang, W.J.; Du, Y.T.; Zhou, Y.B.; Chen, J.; Xu, Z.S.; Ma, Y.Z.; Chen, M.; Min, D.H. Overexpression of *TaCOMT* improves melatonin production and enhances drought tolerance in transgenic *Arabidopsis*. *Int. J. Mol. Sci.* **2019**, *20*, 652. [CrossRef]
31. Zhan, Y.; Qu, Y.; Zhu, L.; Shen, C.; Feng, X.; Yu, C. Transcriptome analysis of tomato (*Solanum lycopersicum* L.) shoots reveals a crosstalk between auxin and strigolactone. *PLoS ONE* **2018**, *13*, e0201124.
32. Li, M.Q.; Hasan, M.K.; Li, C.X.; Ahammed, G.J.; Xia, X.J.; Shi, K.; Zhou, Y.H.; Reiter, R.G.; Yu, J.Q.; Xu, M.X.; Zhou, J. Melatonin mediates selenium-induced tolerance to cadmium stress in tomato plants. *J. Pineal Res.* **2016**, *61*, 291–302. [CrossRef]
33. Ding, F.; Wang, G.; Wang, M.; Zhang, S. Exogenous melatonin improves tolerance to water deficit by promoting cuticle formation in tomato plants. *Molecules* **2018**, *23*, 1605. [CrossRef]
34. Wen, D.; Gong, B.; Sun, S.; Liu, S.; Wang, X.; Wei, M.; Yang, F.; Li, Y.; Shi, Q. Promoting roles of melatonin in adventitious root development of *Solanum lycopersicum* L. by regulating auxin and nitric oxide signaling. *Front Plant Sci.* **2016**, *25*, 718. [CrossRef]
35. Li, C.; Wang, P.; Wei, Z.; Liang, D.; Liu, C.; Yin, L.; Jia, D.; Fu, M.; Ma, F. The mitigation effects of exogenous melatonin on salinity-induced stress in *Malus hupehensis*. *J. Pineal Res.* **2012**, *53*, 298–306. [CrossRef]
36. Apse, M.P.; Aharon, G.S.; Snedden, W.A.; Blumwald, E. Salt tolerance conferred by overexpression of a vacuolar Na$^+$/H$^+$ antiport in *Arabidopis*. *Science* **1999**, *285*, 1256–1258. [CrossRef]
37. Arora, D.; Bhatla, S.C. Melatonin and nitric oxide regulate sunflower seedling growth under salt stress accompanying differential expression of Cu/Zn SOD and Mn SOD. *Free Radical Bio. Med.* **2017**, *106*, 315–328. [CrossRef]
38. He, L.; He, T.; Farrar, S.; Ji, L.B.; Liu, T.Y.; Ma, X. Antioxidants maintain cellular redox homeostasis by elimination of reactive oxygen species. *Cell Physiol. Biochem.* **2017**, *44*, 532–553. [CrossRef]
39. Miller, G.; Suzuki, N.; Ciftci, S.; Mittler, R. Reactive oxygen species homeostasis and signalling during drought and salinity stresses. *Plant Cell Environ.* **2010**, *33*, 453–467.
40. Martinez, V.; Nieves-Cordones, M.; Rodenas, R.; Mestre, T.C.; Garcia-Sanchez, F.; Rubio, F.; Nortes, P.A.; Mittler, R.; Rivero, R.M. Tolerance to stress combination in tomato plants: New insights in the protective role of melatonin. *Molecules* **2018**, *23*, 535. [CrossRef]
41. Carillo, P. GABA shunt in durum wheat. *Front. Plant Sci.* **2018**, *9*, 100. [CrossRef]
42. Ferchichi, S.; Hessini, K.; Dell'Aversana, E.; D'Amelia, L.; Woodrow, P.; Ciarmiello, L.F.; Fuggi, A.; Carillo, P. *Hordeum vulgare* and *Hordeum maritimum* respond to extended salinity stress displaying different temporal accumulation pattern of metabolites. *Funct. Plant Biol.* **2018**, *45*, 1096–1109. [CrossRef]
43. Claussen, W. Proline as a measure of stress in tomato plants. *Plant Sci.* **2005**, *168*, 241–248. [CrossRef]

44. Neill, S.J.; Desikan, R.; Clarke, A.; Hurst, R.D.; Hancock, J.T. Hydrogen peroxide and nitric oxide as signalling molecules in plants. *J. Exp. Bot.* **2002**, *53*, 1237–1247. [CrossRef]

45. Wahid, A.; Gelani, S.; Ashraf, M.; Foolad, M.R. Heat tolerance in plants: An overview. *Environ. Exp. Bot.* **2007**, *61*, 199–223. [CrossRef]

46. Manzer, H.S.; Saud, A.; Mutahhar, Y.; Al-Khaishany, M.Y.; Khan, M.N.; Abdullah, A.; Hayssam, M.A.; Ibrahim, A.A.; Abdulaziz, A.A. Exogenous melatonin counteracts NaCl-induced damage by regulating the antioxidant system, proline and carbohydrates metabolism in tomato seedlings. *Int. J. Mol. Sci.* **2019**, *20*, 353.

47. Yu, Y.; Lv, Y.; Shi, Y.; Li, T.; Chen, Y.; Zhao, D.; Zhao, Z. The role of phyto-melatonin and related metabolites in response to stress. *Molecules* **2018**, *23*, 1887. [CrossRef]

48. Aghdam, M.S.; Luo, Z.; Jannatizadeh, A.; Sheikh-Assadi, M.; Sharafi, Y.; Farmani, B.; Fard, J.R.; Razavi, F. Employing exogenous melatonin applying confers chilling tolerance in tomato fruits by upregulating ZAT2/6/12 giving rise to promoting endogenous polyamines, proline, and nitric oxide accumulation by triggering arginine pathway activity. *Food Chem.* **2019**, *275*, 549–556. [CrossRef]

**Sample Availability:** Not available.

*molecules*

MDPI

*Article*

# Effects of Isosorbide Incorporation into Flexible Polyurethane Foams: Reversible Urethane Linkages and Antioxidant Activity

Se-Ra Shin [1,†], Jing-Yu Liang [1,†], Hoon Ryu [2], Gwang-Seok Song [2] and Dai-Soo Lee [1,*]

[1] Division of Semiconductor and Chemical Engineering, Chonbuk National University, 567 Baekjedaero, Deokjin-gu, Jeonju 54896, Korea; srshin89@jbnu.ac.kr (S.-R.S.); liangjy@naver.com (J.-Y.L.)

[2] Industrial Biotechnology Program, Chemical R&D Center, Samyang Corporation, Daedeok-daero 730, Yuseong-gu, Daejeon 34055, Korea; hoon.ryu@samyang.com (H.R.); gwangseok.song@samyang.com (G.-S.S.)

\* Correspondence: daisoolee@jbnu.ac.kr; Tel.: +82-63-270-2310

† Contributed equally to this work.

Academic Editors: Pavel B. Drasar and Vladimir A. Khripach

Received: 2 March 2019; Accepted: 3 April 2019; Published: 5 April 2019

check for updates

**Abstract:** Isosorbide (ISB), a nontoxic bio-based bicyclic diol composed from two fuzed furans, was incorporated into the preparation of flexible polyurethane foams (FPUFs) for use as a cell opener and to impart antioxidant properties to the resulting foam. A novel method for cell opening was designed based on the anticipated reversibility of the urethane linkages formed by ISB with isocyanate. FPUFs containing various amounts of ISB (up to 5 wt%) were successfully prepared without any noticeable deterioration in the appearance and physical properties of the resulting foams. The air permeability of these resulting FPUFs was increased and this could be further improved by thermal treatment at 160 °C. The urethane units based on ISB enabled cell window opening, as anticipated, through the reversible urethane linkage. The ISB-containing FPUFs also demonstrated better antioxidant activity by impeding discoloration. Thus, ISB, a nontoxic, bio-based diol, can be a valuable raw material (or additive) for eco-friendly FPUFs without seriously compromising the physical properties of these FPUFs.

**Keywords:** isosorbide; reversible urethane linkages; cell opening; antioxidant activity; radical scavenger; flexible polyurethane foam

## 1. Introduction

Polyurethane foams (PUFs) are versatile plastics that have many advantages over other types of foams, such as ease of processing, low density, and excellent physical properties [1]. Among PUFs, flexible PUFs (FPUFs) that have an open cell structure possess excellent air permeability and the physical properties of FPUFs, such as density and resilience, are easily controllable by varying the polyurethane formulation recipe [1–3]. Thus, FPUFs are widely used in many different industries, such as those making cushions for furniture and automobiles, sound-absorbing materials, and packaging materials [4–10]. PUFs are commonly manufactured according to the following steps: (1) Mixing of polymer components with blowing agents; (2) nucleation and growth of cells; (3) gelation and crosslinking; and (4) cell opening and curing [11–13]. The opening of cell windows greatly affects the air permeability of FPUFs and also imparts various physical properties to the FPUFs for a variety of applications. Cell opening is typically induced through a combination of internal and external parameters. Internal parameters include the viscosity of the liquid resin [14,15], urea precipitation [16,17], catalyst balance [13,18,19], the effect of surfactant on bubble nucleation and stability [20,21], and the addition of fillers or additives to facilitate cell opening [22–25]. External parameters enable cell opening after foaming through physical or chemical treatment, such as crushing and reticulation [1,24,26–28]. Furthermore, it is critical that the cell structure and walls corresponding to the polymer matrix must be robust enough to withstand the harsh processes involved in cell opening.

With increasing concerns over environmental issues and the depletion of petroleum-based raw materials, the manufacturing of products based on environmentally-friendly raw material sources, without negatively affecting the performance of the final product, is becoming increasingly important [29,30]. Accordingly, interest has grown dramatically in bio-based materials derived from natural resources that are nontoxic to humans and environmentally friendly [29,31–34]. 1,4:3,6-dianhydrosorbitol or isosorbide (ISB) is one such bio-based resource that can be derived from a natural product. ISB is manufactured via the dehydration of D-sorbitol which can be obtained by hydration of D-glucose [35–39]. ISB is a bicyclic diol composed of tetrahydrofuran rings and hydroxyl groups at carbons 2 and 5. The hydroxyl groups in ISB can be found in two distinct orientations, *exo and endo*, and can be easily modified according to the desired applications [40,41]. Moreover, ISB has been attracting significant interest from multiple fields because of its unique rigid bicyclic structure, nontoxicity, and its ability to improve the heat resistance and mechanical properties of polymers [38,42–44]. For example, ISB can be used to replace bisphenol-A for the manufacture of polycarbonates with properties of high mechanical strength and also to impart ultraviolet (UV)-resistance to polymers [43,45]. Thus, ISB can be a substitute for the role of an aromatic diol due to its bulky and rigid structure.

It is well-known that the urethane units formed by the reaction between the active hydrogen in hydroxyl and isocyanate groups are reversible between 150 and 200 °C, and this feature can induce easier dissociation at the lower temperatures as the steric hindrance of both the hydroxyl groups and isocyanate groups increases [46–50]. In particular, phenolic hydroxyl groups can be used as a general blocking agent by reacting with the isocyanate groups to form a phenolic urethane in order to improve the storage stability of isocyanates from attack of moisture and oxygen [49,50]. At elevated temperatures, this bond can dissociate back to phenolic hydroxyl groups (blocking agent) and isocyanate groups and the regenerated isocyanate groups can participate in further polymerization reactions with hydroxyl or amine groups to form a thermally stable urethane or urea linkages. ISB, which has heterocyclic rings and two secondary hydroxyl groups, is structurally similar to the phenolic hydroxyl group. Thus, we expect that the hydroxyl groups can react reversibly with isocyanate groups to form urethane units that can be dissociated at specific temperatures. Accordingly, in this study, we proposed a novel process for the cell opening of FPUFs via a reversible urethane formation reaction between the hydroxyl groups in ISB and the aromatic isocyanate groups (Scheme 1). Thus, effects of isosorbide incorporation into the formulations of FPUFs on the foam properties were investigated. The core temperature of FPUFs typically rises to around 140–160 °C during foaming, which is sufficient to trigger urethane dissociation for units containing ISB [11,51]. As the high core temperature is reached, the dissociation of ISB-based urethane units causes a decrease in the molecular weight and modulus of polymers, and one can then expect that the relatively thin cell window layers will be broken for cell opening. The free isocyanate groups generated by the reversible urethane linkages can react with adjacent free hydroxyl groups to form urethane linkages. The free isocyanate groups may react with adjacent –NH in urethane or urea groups to form allophanate or biuret groups, respectively [49].

Another interesting observation is that FPUFs based on ISB have been shown to possess better antioxidant activity, as compared with FPUFs without ISB. In general, additives such as UV stabilizers, UV absorbers, and free radical scavengers are added in the formulation of PUFs to prevent discoloration and yellowing through oxidation [52,53]. Thus, imparting innate antioxidant activity is also important for FPUFs and this paper demonstrates that ISB can play the role of antioxidant as well as that of cell opener.

In this study, ISB was incorporated to FPUFs by dissolution in conventional poly (propylene glycol) (PPG), which enabled stable foaming during FPUF formation without deformation, shrinkage, or collapse. We investigated the effects of ISB on the air permeability, resilience, and thermal and mechanical properties of FPUFs. To the best of our knowledge, there are no other studies to date on the effect of ISB as a cell opener and antioxidant in polymers, especially FPUFs. We will thus demonstrate that ISB can be considered an attractive bio-based additive for the manufacturing of FPUFs.

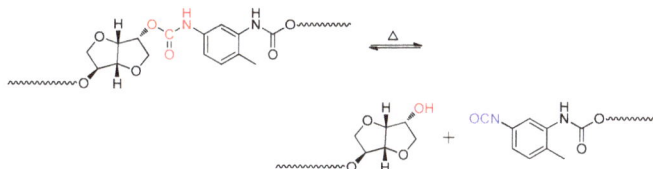

**Scheme 1.** Thermal reversibility of a urethane linkage formed by the reaction between ISB and an aromatic diisocyanate.

## 2. Materials and Methods

### 2.1. Materials

ISB (molecular weight: 146.1 g/mol, hydroxyl value: 767.8 mg KOH/g) was obtained from Samyang Co., Ltd. (Dae-Jeon, Korea). Commercially available poly(propylene glycol) (PPG, TF-3000) with the number average molecular weight of 3000 g/mol and hydroxyl value of 56.1 mg KOH/g was obtained from SKC (Ul-San, Korea). Toluene diisocyanate (TDI-80, isomer ratio 2,4/2,6 = 8/2) was obtained from the OCI Company Ltd. (Gun-San, Korea). TDI-80 is one of popular diisocyanates in PU industries for beds and furniture. Silicone surfactant (Niax silicone L-580) was obtained from Momentive (Waterford, NY, USA). Both amine catalysts A-1 (70% bis[2-dimethylaminoethyl] ether in dipropylene glycol [DPG]) and 33-LV (33% triethylenediamine in DPG) were purchased from Aldrich (Yong-In, Korea). Dibutyltin dilaurate (DBTDL) from Aldrich was used to facilitate gelation. Distilled water was used as the blowing agent. PPG was dehydrated at 80 °C prior to use under vacuum. All chemicals were used as received.

### 2.2. Preparation of FPUFs Based on ISB

ISB was dissolved in PPG at 90 °C and then the ISB-polyol mixture was allowed to cool down to room temperature. The concentration of ISB in the polyol mixture was varied: 1; 2; 3; 4; and 5 wt%. The polyol mixtures of various ISB content were transparent and stable at room temperature (no precipitation or crystallization was observed) (Figure S1). FPUFs were prepared in two steps: Firstly, L-580 (1.20 part per hundred polyol (phr)), A-1 (0.13 phr), 33-LV (0.50 phr), DBTDL (0.16 phr), and distilled water (3.00 phr) were added to the polyol mixture containing various amounts of ISB and were mixed for 30 s; secondly, the stoichiometrically required amount of TDI-80 to react with PPG, ISB, and distilled water was quickly poured into the polyol mixture and vigorously mixed under mechanical stirring at 3000 rpm for 7 s. The isocyanate index was fixed at 100. Then the mixture was poured into a wooden open mold (200 mm × 200 mm × 200 mm) for a free rise. During the foaming, the characteristic times (cream time, rise time, and gel time) were recorded using a stopwatch according to ASTM D7487-13. Cream time (CT) is the time between the start of mixing and the point at which fine bubbles begin to appear. Rise time (RT) is the time at which the foam stops expanding, as observed visually. Gel time (GT) time at which long strings of tacky material can be pulled away from the surface of the foam when the surface is touched by the edge of a tongue depressor or similar implement. The FPUFs were demolded and cured at 110 °C for one day before characterization. After curing, visible deformations (shrinkage and collapse) were not observed (Figure S2) and the additional processes for cell opening, such as crushing and reticulation, were not performed. The samples were named PUF-IX, where X was the content of ISB in the conventional PPG mixture. Table 1 summarizes the formulation for the FPUFs with various amounts of ISB, which were designed to manufacture general purpose

FPUFs for bed and furniture with density values of about 30 kg/m$^3$. The bio-content in FPUFs with various ISB content is shown in Table S1.

Table 1. Sample code and formulation for FPUFs with various amounts of ISB.

| Sample Code | PUF-I0 | PUF-I1 | PUF-I2 | PUF-I3 | PUF-I4 | PUF-I5 |
|---|---|---|---|---|---|---|
| | (Composition by wt.) | | | | | |
| *Polyol part* | | | | | | |
| TF-3000 | 100 | 99.0 | 98.0 | 97.0 | 96.0 | 95.0 |
| ISB | - | 1.00 | 2.00 | 3.00 | 4.00 | 5.00 |
| L-580 | 1.20 | 1.20 | 1.20 | 1.20 | 1.20 | 1.20 |
| A-1 | 0.13 | 0.13 | 0.13 | 0.13 | 0.13 | 0.13 |
| 33-LV | 0.50 | 0.50 | 0.50 | 0.50 | 0.50 | 0.50 |
| DBTDL | 0.16 | 0.16 | 0.16 | 0.16 | 0.16 | 0.16 |
| Distilled water | 3.00 | 3.00 | 3.00 | 3.00 | 3.00 | 3.00 |
| *Isocyanate part* | | | | | | |
| TDI-80 | 37.7 | 38.8 | 39.9 | 41.0 | 42.1 | 43.2 |
| Isocyanate index | 100 | 100 | 100 | 100 | 100 | 100 |

*2.3. Preparation of Polyurethane (PU) Films Based on ISB*

To confirm the reversible feature of the urethane linkages formed between the isocyanate groups and hydroxyl groups in ISB, PU films of differing ISB content were prepared using the same components as the FPUFs, but without a silicone surfactant, amine catalyst, and blowing agent. The ISB/PPG (TF-3000) mixture was mixed with the stoichiometrically required amount of TDI-80. DBTDL at 0.1 wt% with respect to the polyol weight was added to promote gelation. Then, the mixture was degassed under vacuum and poured into a glass mold. Finally, the PU films were cured at 110 °C for 24 h. The samples were named PU-IX, where X was the content of ISB (wt%) with respect to the total PPG weight. The sample code and formulation for PU films with various ISB content are shown in Table S2.

*2.4. Free Radical Scavenging of PU Films Based on ISB*

The free radical scavenging activity of the PU films was evaluated according to the 2,2-diphenyl-1-picrylhydrazyl (DPPH) method [54–56]. Without exposure to light, 100 mg of the PU film containing ISB (5 wt%) was immersed in 3 mL of 0.3 mM DPPH solution (in methanol). The DPPH solution containing a PU film without ISB was also prepared as a control for comparison. The absorbance of the DPPH solution was measured at 515 nm on a Jasco V-670 (Easton, MD, USA) UV-vis spectrometer. The decrease in the absorbance at this wavelength was monitored hourly. The percentage of free radical scavenging activity was calculated using the following Equation (1):

$$\text{Free Radical Scavenging Activity (\%)} = [(A_b - A_s)/A_b] \times 100 \qquad (1)$$

where $A_b$ is the absorbance of blank DPPH solution without PU film and $A_s$ is the absorbance of the DPPH solution containing PU films.

*2.5. Characterization*

Reversibility of urethane linkages was monitored using a Jasco Fourier-transform infrared (FTIR) spectrometer (FTIR 2000, Easton, MD, USA) equipped with a heating block, thermocouple, and digital thermal controller. The PU films were dissolved in dimethylformamide and the solution was coated on a KBr window. The solvent was then removed with mild heating (60 °C). FTIR spectra were recorded in the wave number range from 4000 to 500 cm$^{-1}$ at a resolution of 4 cm$^{-1}$. SEM was performed on a Jeol JSM 6400 (Akishima, Tokyo, Japan) to examine the cell morphology at an accelerating voltage of 20 kV. The samples were prepared by coating with gold to avoid charging of the electrons before measurement. The average cell size and thickness of cell walls were determined by using analysis software. About 60–70 cells per each sample were counted to determine the average cell size and thickness of cell walls. Air permeability of the FPUFs was evaluated using a Hallam F0023 foam porosity tester (Victoria, Australia), according to ASTM D 3574. The test specimens were cut into 50 mm × 50 mm × 25 mm (width × length × height) pieces and placed in the test cavity. The pressure differential was controlled at 125 Pa and it was maintained for 10 s. The tensile properties of the FPUFs, such as tensile stress and elongation at break, were measured using a universal testing machine (UTM, Z020, Zwick/Roell, Ulm, Germany). Dog bone specimens (10 mm thick) were used for the measurements and the crosshead speed were set at 500 mm/min. The tensile tests of four specimens per sample were evaluated and averaged. Tear strength was evaluated using a UTM and the procedure for tear strength testing was identical to that of the tensile test. Compression force deflection (CFD) was measured using UTM (Z020, Zwick/Roell, Ulm, Germany) according to ASTM D 3574. Three specimens per samples were measured and averaged. The resilience of FPUFs was evaluated by employing a ball rebounding tester according to ASTM D 3574. The center of the test specimens (100 mm × 100 mm × 50 mm; width × length × height) was located at the bottom center of the tube. Then, a 16.3 g steel ball (16 mm diameter) was dropped from a height of 500 mm and the maximum rebound height was recorded. The median of three specimens per sample was obtained. The thermal decomposition behavior of the FPUFs was investigated by employing TGA from TA instruments (Q50 machine, New Castle, DE, USA). Small pieces of sample were placed on a platinum pan and heated from room temperature to 800 °C at a heating rate of 20 °C/min under nitrogen. TGA measurements were performed thrice per sample and representative data were used for analysis. Dynamic mechanical property measurements of the FPUFs were carried out on a DMA from TA instruments (Q800, New Castle, DE, USA) in tension mode from −100 °C to 150 °C at a heating rate of 5 °C/min (frequency of 1 Hz and amplitude of 15%).

## 3. Results and Discussion

### 3.1. Physical Properties of FPUFs Based on ISB

For all the FPUFs in this study, the same amounts of catalyst and blowing agent were added during formulation. Thus, the changes in reactivity and density are entirely dependent on ISB content and the effects of ISB. ISB was successfully incorporated into FPUFs, and these FPUFs formed open cellular structures without any deterioration (shrinkage or collapse). Table 2 shows the reactivity and density of FPUFs formed with differing amounts of ISB. The CT was found to increase with increasing ISB content. In the formulations of FPUFs, the amount of TDI-80 increased with increasing ISB content due to the increased hydroxyl value of polyols, while the catalyst content was fixed. Thus, the concentrations of the catalysts in the formulations were lowered with increasing the ISB content. The increase of CT with increasing ISB content is attributable to the decrease of catalysts concentrations as well as the increased viscosity of the polyol/ISB mixture (Table S3). This increased difficulty in reactive contact of the reagent molecules and hampered the blowing reaction. Accordingly, FPUFs based on ISB exhibited a slower RT, as compared with the control FPUF (PUF-I0). FPUFs with less ISB content (PUF-I1 and PUF-I2) showed an increased GT; however, GT decreased with increasing ISB content. This was mainly due to the increased number of hydroxyls in the polyol/ISB mixture because there are more hydroxyl groups that are capable of reacting with isocyanate groups in the overall mixture. Thus, more urethane bonds could be formed by the incorporation of ISB. Furthermore, the hydroxyl groups of the short diol, ISB, can approach isocyanate groups more easily than those hydroxyl groups of conventional PPG, leading to a decrease in GT. Density was an important parameter in the physical properties of FPUFs and it strongly depended on the amount of blowing agent [13]. Since the same amount of distilled water was used for the preparation of all FPUFs, all the FPUFs had a similar density ($30 \pm 1$ kg/m$^3$). This indicated that the incorporation of ISB did not influence the density of FPUF.

**Table 2.** Characteristics of FPUFs with differing amounts of ISB.

| Sample Code | PUF-I0 | PUF-I1 | PUF-I2 | PUF-I3 | PUF-I4 | PUF-I5 |
|---|---|---|---|---|---|---|
| CT (s) | 11 | 11 | 13 | 14 | 16 | 17 |
| RT (s) | 72 | 79 | 79 | 76 | 76 | 76 |
| GT (s) | 93 | 101 | 98 | 90 | 86 | 76 |
| Density (kg/m$^3$) | 31 | 31 | 30 | 29 | 30 | 31 |

Figure 1 shows the scanning electron microscopy (SEM) images of the various FPUFs studied. All the FPUFs showed spherical or polyhedral shapes. Broken and contorted cells were not observed; thus, incorporation of ISB did not lead to such negative effects. Average cell size, average thickness of cell walls, and the number of cells per unit area of ISB-containing FPUFs are summarized in Table S4. The incorporation of ISB to FPUFs led to a slight decrease in average thickness of cell walls due to the slower CT. That is, slower CT implies the slow increase in mixture viscosity allowing the liquid layer between the bubbles to thin during initial cell formation. On the other hand, the average cell size of FPUFs containing ISB was reduced in comparison with that of control FPUF (PUF-I0) due to the decreased GT. Thus, it was observed that the number of cells of ISB-containing FPUFs per unit area were larger than that of control FPUF. The fast formation of urethane or urea linkage may restrict the growth and coalescence of bubbles, and this leads to the formation of smaller cell sizes. Some of the cell windows were also found partially open but most of them were closed. In this study, the FPUFs did not allow any physical and chemical treatment after their preparation to confirm the effect of ISB as a cell opener. The air permeability of FPUFs was investigated to quantitatively evaluate the degree of cell opening and the results are shown in Figure 2. The air permeability of the FPUFs increased with increasing ISB content. The maximum air permeability was found for 2 wt% ISB content (0.021 m$^3$/min). This was nearly three times that of the control FPUF (0.006 m$^3$/min). Further increases in ISB content led to decreased air permeability due increased formation of closed cells. In general, closed cells can be formed when the gelation dominates the blowing reaction during formation of FPUFs. As shown in Table 2, the incorporation of ISB slowed the blowing reaction and accelerated the gelation reaction. Accordingly, the air permeability of FPUFs with high ISB content became reduced.

ISB contains a fuzed bicyclic furan structure; thus, it can impart stiff properties and lead to a high glass-transition temperature [38]. Another factor for concern is that FPUFs based on ISB might exhibit a lower crosslinking density. Because ISB has two hydroxyl groups, the incorporation of ISB into FPUFs leads to a longer distance between the physical crosslinking points (Figure S3). Therefore, the addition of ISB may have both a reinforcing effect (from the fused bicyclic furan) and a plasticizing effect (by lowering the crosslinking density). Figure 3a shows the tensile properties of the FPUFs with various ISB content. The tensile strength of the FPUFs based on ISB was enhanced compared with the control FPUF (103.0 kPa). Tensile strength was found to increase with increasing ISB content and the maximum was 155.6 kPa for PUF-I5. When 2 wt% ISB was incorporated into FPUF, a slight decrease in the tensile strength (108.4 kPa) was observed, which was attributed to the higher air permeability. In general, the mechanical properties of FPUFs weaken as the open cell content increases [4,57]. FPUFs based on ISB also showed an increased elongation at break compared with the control FPUF, which was attributed to decrease in crosslinking density of the polymer network. When a high content of ISB was incorporated (PUF-I4 and PUF-I5), a decrease in the elongation was observed at break due to the increased stiffness.

Figure 3b shows the tear strength of the various FPUFs. The tear strength of FPUFs based on ISB was comparable or slightly improved over that of the control FPUF. The observed decrease and thus weakening in tear strength for PUF-I2 was caused by the increase in open cell windows. Furthermore, ISB-containing FPUFs prepared in this study exhibited the comparable or superior tensile and tear strength to reference FPUFs in the literature (100–150 kPa for tensile strength and 500–880 N/m for tear strength) [1–3,27].

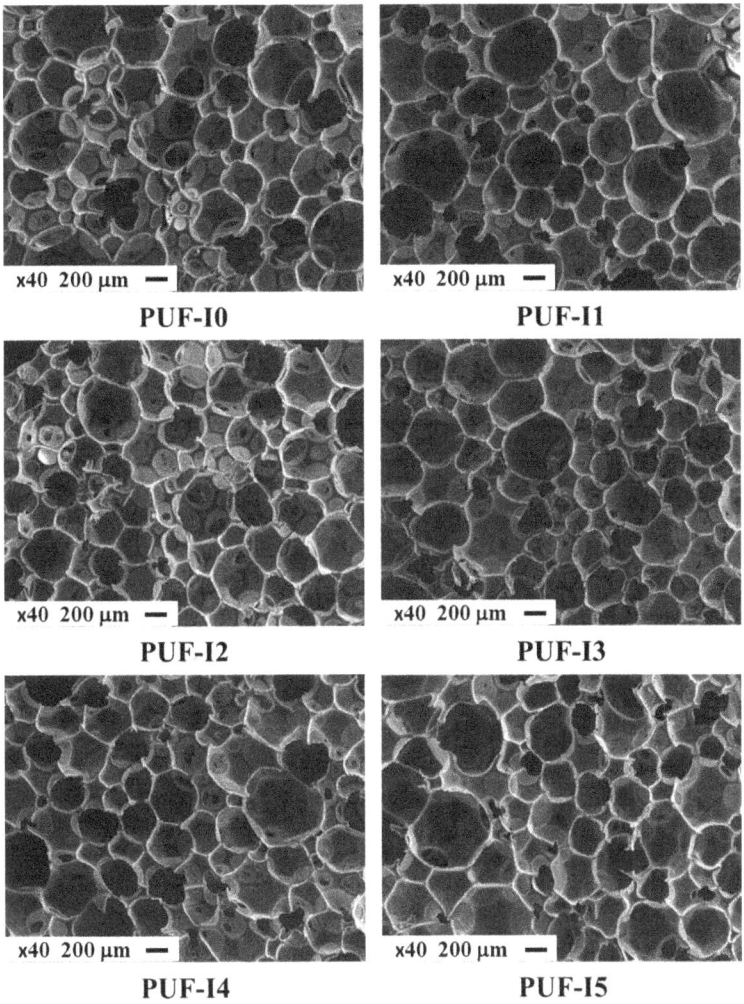

**Figure 1.** SEM images of the various FPUFs studied.

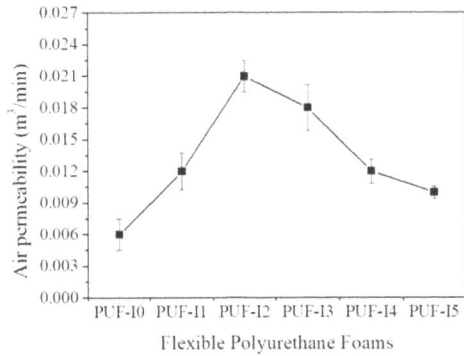

**Figure 2.** Air permeability of the various FPUFs studied.

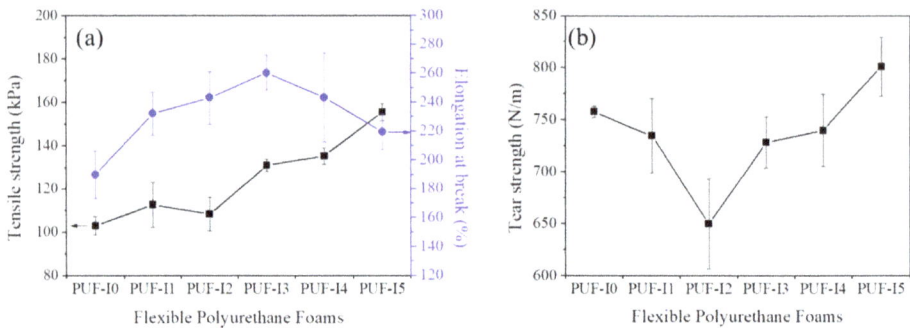

**Figure 3.** (a) Tensile strength and elongation at break; and (b) tear strength of the various FPUFs.

Figure 4 shows the CFD of the FPUFs with various ISB content. CFDs of FPUFs with lower ISB content (PUF-I1 and PUF-I2) were lower than that of PUF-I0 due to the increased air permeability. It is speculated that the increase of cell opening led to the decrease in resistance to external compression force. Further increase of ISB content led to an increase of CFD, which is attributed to the increased stiffness by the incorporation of ISB as well as decreased air permeability (increased closed cell content). The slightly enhanced CFD value of PUF-I5 (3.61 kPa) was observed compared with PUF-I0. However, CFD of ISB-based FPUFs prepared in this study showed a lower value compared with that of reference FPUF in the literature (>4~5 kPa) [58].

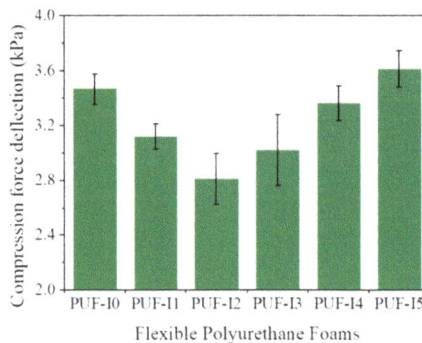

**Figure 4.** CFD of FPUFs with various ISB content.

The resilience of FPUFs is another important parameter for specific FPUF applications, such as automotive seats and furniture. The resilience of FPUF is strongly affected by the molecular weight and structure of soft and hard domains, crosslinking density, fillers, and the proportion of open cells [59–61]. The resilience of FPUF is generally evaluated by the ball rebounding test and loss factor (tan δ) in dynamic mechanical analyzer (DMA) measurement. The maximum ball rebound height of the prepared FPUFs is presented in Figure 5a. All the ISB containing FPUFs exhibited a lower rebounding height than the control FPUF (PUF-I0). This was attributed to the decreased crosslinking density from ISB incorporation [59]. However, ball rebounding height of ISB-containing FPUFs did not show significant trend with ISB content. Figure 5b–d show the DMA results (storage modulus, loss modulus, and tan δ curves) of the various FPUFs. The glass-transition temperature ($T_g$) was determined as the maximum point in the tan δ curve, and the results are summarized in Table 3. The storage moduli of the FPUFs with less ISB content (PUF-I1 and PUF-I2) at room temperature were lower than the control FPUF. This was mainly due to the loosening of the crosslinking network caused by the lengthened physical crosslinking points; thus, the $T_g$s of ISB-containing FPUFs were reduced (Table 3). On the other hand, further increase in ISB content led to an increase in the storage modulus and $T_g$ again, resulting from the increased stiffness by the addition of ISB with rigid bicyclic ring. The peak value and width of the tan δ curves is another meaningful indicator of resilient performance of FPUFs [25,61]. As shown in Figure 5d, the tan δ values of the ISB-containing FPUFs were higher compared with the PUF-I0 at room temperature; therefore, a decreased resilience can be expected for the ISB-containing FPUFs. In common with ball rebounding test, the significant change in resilience performance of ISB-containing FPUFs was not found to depend on ISB content. It indicates that resilient performance of FPUFs was reduced by the incorporation of ISB, but no significantly affect on ISB content.

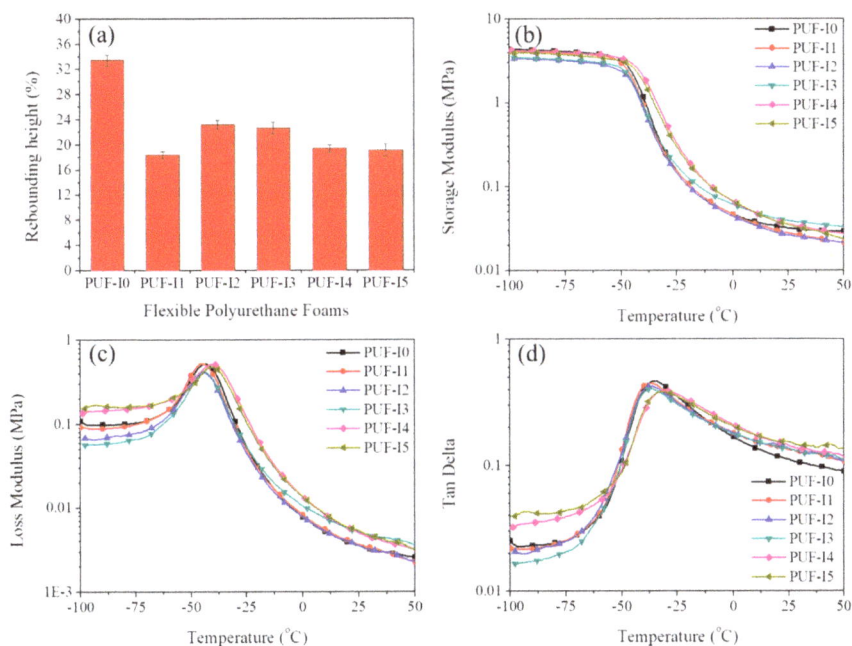

**Figure 5.** (a) Maximum rebounding height (%); (b) temperature dependence of storage modulus; (c) loss modulus; and (d) tan δ curves of the various FPUFs.

**Table 3.** Glass transition temperature ($T_g$) of the various FPUFs determined by DMA.

| Sample Code | PUF-I0 | PUF-I1 | PUF-I2 | PUF-I3 | PUF-I4 | PUF-I5 |
|---|---|---|---|---|---|---|
| $T_g$ (°C) | −36.7 | −37.8 | −37.7 | −37.4 | −31.4 | −31.2 |

### 3.2. Cell Opening by the Reversibility of Urethane Units Based on ISB in FPUFs

We studied the effect of ISB as a thermally triggered reversible cell opening agent (Scheme 1). In general, the core temperature of the FPUF during foaming of slab stocks in the plants approaches 160 °C due to the exothermic heat of reaction [11]. This temperature is sufficient for urethane bonds formed by ISB in the polymer network to form and break reversibly. At that moment, the modulus of polymers, especially for thin cell windows, weakens and then the weak cell windows can be opened. One important consideration for FPUFs is that the solid matrix that forms the cell strut and wall should be strong enough to withstand the harsh cell opening conditions. If the solid matrix is too weak, the cell walls break and the overall structure can even collapse, leading to serious deterioration of the mechanical properties of the FPUFs. In this experimental study, the foam dimensions prepared in the laboratory were too small for the temperature to reach 160 °C during the foaming. While the air permeability of the ISB-containing FPUFs was improved, as compared with the control FPUF (Figure 2), the ISB-containing FPUFs were thermally treated in a convection oven (160 °C) to confirm the desired effect of ISB as a cell opener via thermal reversibility of the urethane linkages. Before thermal treatment, the reversibility of the urethane linkages formed by ISB and isocyanate groups was confirmed using temperature-variable FTIR. It is difficult to evaluate the reversibility of urethane bonds in a foam, thus we prepared PU films using the same raw materials but without surfactant, catalyst, and distilled water (see experimental section for details). The chemical compositions of PU films were different from those of FPUFs since PU films were prepared in the absence of distilled water; thus they did not contain urea linkages. According to Delebe and Rolph, the urea linkages are more stable than urethane linkages thermally [47,48]. Temperature-dependent FTIR spectra of PUF-I0 and PUF-I5 are shown in Figure S4. When the temperature was increased above 160 °C, the peak intensity of urea carbonyl groups (C=O) at 1690 cm$^{-1}$ decreased significantly in both FTIR spectra of PUF-I0 and PUF-I5, as shown in Figure S4c,d. Accordingly, peaks due to the isocyanate groups could also be observed at 2270 cm$^{-1}$ (Figure S4e,f). The generation of isocyanate groups for PUF-I0 was attributed to the thermal reversibility of urea bonds, while that for PUF-I5 was due to the thermal reversibility of both urethane and urea bonds. The isocyanate peak in FTIR of PUF-I0 was observed at higher temperature (180 °C) compared with PUF-I5 (160 °C), implying that the reversibility of ISB-based urethane linkages appeared at lower temperature. Consequently, the PU films without urea linkages would be the better choice to confirm the reversibility of ISB-based urethane linkages. Figure 6 shows the FTIR spectra of PU films with or without ISB at different temperatures. To facilitate comparison of the temperature-dependent changes, all the FTIR spectra were normalized against the intensity of the C–O peak in the soft segments (1110 cm$^{-1}$) [5,62,63]. The absorption peak of the isocyanate group can be observed at 2270 cm$^{-1}$. An expansion of the FTIR spectra near the characteristic peak of the isocyanate group for the PU films with differing ISB content is shown in Figure S5. No isocyanate peak was observed across the entire temperature range for PU films without ISB (PU-I0). However, the PU film with 5 wt% ISB showed a clear absorption peak near 2270 cm$^{-1}$, corresponding with the isocyanate group when the temperature exceeded 160 °C (Figure S5). Furthermore, while PU-I0 showed a slight decrease in peak intensity at 1730 cm$^{-1}$ (assigned to the absorption of carbonyl of urethane groups) as the temperature changed, PU-I5 showed a large decrease in the peak intensity with increasing temperature. The decrease of peak intensity at 1730 cm$^{-1}$ is attributable to the reversible urethane groups. Therefore, the reversibility of urethane groups at elevated temperatures (at least 160 °C) was confirmed for ISB-containing PU films and cell opening thermal treatment should be performed above 160 °C.

The thermal reversibility of ISB-based urethane units was also confirmed by thermogravimetric analysis (TGA) measurements. Figure 7 shows the thermogravimetry (TG) and derivative TG (DTG) thermograms measured under nitrogen atmosphere. The TG and DTG curves revealed two main decomposition stages. The first stage occurs at 270 °C and corresponds to the dissociation and break down of the hard segments. The second stage occurs at 360 °C and is associated with the decomposition of the soft segments; that is, conventional PPG [30,64]. The characteristic temperatures of various FPUFs obtained from TGA measurement were summarized in Table 4. The two maximum decomposition temperatures did not show significant changes depending on ISB content. All the FPUFs exhibited the almost similar maximum decomposition temperature of soft and hard segments. One remarkable observation was that the shoulder at about 240 °C in the decomposition of the hard segments became increasingly prominent as the ISB content increased. This shoulder was not observed in PUF-I0, but it could be evidently observed for the FPUFs containing ISB. To determine what this shoulder peak indicated, we scrutinized our FTIR data. Both hydroxyl groups and isocyanate groups were completely consumed during the preparation of the FPUFs (Figure S6). No peaks corresponding to the free hydroxyl (3400–3600 cm$^{-1}$) or the isocyanate functional groups (2270 cm$^{-1}$) were observed in the FTIR spectra of any of the FPUFs studied. Therefore, the shoulder peak is strongly believed to be the decomposition of hard segments specific to the reversible urethane linkages formed between the hydroxyl groups of ISB and the isocyanate groups. Accordingly, $T_{5\%}$ and $T_{10\%}$ were reduced with increasing ISB content due to the thermal decomposition of ISB, resulting from the reversibility of ISB-based urethane linkages at elevated temperature.

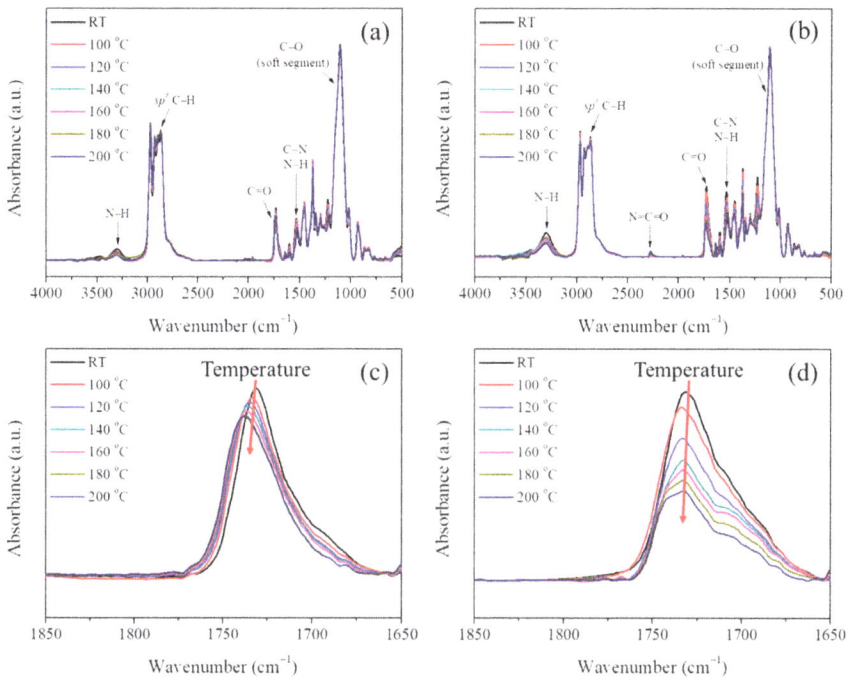

**Figure 6.** Temperature-dependent FTIR spectra of PU films: (**a**) PU-I0; (**b**) PU-I5; (**c**) PU-I0 expanded at 1650–1850 cm$^{-1}$; and (**d**) PU-I5 expanded at 1650–1850 cm$^{-1}$.

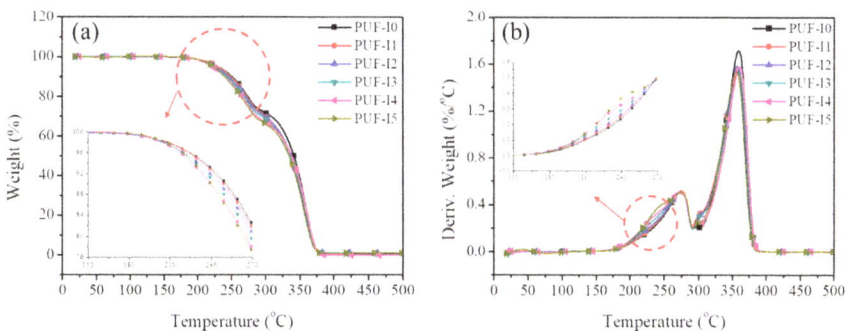

**Figure 7.** TG (**a**) and DTG (**b**) thermograms of FPUFs containing various amounts of ISB in nitrogen gas atmosphere.

SEM images of the FPUFs after thermal treatment at 160 °C are presented in Figure 8. The average cell size and average thickness of cell walls are summarized in Table S4. The average thickness of cell walls of FPUFs after thermal treatment was not significantly changed compared with those of FPUFs before thermal treatment. However, the average cell size was slightly reduced after thermal treatment. When compared with the SEM images of FPUFs before thermal treatment (Figure 1), almost all of the cell windows in the ISB-containing FPUFs were opened without serious destruction of the cell wall and of the overall structure. Furthermore, the opening of the cell windows occurred more noticeably for the FPUFs with higher ISB content. This indicates that cell opening can occur through the designed reversible urethane linkages formed by the hydroxyl groups of ISB and the isocyanate groups. Figure 9 shows the air permeability of the studied FPUFs after thermal treatment at 160 °C. The air permeability data revealed that the air permeability of FPUFs based on ISB increased significantly, in good agreement with the SEM images. PUF-I0 also exhibited increased air permeability after thermal treatment ($0.028$ m$^3$/min), indicating that thermal treatment itself causes cell opening for even FPUFs without ISB. However, the ISB-containing FPUFs showed better air permeability as compared with PUF-I0, suggesting that the designed reversible urethane linkages based on ISB can additionally weaken the modulus of the polymer to further promote the opening of cell windows. The maximum air permeability was obtained for the FPUF with 3 wt% ISB after the thermal treatment.

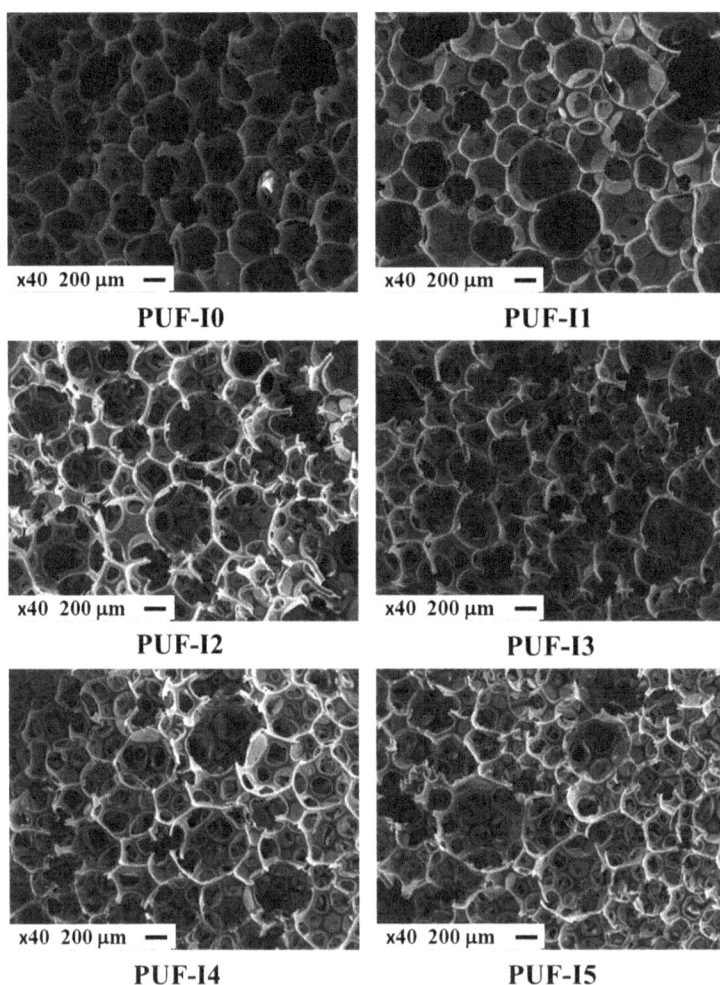

**Figure 8.** SEM images of the FPUFs after thermal treatment at 160 °C.

**Table 4.** Characteristic decomposition temperatures of various FPUF.

| Sample Code | PUF-I0 | PUF-I1 | PUF-I2 | PUF-I3 | PUF-I4 | PUF-I5 |
|---|---|---|---|---|---|---|
| $T_{5\%}$ [a] (°C) | 230.2 | 228.5 | 225.3 | 223.7 | 221.3 | 221.3 |
| $T_{10\%}$ [b] (°C) | 250.2 | 248.6 | 246.2 | 243.0 | 239.5 | 238.2 |
| $T_{max1}$ [c] (°C) | 273.3 | 274.3 | 274.3 | 274.3 | 273.3 | 274.3 |
| $T_{max2}$ [d] (°C) | 359.7 | 357.8 | 357.8 | 358.7 | 360.5 | 355.8 |

[a] $T_{5\%}$ indicates the decomposition temperature at 5 wt% of weight loss; [b] $T_{10\%}$ indicates the decomposition temperature at 10 wt% of weight loss; [c] $T_{max1}$ indicates the maximum decomposition temperature of first decomposition stage; and [d] $T_{max2}$ indicates the maximum decomposition temperature of second decomposition stage.

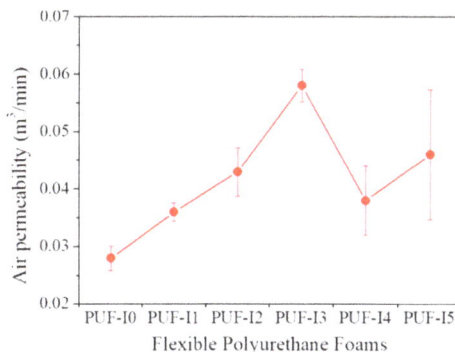

**Figure 9.** Air permeability of the FPUFs with various ISB content after thermal treatment at 160 °C.

## 3.3. Antioxidant Activity of ISB-containing FPUFs

Interestingly, we found that the difference in yellowing was significant for the ISB-containing FPUFs after thermal treatment at 160 °C. In general, polymers or plastics containing aromatic rings exhibit yellowing by atmospheric oxygen, light, and heat, causing change from the original color to yellow (yellowing). As is shown in Figure 10, all the FPUFs were bright beige before the thermal treatment. After thermal treatment, the observed discoloration was dependent on the ISB content. In other words, the control FPUF without ISB showed much yellowing after the thermal treatment at 160 °C; however, the levels of yellowing decreased as the ISB content in the FPUFs increased. Additionally, when PUF-I0 and PUF-I5 were left under ambient conditions (in air) for more than 30 days, the degree of yellowing between the two foams was noticeably different (Figure S7). The yellow index (YI) of the FPUF samples was investigated according to ASTM E313 and the results are presented in Table 5. As the content of ISB increased, the YI value of FPUFs significantly decreased. The YI value decreased from 46.40 for PUF-I0 to 19.72 for PUF-I5. Thus, the FPUFs containing ISB were demonstrated to prevent or slow FPUF oxidation caused by external factors (oxygen, light, and heat) and showed better antioxidant properties, as compared with the control FPUF without ISB. Further studies on the effects of ISB in comparison with the commercial anti-yellowing agents for FPUFs are necessary for the practical applications in industries.

**Table 5.** YI of the FPUFs after thermal treatment at 160 °C.

| Sample Code | PUF-I0 | PUF-I1 | PUF-I2 | PUF-I3 | PUF-I4 | PUF-I5 |
|---|---|---|---|---|---|---|
| YI | 46.4 | 45.3 | 35.0 | 29.2 | 23.6 | 19.7 |

**Figure 10.** Color comparison of FPUF samples (**a**) before and (**b**) after thermal treatment at 160 °C.

In general, antioxidants can possess a variety of functional groups depending on the specific starting material, which includes phenolic hydroxyls, amines, and phosphoryl groups. Specific to polymers, phenolic antioxidants or hindered amine light stabilizers are widely used as primary antioxidants to stabilize polymers or plastics from the UV-mediated oxidation by absorbing UV light or scavenging the resulting free radicals [54,65,66]. The antioxidant activity of ISB can be inferred from the following: At least some ISB may be autoxidized on exposure to atmospheric oxygen, as is shown in Scheme 2 [67]. Matsubara et al. in their study reported that tetrahydrofuran (THF) and tetrahydropyran (THP) groups can be easily oxidized at the α-position to the ring oxygen atom to form peroxides [68]. Therefore, it is postulated that ISB may prevent degradative polymer oxidation through oxygen capture or free radical scavenging. Another possibility is that the hydroxyl groups on ISB can be oxidized to form a ketone, although the oxidation of the hydroxyl group on ISB will have low levels of conversion (4%) in the absence of a catalyst (Scheme 3) [69]. The two hydroxyl groups on ISB can both be oxidized and different amounts of Gibbs free energy are required for the oxidation of these hydroxyls (compared to generic secondary alcohols), due to the higher steric hindrance of *endo*-hydroxyl groups. This means that ISB itself will demonstrate lower antioxidant activity outside of a FPUF. In the 2,2-diphenyl-1-picrylhydrazyl (DPPH) test of ISB, the free radical scavenging activity of the pure ISB solution (0.3 mM in methanol) showed a lower value (5.1%) than the equivalent value for ISB-containing PU (Figure S8).

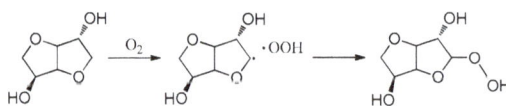

**Scheme 2.** Scheme for furan ring autoxidation in ISB [67].

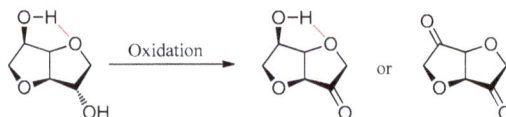

**Scheme 3.** Scheme for possible hydroxyl group oxidation in ISB [69].

As shown in Schemes 2 and 3, free radicals formed in ISB molecules can act as a radical scavenger to terminate the chain reaction of free radicals and stabilize the polymers from oxidation. Accordingly, free radical scavenging activity of ISB-containing PU films was evaluated by using the DPPH method, which is widely used for evaluation of the antioxidant activities of polymers [55,56]. Figure 11 shows free radical scavenging activity of PU samples with or without ISB, and this test was performed by using PU films containing ISB instead of FPUFs as the foams absorbed the DPPH solution. The representative

PU films with 0 and 5 wt% ISB of polyols (PU-I0 and PU-I5, respectively), were tested. The blank DPPH solution without sample showed almost constant absorbance over time while the DPPH solution containing PU films showed a decrease in absorbance over time at 515 nm. Because conventional PPG for FPUF contains an antioxidant in general, the absorbance of the ISB-free PU film (PU-I0) decreased. However, a far greater decrease in absorbance was observed for PU-I5. After immersion time of 4 h, the free radical scavenging activity calculated using Equation (1) was 68.7% and 82.3% for the PU film free of and the PU film containing ISB, respectively, which indicated that the ISB-containing PU film possesses better antioxidant activity. For future work, more experiments and analyses are needed to elucidate the antioxidant mechanism of ISB.

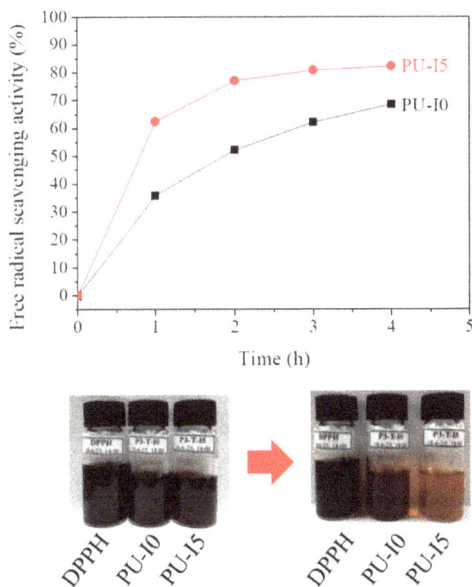

**Figure 11.** Free radical scavenging activity of PU films with or without ISB.

## 4. Conclusions

The rigid bicyclic diol, ISB, could be successfully incorporated into the FPUFs. ISB was miscible with conventional PPG up to 5 wt% of polyols, and the effects of ISB as a cell opener on the FPUFs were investigated. The maximum improvement in air permeability of FPUFs was observed in FPUF with 2 wt% of ISB in polyol. The slight decrease in air permeability of ISB-based FPUFs with further increase of ISB content above 2wt% was attributed to the increase of gelation rate, hindering the cell opening of FPUFs. The FPUFs based on ISB also exhibited comparable or superior mechanical properties (tensile strength and tear strength) in comparison with the control FPUF without ISB (PUF-I0) due to the inherent nature of ISB having a rigid and stiff bicyclic structure. Cell opening of ISB-containing FPUFs was studied additionally by heating to 160 °C, the core temperature of FPUFs during the foaming in the plants for the slab stock production. The air permeability of the FPUFs after the thermal treatment was significantly improved; in particular, the ISB-containing FPUFs showed 100% increases in air permeability compared with the control FPUF. Thus, the reversibility of the urethane linkages can be helpful for decreasing the modulus of the polymer at elevated temperatures and for promoting the cell opening.

Interestingly, the FPUFs containing ISB exhibited excellent antioxidant activity. In the DPPH test, the ISB-containing PU film showed increased free radical scavenging activity (up to 82.3%) compared with the ISB-free control (62.5%), indicating the better antioxidant activity of the ISB-based PU polymer.

*Molecules* **2019**, *24*, 1347

In addition, the yellow index of PUF-I5 was significantly lower than that of PUF-I0 after thermal treatment at 160 °C. Consequently, ISB is a promising bio-based raw material or additive for FPUFs as both an antioxidant and a cell opener without serious deterioration on the physical properties of FPUFs.

**Supplementary Materials:** Figure S1: Photographs of TF-3000 (PPG)/ISB mixture samples containing different amount of ISB. The last numbers of the sample codes denote wt% of ISB in the samples, Figure S2: Photographs of FPUFs investigated in this study, Figure S3: Schematic illustration of crosslinked network structures of (a) PU without ISB and (b) those with ISB, Figure S4: Temperature-dependent FTIR spectra of PUF-I0 and PUF-I5: (a) PUF-I0; (b) PUF-I5; (c) PUF-I0 expanded at 1660–1850 cm$^{-1}$; (d) PUF-I5 expanded at 1660–1850 cm$^{-1}$; (e) PUF-I0 expanded at 2150–2400 cm$^{-1}$; (f) PUF-I5 expanded at 2150–2400 cm$^{-1}$, Figure S5: Temperature-dependent FTIR spectra of PU films based on different concentrations of ISB, Figure S6: FTIR spectra of different FPUFs after curing at room temperature for 1 h, Figure S7: Photographs of PUF-I0 (left) and PUF-I5 (right) left at room temperature for 30 days, Figure S8: Free radical scavenging activity of ISB solution in methanol (0.3 mM) by DPPH method at room temperature, Table S1: Bio-based content of FPUFs investigated, Table S2: Sample code and formulation for PU films with various ISB content, Table S3: Shear viscosity of PPG/ISB mixture at 25 °C, Table S4: Average cell size, average thickness of cell walls, and the number of cells per unit area of FPUFs with various ISB content before and after thermal treatment.

**Author Contributions:** S.-R.S. and D.-S.L. contributed to the manuscript via literature survey and experimental design, data analysis and writing; S.-R.S., and J.-Y.L. performed the experiments and analyzed the data; H.R. and G.-S.S. investigated the potential applications.

**Acknowledgments:** This material is based upon work supported by the Ministry of Trade, Industry & Energy (MOTIE, Korea) under Industrial Technology Innovation Program. (No. 10049677, Bio-Isosynatates and Alternative Biobased Materials for Polyurethane Using Green Carbon).

**Conflicts of Interest:** The authors declare no conflict of interest.

# References

1. Szycher, M. *Szycher's Handbook of Polyurethanes*; CRC Press: New York, NY, USA, 2012; pp. 183–192.
2. Latinwo, G.K.; Aribike, D.S.; Susu, A.A.; Kareem, S.A. Effects of different filler treatments on the morphology and mechanical properties of flexible polyurethane foam composites. *Nat. Sci.* **2010**, *8*, 23–31. [CrossRef]
3. Konig, A.; Fehrenbacher, U.; Hirth, T.; Kroke, E. Flexible Polyurethane Foam with the Flame-retardant Melamine. *J. Cell. Plast.* **2008**, *44*, 469–480. [CrossRef]
4. Zhang, C.; Li, J.; Hu, Z.; Zhu, F.; Huang, Y. Correlation between the acoustic and porous cell morphology of polyurethane foam: Effect of interconnected porosity. *Mater. Des.* **2012**, *41*, 319–325. [CrossRef]
5. Kattiyaboot, T.; Thongpin, C. Effect of Natural Oil Based Polyols on the Properties of Flexible Polyurethane Foams Blown by Distilled Water. *Energy Procedia* **2016**, *89*, 177–185. [CrossRef]
6. Verdejo, R.; Stämpfli, R.; Alvarez-Lainez, M.; Mourad, S.; Rodriguez-Perez, M.A.; Brühwiler, P.A.; Shaffer, M. Enhanced acoustic damping in flexible polyurethane foams filled with carbon nanotubes. *Compos. Sci. Technol.* **2009**, *69*, 1564–1569. [CrossRef]
7. Zaretsky, E.; Asaf, Z.; Ran, E.; Aizik, F. Impact response of high density flexible polyurethane foam. *Int. J. Impact Eng.* **2012**, *39*, 1–7. [CrossRef]
8. Gwon, J.G.; Kim, S.K.; Kim, J.H. Sound absorption behavior of flexible polyurethane foams with distinct cellular structures. *Mater. Des.* **2016**, *89*, 448–454. [CrossRef]
9. Lin, Y.; Hsieh, F. Water-blown flexible polyurethane foam extended with biomass materials. *J. Appl. Polym. Sci.* **1997**, *65*, 695–703. [CrossRef]
10. Hodlur, R.; Rabinal, M. Self assembled graphene layers on polyurethane foam as a highly pressure sensitive conducting composite. *Compos. Sci. Technol.* **2014**, *90*, 160–165. [CrossRef]
11. Zhang, X.; Davis, H.; Macosko, C. A new cell opening mechanism in flexible polyurethane foam. *J. Cell. Plast.* **1999**, *35*, 458–476. [CrossRef]
12. Yasunaga, K.; Neff, R.; Zhang, X.; Macosko, C. Study of cell opening in flexible polyurethane foam. *J. Cell. Plast.* **1996**, *32*, 427–448. [CrossRef]
13. Herrington, R.; Hock, K. *Flexible Polyurethane Foams*; Dow Chemical Company: Midland, TX, USA, 1997.
14. Neff, R.A.; Macosko, C.W. Simultaneous measurement of viscoelastic changes and cell opening during processing of flexible polyurethane foam. *Rheol. Acta* **1996**, *35*, 656–666. [CrossRef]

15. Singh, A.P.; Bhattacharya, M. Viscoelastic changes and cell opening of reacting polyurethane foams from soy oil. *Polym. Eng. Sci.* **2004**, *44*, 1977–1986. [CrossRef]

16. Li, W.; Ryan, A.J.; Meier, I.K. Effect of chain extenders on the morphology development in flexible polyurethane foam. *Macromolecules* **2002**, *35*, 6306–6312. [CrossRef]

17. Aneja, A.; Wilkes, G.L. Exploring macro-and microlevel connectivity of the urea phase in slabstock flexible polyurethane foam formulations using lithium chloride as a probe. *Polymer* **2002**, *43*, 5551–5561. [CrossRef]

18. Park, J.H.; Minn, K.S.; Lee, H.R.; Yang, S.H.; Yu, C.B.; Pak, S.Y.; Oh, C.S.; Song, Y.S.; Kang, Y.J.; Youn, J.R. Cell openness manipulation of low density polyurethane foam for efficient sound absorption. *J. Sound Vibrat.* **2017**, *406*, 224–236. [CrossRef]

19. Dworakowska, S.; Bogdał, D.; Zaccheria, F.; Ravasio, N. The role of catalysis in the synthesis of polyurethane foams based on renewable raw materials. *Catal. Today* **2014**, *223*, 148–156. [CrossRef]

20. Zhang, X.; Macosko, C.; Davis, H.; Nikolov, A.; Wasan, D. Role of silicone surfactant in flexible polyurethane foam. *J. Colloid Interface Sci.* **1999**, *215*, 270–279. [CrossRef] [PubMed]

21. Rossmy, G.; Kollmeier, H.J.; Lidy, W.; Schator, H.; Wiemann, M. Mechanism of the Stabilization of Flexible Polyether Polyurethane Foams by Silicone-Based Surfactants. *J. Cell. Plast.* **2016**, *17*, 319–327. [CrossRef]

22. Rath, A.; Apichatachutapan, W.; Gummaraju, R.; Neff, R.; Heyman, D. Effect of average particle size and distribution on the performance of copolymer polyols. *J. Cell. Plast.* **2003**, *39*, 387–415. [CrossRef]

23. Ahn, W.; Lee, J.-M. Open-Cell Rigid Polyurethane Foam Using Lithium Salt of 12-Hydroxystearic acid as a Cell Opening Agent. *Polym. Korea* **2018**, *42*, 919–924. [CrossRef]

24. Dounis, D.V.; Wilkes, G.L. Structure-property relationships of flexible polyurethane foams. *Polymer* **1997**, *38*, 2819–2828. [CrossRef]

25. Zou, J.; Lei, Y.; Liang, M.; Zou, H. Effect of nano-montmorillonite as cell opener on cell morphology and resilient performance of slow-resilience flexible polyurethane foams. *J. Polym. Res.* **2015**, *22*, 201. [CrossRef]

26. Vaughan, B.R.; Wilkes, G.L.; Dounis, D.V.; McLaughlin, C. Effect of vegetable-based polyols in unimodal glass-transition polyurethane slabstock viscoelastic foams and some guidance for the control of their structure-property behavior. I. *J. Appl. Polym. Sci.* **2011**, *119*, 2683–2697. [CrossRef]

27. Javni, I.; Song, K.; Lin, J.; Petrovic, Z.S. Structure and properties of flexible polyurethane foams with nano- and micro-fillers. *J. Cell. Plast.* **2011**, *47*, 357–372. [CrossRef]

28. Kaushiva, B.D.; Dounis, D.V.; Wilkes, G.L. Influences of copolymer polyol on structural and viscoelastic properties in molded flexible polyurethane foams. *J. Appl. Polym. Sci.* **2000**, *78*, 766–786. [CrossRef]

29. Sonnenschein, M.F.; Wendt, B.L. Design and formulation of soybean oil derived flexible polyurethane foams and their underlying polymer structure/property relationships. *Polymer* **2013**, *54*, 2511–2520. [CrossRef]

30. Ugarte, L.; Saralegi, A.; Fernández, R.; Martín, L.; Corcuera, M.A.; Eceiza, A. Flexible polyurethane foams based on 100% renewably sourced polyols. *Ind. Crop. Prod.* **2014**, *62*, 545–551. [CrossRef]

31. Bernardini, J.; Cinelli, P.; Anguillesi, I.; Coltelli, M.-B.; Lazzeri, A. Flexible polyurethane foams green production employing lignin or oxypropylated lignin. *Eur. Polym. J.* **2015**, *64*, 147–156. [CrossRef]

32. Prociak, A.; Malewska, E.; Kurańska, M.; Bąk, S.; Budny, P. Flexible polyurethane foams synthesized with palm oil-based bio-polyols obtained with the use of different oxirane ring opener. *Ind. Crop. Prod.* **2018**, *115*, 69–77. [CrossRef]

33. Lan, Z.; Daga, R.; Whitehouse, R.; McCarthy, S.; Schmidt, D. Structure–properties relations in flexible polyurethane foams containing a novel bio-based crosslinker. *Polymer* **2014**, *55*, 2635–2644. [CrossRef]

34. Rashmi, B.J.; Rusu, D.; Prashantha, K.; Lacrampe, M.F.; Krawczak, P. Development of water-blown bio-based thermoplastic polyurethane foams using bio-derived chain extender. *J. Appl. Polym. Sci.* **2013**, *128*, 292–303. [CrossRef]

35. Oltmanns, J.U.; Palkovits, S.; Palkovits, R. Kinetic investigation of sorbitol and xylitol dehydration catalyzed by silicotungstic acid in water. *Appl. Catal. A Gen.* **2013**, *456*, 168–173. [CrossRef]

36. Zou, J.; Cao, D.; Tao, W.; Zhang, S.; Cui, L.; Zeng, F.; Cai, W. Sorbitol dehydration into isosorbide over a cellulose-derived solid acid catalyst. *RSC Adv.* **2016**, *6*, 49528–49536. [CrossRef]

37. Fleche, G.; Huchette, M. Isosorbide. Preparation, properties and chemistry. *Starch-Stärke* **1986**, *38*, 26–30. [CrossRef]

38. Fenouillot, F.; Rousseau, A.; Colomines, G.; Saint-Loup, R.; Pascault, J.P. Polymers from renewable 1,4:3,6-dianhydrohexitols (isosorbide, isomannide and isoidide): A review. *Prog. Polym. Sci.* **2010**, *35*, 578–622. [CrossRef]

39. Dussenne, C.; Delaunay, T.; Wiatz, V.; Wyart, H.; Suisse, I.; Sauthier, M. Synthesis of isosorbide: An overview of challenging reactions. *Green Chem.* **2017**, *19*, 5332–5344. [CrossRef]

40. Kasmi, N.; Roso, M.; Hammami, N.; Majdoub, M.; Boaretti, C.; Sgarbossa, P.; Vianello, C.; Maschio, G.; Modesti, M.; Lorenzetti, A. Microwave-assisted synthesis of isosorbide-derived diols for the preparation of thermally stable thermoplastic polyurethane. *Des. Monomers Polym.* **2017**, *20*, 547–563. [CrossRef]

41. Smiga-Matuszowicz, M.; Janicki, B.; Jaszcz, K.; Lukaszczyk, J.; Kaczmarek, M.; Lesiak, M.; Sieron, A.L.; Simka, W.; Mierzwinski, M.; Kusz, D. Novel bioactive polyester scaffolds prepared from unsaturated resins based on isosorbide and succinic acid. *Mater. Sci. Eng. C* **2014**, *45*, 64–71. [CrossRef] [PubMed]

42. Liu, W.; Xie, T.; Qiu, R. Biobased Thermosets Prepared from Rigid Isosorbide and Flexible Soybean Oil Derivatives. *ACS Sustain. Chem. Eng.* **2016**, *5*, 774–783. [CrossRef]

43. Park, S.-A.; Choi, J.; Ju, S.; Jegal, J.; Lee, K.M.; Hwang, S.Y.; Oh, D.X.; Park, J. Copolycarbonates of bio-based rigid isosorbide and flexible 1,4-cyclohexanedimethanol: Merits over bisphenol-A based polycarbonates. *Polymer* **2017**, *116*, 153–159. [CrossRef]

44. Besse, V.; Auvergne, R.; Carlotti, S.; Boutevin, G.; Otazaghine, B.; Caillol, S.; Pascault, J.-P.; Boutevin, B. Synthesis of isosorbide based polyurethanes: An isocyanate free method. *React. Funct. Polym.* **2013**, *73*, 588–594. [CrossRef]

45. Noordover, B.A.; van Staalduinen, V.G.; Duchateau, R.; Koning, C.E.; van Benthem, R.A.; Mak, M.; Heise, A.; Frissen, A.E.; van Haveren, J. Co-and terpolyesters based on isosorbide and succinic acid for coating applications: Synthesis and characterization. *Biomacromolecules* **2006**, *7*, 3406–3416. [CrossRef] [PubMed]

46. Ravey, M.; Pearce, E.M. Flexible polyurethane foam. I. Thermal decomposition of a polyether-based, water-blown commercial type of flexible polyurethane foam. *J. Appl. Polym. Sci.* **1997**, *63*, 47–74. [CrossRef]

47. Delebecq, E.; Pascault, J.P.; Boutevin, B.; Ganachaud, F. On the versatility of urethane/urea bonds: Reversibility, blocked isocyanate, and non-isocyanate polyurethane. *Chem. Rev.* **2013**, *113*, 80–118. [CrossRef] [PubMed]

48. Rolph, M.S.; Markowska, A.L.J.; Warriner, C.N.; O'Reilly, R.K. Blocked isocyanates: From analytical and experimental considerations to non-polyurethane applications. *Polym. Chem.* **2016**, *7*, 7351–7364. [CrossRef]

49. Nasar, A.S.; Kalaimani, S. Synthesis and studies on forward and reverse reactions of phenol-blocked polyisocyanates: An insight into blocked isocyanates. *RSC Adv.* **2016**, *6*, 76802–76812. [CrossRef]

50. Kalaimani, S.; Nasar, A.S. Catalysis of deblocking and cure reactions of easily cleavable phenol blocked polyisocyanates with poly(polytetrahydrofuran carbonate) diol. *Eur. Polym. J.* **2017**, *91*, 221–231. [CrossRef]

51. Elwell, M.J.; Mortimer, S.; Ryan, A.J. A synchrotron SAXS study of structure development kinetics during the reactive processing of flexible polyurethane foam. *Macromolecules* **1994**, *27*, 5428–5439. [CrossRef]

52. Blair, G.R.; McEvoy, J.; de Priamus, M.R.; Dawe, B.; Pask, R.; Wright, C. *The Effect of Visible Light on the Variability of Flexible Foam Compression Sets*; Centre of the Polyurethane Industry, American Chemistry Council: Washington, DC, USA, 2007.

53. Newman, C.R.; Forciniti, D. Modeling the ultraviolet photodegradation of rigid polyurethane foams. *Ind. Eng. Chem. Res.* **2001**, *40*, 3346–3352. [CrossRef]

54. Jung, B.-O.; Chung, S.-J.; Lee, S.B. Preparation and characterization of eugenol-grafted chitosan hydrogels and their antioxidant activities. *J. Appl. Polym. Sci.* **2006**, *99*, 3500–3506. [CrossRef]

55. Modjinou, T.; Versace, D.-L.; Abbad-Andaloussi, S.; Langlois, V.; Renard, E. Antibacterial and antioxidant photoinitiated epoxy co-networks of resorcinol and eugenol derivatives. *Mater. Today Commun.* **2017**, *12*, 19–28. [CrossRef]

56. Brand-Williams, W.; Cuvelier, M.-E.; Berset, C. Use of a free radical method to evaluate antioxidant activity. *LWT-Food Sci. Technol.* **1995**, *28*, 25–30. [CrossRef]

57. Suminokura, T.; Sasaki, T.; Aoki, N. Effect of the cell membrane on mechanical properties of flexible polyurethane foams. *Jpn. J. Appl. Phys.* **1968**, *7*, 330. [CrossRef]

58. Gabbard, J.D. Flexible Water-Blown Polyurethane Foams. U.S. Patent 5,624,968, 29 April 1997.

59. Wang, W.; Gong, W.; Zheng, B. Preparation of low-density polyethylene foams with high rebound resilience by blending with polyethylene-octylene elastomer. *Polym. Eng. Sci.* **2013**, *53*, 2527–2534. [CrossRef]

60. Li, F.; Hou, J.; Zhu, W.; Zhang, X.; Xu, M.; Luo, X.; Ma, D.; Kim, B.K. Crystallinity and morphology of segmented polyurethanes with different soft-segment length. *J. Appl. Polym. Sci.* **1996**, *62*, 631–638. [CrossRef]

61. Indennidate, L.; Cannoletta, D.; Lionetto, F.; Greco, A.; Maffezzoli, A. Nanofilled polyols for viscoelastic polyurethane foams. *Polym. Int.* **2010**, *59*, 486–491. [CrossRef]

62. Yuan, Q.; Zhou, T.; Li, L.; Zhang, J.; Liu, X.; Ke, X.; Zhang, A. Hydrogen bond breaking of TPU upon heating: Understanding from the viewpoints of molecular movements and enthalpy. *RSC Adv.* **2015**, *5*, 31153–31165. [CrossRef]

63. Deng, Y.; Li, S.; Zhao, J.; Zhang, Z.; Zhang, J.; Yang, W. Crystallizable and tough aliphatic thermoplastic poly (ether urethane) s synthesized through a non-isocyanate route. *RSC Adv.* **2014**, *4*, 43406–43414. [CrossRef]

64. Allan, D.; Daly, J.; Liggat, J.J. Thermal volatilisation analysis of TDI-based flexible polyurethane foam. *Polym. Degrad. Stab.* **2013**, *98*, 535–541. [CrossRef]

65. Geoffrey, P. *Plastics Additives: An AZ Reference*; Chapman & Hall Publisher: London, UK, 1998; pp. 55–107.

66. Bolgar, M.; Hubball, J.; Groeger, J.; Meronek, S. *Handbook for the Chemical Analysis of Plastic and Polymer Additives*; CRC Press: New York, NY, USA, 2007; p. 61.

67. Hagberg, E.; Rockafellow, E.M.; Smith, B.; Stensrud, K.F. Hydrogenation of Isohexide Products for Improved Color. U.S. Patent 9,321,784, 26 April 2016.

68. Matsubara, H.; Suzuki, S.; Hirano, S. An ab initio and DFT study of the autoxidation of THF and THP. *Org. Biomol. Chem.* **2015**, *13*, 4686–4692. [CrossRef]

69. Gross, J.; Tauber, K.; Fuchs, M.; Schmidt, N.G.; Rajagopalan, A.; Faber, K.; Fabian, W.M.; Pfeffer, J.; Haas, T.; Kroutil, W. Aerobic oxidation of isosorbide and isomannide employing TEMPO/laccase. *Green Chem.* **2014**, *16*, 2117–2121. [CrossRef]

**Sample Availability:** Samples of the compounds are not available from the authors.

*molecules*

MDPI

*Article*

# Synthesis and In Vitro Evaluation of Caffeoylquinic Acid Derivatives as Potential Hypolipidemic Agents

Yu Tian [1,†] , Xiao-Xue Cao [1,†], Hai Shang [1], Chong-Ming Wu [1], Xi Zhang [2], Peng Guo [1,*], Xiao-Po Zhang [3,*] and Xu-Dong Xu [1,*]

1    Beijing Key Laboratory of Innovative Drug Discovery of Traditional Chinese Medicine (Natural Medicine) and Translational Medicine; Key Laboratory of Bioactive Substances and Resources Utilization of Chinese Herbal Medicine, Ministry of Education; Key Laboratory of Efficacy Evaluation of Chinese Medicine against Glycolipid Metabolic Disorders, State Administration of Traditional Chinese Medicine; Zhong Guan Cun Open Laboratory of the Research and Development of Natural Medicine and Health Products; Key Laboratory of New Drug Discovery Based on Classic Chinese Medicine Prescription; Key Laboratory of New Drug Discovery Based on Classic Chinese Academy of Medical Sciences; Institute of Medicinal Plant Development, Chinese Academy of Medical Sciences & Peking Union Medical College, Beijing 100193, China; ytian@implad.ac.cn (Y.T.); snow2018cxx@163.com (X.-X.C.); hshang@implad.ac.cn (H.S.); cmwu@implad.ac.cn (C.-M.W.)
2    Center of Research and Development on Life Sciences and Environment Sciences, Harbin University of Commerce, Harbin 150076, China; 18800467885@163.com
3    School of Pharmacy, Hainan Medical University, Haikou 571199, China
*    Correspondence: pguo@implad.ac.cn (P.G.); z_xp1412@163.com (X.-P.Z.); xdxu@implad.ac.cn (X.-D.X.); Tel.: +010-5783-3235 (P.G.); +0898-6689-3826 (X.-P.Z.); +010-5783-3013 (X.-D.X.)
†    These authors contributed equally to this work.

Academic Editors: Pavel B. Drasar and Vladimir A. Khripach
Received: 14 January 2019; Accepted: 5 March 2019; Published: 9 March 2019

check for updates

**Abstract:** A series of novel caffeoylquinic acid derivatives of chlorogenic acid have been designed and synthesized. Biological evaluation indicated that several synthesized derivatives exhibited moderate to good lipid-lowering effects on oleic acid-elicited lipid accumulation in HepG2 liver cells. Particularly, derivatives **3d**, **3g**, **4c** and **4d** exhibited more potent lipid-lowering effect than the positive control simvastatin and chlorogenic acid. Further studies on the mechanism of **3d**, **3g**, **4c** and **4d** revealed that the lipid-lowering effects were related to their regulation of TG levels and merit further investigation.

**Keywords:** caffeoylquinic acids; chlorogenic acid; derivatives; lipid-lowering effects; oleic acid-elicited; HepG2 cells

## 1. Introduction

Hyperlipidemia is among the key risk factors for cardio-vascular diseases (CVD), inducing essential hypertension (EHT), coronary heart disease (CHD), heart failure, atherosclerosis (AS), and sudden cardiac death (SCD). Several studies [1–5] have provided evidence that low-density lipoprotein cholesterol (LDL-c) has been considered as the main risk factor of CHD and AS [6,7], and the increased circulating levels of triglycerides (TG) show a significant influence on CHD and AS [8,9]. Therefore, it is quite beneficial to research and explore anti-hypolipidemic agents that could modulate the dysregulation of lipid metabolism and decrease the elevated levels of serum TG [10].

*Pandanus tectorius* (L.) Parkins. (PTPs, Figure 1), distributed in South China, is a traditional folk medicine for the treatment of leprosy, bronchitis, measles, dermatitis, and diabetes [11,12]. Chlorogenic acid (3-*O*-caffeoylquinic acid, 1, Figure 1) is one of the secondary phenolic metabolites, which is one of the the primary components found in PTPs, and was reported to have potential antioxidant, antimicrobial, anti-inflammatory, and anti-hpyerglycemic effects [13–16]. In particular, chlorogenic acids and their analogues, which contain one or more caffeoylquinic acid scaffolds, also efficiently improve hypercholesterolemia and hyperglycemia. In addition, as a kind of rich natural product, chlorogenic acids and their analogues are widely distributed in a variety of plants that show hypolipidemic functions, such as *Hypericum salsugineum*, *L. czeczottianus*, *L. nissolia*, *Hypericum androsaemum* L., and *Ipomea batatas* L. [17–20]. Moreover, our previous studies revealed that some of the natural caffeoylquinic acids were confirmed to possess obvious anti-hyperlipidemia activity and could significantly inhibit lipid accumulation and TG levels in the liver [10,12]. Recently, we also found that the chlorogenic acid derivatives (chlorogenic acid diketal, 2, Figure 1) could still significantly decrease lipid accumulation in oleic acid-elicited HepG2 cells, in spite of the ketal substituent on the 1,4,5-position hydroxyl and carboxyl group. However, despite the promising and diverse pharmacological properties of caffeoylquinic acid analogues, synthesis and investigation of the structure–activity relationships of caffeoylquinic acid derivatives' lipid-decreasing effect has not been discussed yet.

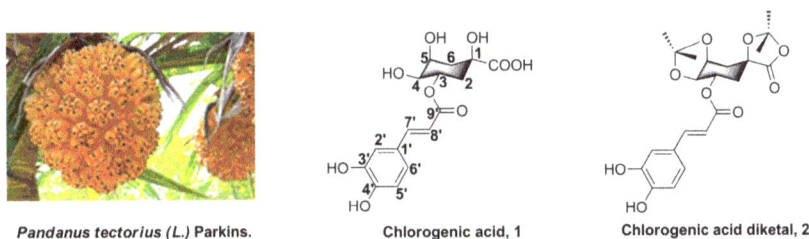

*Pandanus tectorius (L.)* Parkins.　　　　Chlorogenic acid, 1　　　　Chlorogenic acid diketal, 2

**Figure 1.** The plant *Pandanus tectorius* (L.) Parkins. and the structures of chlorogenic acid (**1**) and its analogues Chlorogenic acid diketal (**2**).

In this present research, we describe the synthesis of fourteen novel caffeoylquinic acid derivatives with various amines (n-butylamine, isobutylamine, n-octylamine, propargylamine, benzylamine, piperidine, and cyclohexylamine) which were appended to the 1-position carboxyl group of chlorogenic acid. Meanwhile, in view of the remaining chlorogenic acid diketal attenuating lipid accumulation activity, we also designed the amide derivatives with 4, 5-position hydroxyl groups substituted by ketal (isopropylidene). Therefore, the purpose of this study was to synthesize novel caffeoylquinic acid derivatives (**3a–g**, **4a–g**, Figure 2) and investigate the effects of derivatives on oleic acid-elicited lipid accumulation in HepG2 cells. The underlying lipid-lowering effect mechanisms of derivatives were also elucidated using intracellular TG quantification.

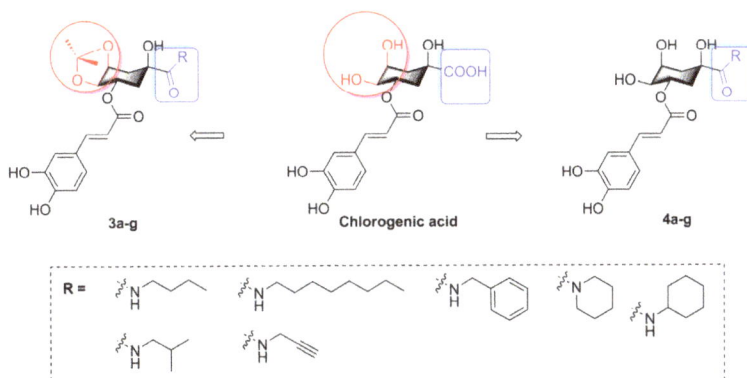

**Figure 2.** The structures of chlorogenic acid derivatives.

## 2. Results and Discussion

### 2.1. Chemistry

The synthesis of derivatives **3a–g** and **4a–g** is outlined in Scheme 1. The naturally abundant chlorogenic acid (**1**) was treated with dimethyoxypropane (DMP) and p-toluenesulfonic acid (TsOH) in dry acetone to obtain chlorogenic acid ketal derivative **5**. Compounds **3a–g** were attained via amidation with various amines (n-butylamine, isobutylamine, n-octylamine, propargylamine, benzylamine, piperidine, and cyclohexylamine) of 1-position carboxyl group of compound **5** in BOP (benzotriazol-1-yl-oxytris (dimethylamino) phosphoniumhexa-fluorophosphate) and DIEA (dimethyltriethylamine) conditions, and then followed by deprotection at 4,5-position ketal in the presence of a TFA/DCM (trifluoroacetic acid/ dichloromethane) solution, to gain compounds **4a–g**.

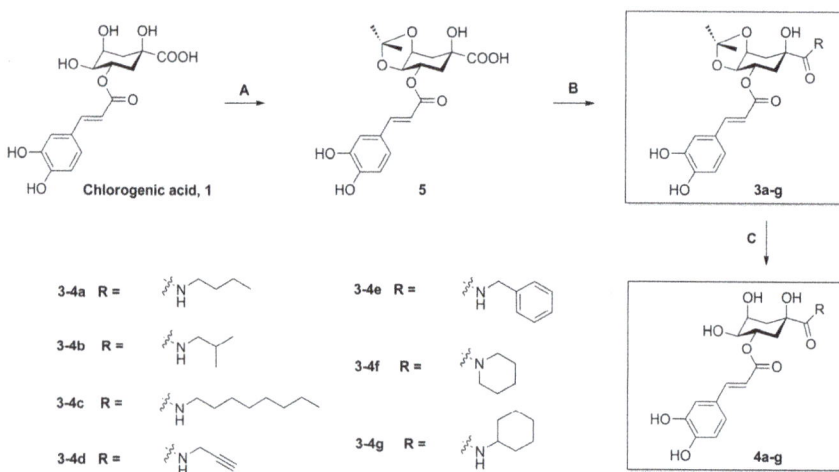

**Scheme 1.** The synthesis of compounds **3a–g** and **4a–g**. Reagents and conditions: (**A**) DMP, TsOH, acetone; (**B**) BOP, DIEA, amines; (**c**) TFA, DCM, $H_2O$ (9:1:1).

### 2.2. Biological Results and Discussion

#### 2.2.1. Lipid-Lowering Effects of Chlorogenic Acid and Compounds **3a–g** and **4a–g** on Oleic Acid-Elicited Lipid Accumulation in HepG2 Liver Cells

The lipid-lowering effects of chlorogenic acid and its derivatives against oleic acid-elicited lipid accumulation in HepG2 liver cells were detected using the Oil red O staining assay. The MTS assay indicated that 10 μmol/L chlorogenic acid derivatives **3a–3g** (Figure 3a) and **4a–4g** (Figure 3b) displayed no cytotoxicity, within 24 h, to HepG2 liver cells. Oil red O staining results showed that some derivatives exhibited moderate to good regulation effects on oleic acid-elicited lipid accumulation in HepG2 liver cells. As shown in Figure 3c–d, the preliminary test of compounds **3a–g** and **4a–g** at 10 μmol/L revealed that **3d**, **3f**, **3g**, **4a**, and **4c–g** better moderated the lipid-lowering effects of HepG2 cells. Treatment of HepG2 liver cells with compounds **3d**, **3g**, **4a**, and **4c–g** exhibited better regulating effects than chlorogenic acid (CA). The absorbance decreased from 0.261 (with oleic acid treatment alone) to 0.249, 0.255, and 0.249 after treatment with 10 μmol/L compounds **3d**, **3f**, and **3g** in Figure 3c, respectively. In Figure 3d, the absorbance decreased from 0.282 (oleic acid alone) to 0.271, 0.267, 0.270, 0.273, 0.271, and 0.272 after treatment with 10 μmol/L derivatives **4a** and **4c–g** in Figure 3b, respectively. The Oil red O staining images are shown in Figure 3e. Among them, derivatives **3d**, **3g**, **4c**, and **4d** exhibited more potent regulation effects than the others, after treatment for 24 h, which suggests that

**3d**, **3g**, **4c**, and **4d** deserve further evaluation as potential hypolipidemic agents for regulation on oleic acid-elicited lipid accumulation in HepG2 liver cells.

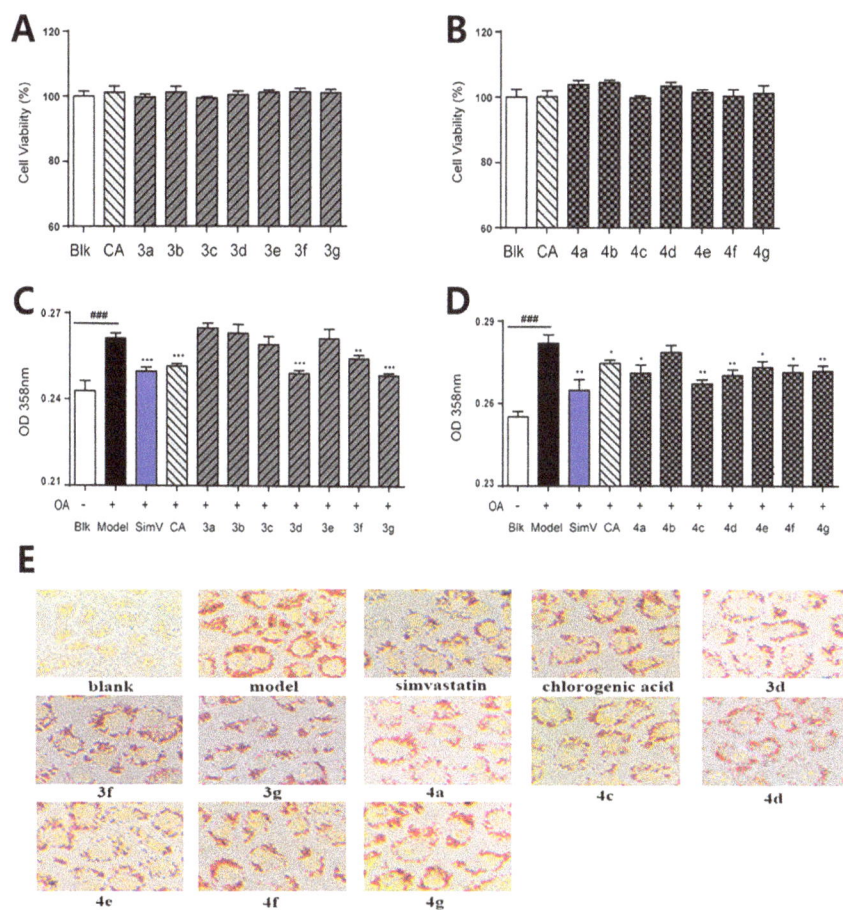

**Figure 3.** The nontoxic effects of chlorogenic acid and its derivatives on cell viability. (**A**) Cells treated with 10 µmol/L compounds **3a**–**3g**, for 24 h, respectively, and (**B**) compounds **4a**–**g,** for 24 h, respectively. The regulation effects of chlorogenic acid and its derivatives on oleic acid-elicited lipid accumulation in HepG2 liver cells. HepG2 cells were co-incubated in serum-free DMEM containing 100 µmol/L oleic acid (OA) (Model) and co-treated with 10 µmol/L of simvastatin (SimV) or chlorogenic acid (CA), (**C**) compounds **3a**–**g**, for 24 h, respectively, and (**D**) compounds **4a**–**g**, for 24 h, respectively. (**E**) Oil red O staining images. The data are expressed as the mean ± SEM (*n* = 8) from three independent experiments. ### *p* < 0.001 versus blank group (Blk); * *p* < 0.05 versus OA-treated cells; ** *p* < 0.01 versus OA-treated cells; *** *p* < 0.001 versus OA-treated cells.

The primary structure-activity relationships (SARs) suggested that, compared to the positive groups, simvastatin (SimV) and chlorogenic acid (CA), the amide derivatives of chlorogenic acid with 4, 5-position hydroxyl groups substituted by ketal had weaker potency than those with 4, 5-position hydroxyl groups exposed. Beyond that, the derivatives **3**–**4d**, **3**–**4f**, and **3**–**4g** with propynylamine, piperidine and cyclohexylamine groups could better regulate lipid accumulation after oleic acid treatment, indicating that the introduction of propargyl, piperidyl, and cyclohexyl groups could ameliorate lipid accumulation. Among them, derivative **3**–**4d** exhibited more potent lipid-lowering effects compared to the other compounds, which means that the propargyl group is a favourable substituent for regulation of lipid accumulation in HepG2 liver cells. However, derivatives **3**–**4b**, including isobutyl group, were inert, suggesting that the isobutyl group was negative factor in the derivatives. In addition, the above derivatives show lipid-lowering effects in Figure 3a–b, illustrating that the 1-position carboxyl group and 4, 5-position hydroxyl groups of chlorogenic acid were not essential groups for regulating lipid effects and different groups on the 1-position carboxyl group influenced lipid-lowering activities obviously.

### 2.2.2. Inhibitory Effects of Compounds **3d**, **3g**, **4c**, and **4d** Toward Triglycerides

On account of the preferable potency and typical structure, derivatives **3d**, **3g**, **4c**, and **4d** were chosen for further investigation with simvastatin and chlorogenic acid. We measured the levels of TG in HepG2 liver cells and found that consumptions of **3d**, **3g**, **4c**, and **4d** markedly reduced the content of TG in liver cells (49.63%, 48.56%, 49.00%, and 48.72% by 10 μmol/L derivatives, respectively). The efficacy of **3d**, **3g**, **4c**, and **4d** in decreasing TG levels was a little weaker than that of simvastatin, but better than that of chlorogenic acid, in decreasing TG levels at 10 μmol/L (Figure 4). Taken together, these results indicate that derivatives **3d**, **3g**, **4c**, and **4d** inhibited TG levels markedly on oleic acid-elicited lipid accumulation in HepG2 liver cells.

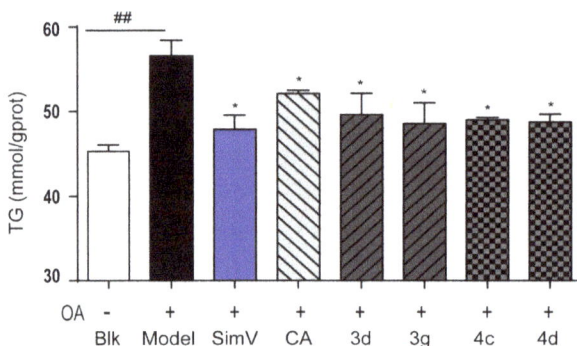

**Figure 4.** Inhibitory effects of compounds toward triglycerides in HepG2 liver cells. Positive control: simvastatin (SimV, 47.89%) and chlorogenic acid (CA, 51.51%); blank (Blk, 45.30%): DMEM; OA (56.61%): oleic acid. Intracellular levels of triglycerides were measured by kits according to the manufacturer's instructions. Bars depict the means $\pm$ SEM in triplicate. ## $p < 0.01$ versus blank group (Blk); * $p < 0.05$ versus OA-treated cells.

## 3. Experimental Section

### 3.1. General Information

All the reagents were used without further purification unless otherwise specified. Solvents were dried and redistilled prior to use in the usual manner. Analytical TLC was performed using silica gel HF254. Preparative column chromatography was performed with silica gel H. $^1$H and $^{13}$C-NMR spectra were recorded on a Bruker Advance III 600 MHz spectrometer. HRMS was obtained on a Thermofisher LTQ-Obitrap XL. Chlorogenic acid, DMP (Dimethyoxypropane), TsOH (p-Toluenesulfonic acid), BOP (Benzotriazol-1-yl-oxytris (dimethylamino)phosphoniumhexa-fluorophosphate), DIEA (Dimethyltriethylamine), TFA (trifluoroacetic acid), and DCM (dichloromethane) were purchased from the Energy Chemical Company. Dulbecco's modified Eagle medium (DMEM) fetal bovine serum was purchased from Corning Inc (Corning, CA, USA). Penicillin and streptomycin were procured from Hyclone (South Logan, UT, USA). Simvastatin, oil-red O, oleic acid (OA), and dimethyl Sulphoxide DMSO were purchased from Sigma-Aldrich (St. Louis, MO, USA). The kits for Triglyceride (TG) were purchased from Jian Cheng Biotechnology Company (Nanjing, China). The Total RNA extraction reagent Trizol (Ambion, Austin, TX, USA), the PrimeScript RT reagent kit, and the SYBR-Green PCR kit were purchased from Transgene Biotech, Inc. (Beijing, China).

### 3.2. Chemistry

#### 3.2.1. Procedure for the Synthesis of Intermediate 5

To a suspension of chlorogenic acid (7.0 g, 19.8 mmol) in dry acetone (60 mL) and DMP (40 mL), catalytic amount of TsOH (50 mg, 0.26 mmol) was added. Then the reaction mixture was stirred at room temperature for 30 min. Reaction was monitored by TLC. The crude mixture was neutralized with $Na_2CO_3$ powder to pH 6. Then the suspension was filtered out and the filtrate was evaporated and the crude product was subjected to column chromatography (eluent: PE-EtOAc, 1:1) to offer a pure light-yellow solid compound **5** (6.8 g, 87% yield).

#### 3.2.2. General Procedure for the Synthesis of Compounds 3a–g

To a mixture of compound **5** (2.4 g, 6.1 mmol) and BOP (2.7 g, 6.1 mmol) in dry THF (100 mL), organic base DIEA (1.6 g, 12.2 mmol) was added and stirred at room temperature under $N_2$ air. Then various amines (n-butylamine, isobutylamine, n-octylamine, propargylamine, benzylamine, piperidine, and cyclohexylamine) (6.7 mmol) were dropped and reacted respectively for 4–16 h. When complete, the reaction solvent was evaporated and the crude product was subjected to column chromatography (eluent: DCM-CH$_3$OH, 10:1) to gain pure compound **3a–g** as a light-yellow solid.

*1-N-n-butyl-4, 5-di-O-Isopropylidene-Chlorogenic Acid Amide* (**3a**): Yellowish powder, 81% yield; 1H-NMR (600 MHz, DMSO) $\delta$: 9.58(s, 1H, Ph-OH), 9.14(s, 1H, Ph-OH), 7.71 (t, $J$ = 6.0 Hz, 1H, NH), 7.48 (d, $J$ = 15.8 Hz, 1H, H-7′), 7.05 (d, $J$ = 1.7 Hz, 1H, H-2′), 7.00 (dd, $J$ = 8.3 Hz, 1.7 Hz, 1H, H-6′), 6.77 (d, $J$ = 8.2 Hz, 1H, H-5′), 6.24 (d, $J$ = 15.9 Hz, 1H, H-8′), 5.41 (s, 1H, H-1), 5.37–5.32 (m, 1H, H-3), 4.43–4.41 (m, 1H, H-4), 4.13–4.11 (m, 1H, H-5), 3.11–3.01 (m, 2H, NH-C$\underline{H}_2$), 2.26–2.22 (m, 1H, H-2), 1.99–1.94 (m, 1H, H-2), 1.80–1.73 (m, 2H, H-6), 1.42 (s, 3H, CH$_3$-C-O), 1.39–1.36 (m, 2H, NH-CH$_2$-C$\underline{H}_2$), 1.26–1.22 (m, 5H, C$\underline{H}_3$-C-O, C$\underline{H}_2$-CH$_3$), 0.84 (t, $J$ =7.3 Hz, 3H, CH$_2$-C$\underline{H}_3$); $^{13}$C-NMR (150 MHz, DMSO) $\delta$: 175.0, 165.9, 148.4, 145.6, 145.3, 125.6, 121.4, 115.8, 114.9, 114.1, 108.0, 76.9, 74.1, 73.5, 70.9, 38.2, 37.0, 34.5, 31.2, 28.0, 25.9, 19.4, 13.7; HRMS (ESI): Calcd for [M + Na]$^+$ C$_{23}$H$_{31}$NNaO$_8$: 472.1947, found 472.1948.

*1-N-isobutyl-4, 5-di-O-Isopropylidene-Chlorogenic Acid Amide* (**3b**): Yellowish powder, 77% yield; $^1$H-NMR (600 MHz, DMSO) $\delta$: 7.70 (t, $J$ = 6.1 Hz, 1H, NH), 7.47 (d, $J$ =15.8 Hz, 1H, H-7′), 7.04 (d, $J$ = 2.0 Hz, 1H, H-2′), 7.00 (dd, $J$ = 8.2 Hz, 2.0 Hz, 1H, H-6′), 6.76 (d, $J$ = 8.2 Hz, 1H, H-5′), 6.24 (d, $J$ = 15.9 Hz, 1H, H-8′), 5.47 (br, 1H, H-1), 5.37–5.33 (m, 1H, H-3), 4.43–4.41 (m, 1H, H-4), 4.14–4.11 (m, 1H, H-5), 2.94–2.86 (m, 2H, NH-C$\underline{H}_2$), 2.26–2.22 (m, 1H, H-2), 1.99–1.96 (m, 1H, H-2), 1.80–1.75 (m, 2H, H-6), 1.74–1.69 (m, 1H, NH-CH$_2$-C$\underline{H}$), 1.42 (s, 3H, CH$_3$-C-O), 1.26 (s, 3H, CH$_3$-C-O), 0.82–0.80 (m, 6H, 2×C$\underline{H}_3$); $^{13}$C-NMR (150 MHz, DMSO) $\delta$: 175.1, 165.9, 148.5, 145.6, 145.4, 125.6, 121.4, 115.8, 114.9, 114.1, 108.0, 76.9, 74.2, 73.6, 70.9, 45.9, 37.1, 34.4, 28.1, 26.0, 20.0; HRMS (ESI): Calcd for [M + Na]$^+$ C$_{23}$H$_{31}$NNaO$_8$: 472.1947, found 472.1946.

*1-N-n-octyl-4, 5-di-O-Isopropylidene-Chlorogenic Acid Amide* (**3c**): Yellowish powder, 82% yield; $^1$H-NMR (600 MHz, DMSO) $\delta$: 7.74 (t, $J$ = 5.9 Hz, 1H, NH), 7.47 (d, $J$ =15.8 Hz, 1H, H-7′), 7.04 (d, $J$ = 2.0 Hz, 1H, H-2′), 7.00 (dd, $J$ = 8.2 Hz, 2.0 Hz, 1H, H-6′), 6.76 (d, $J$ = 8.2 Hz, 1H, H-5′), 6.24 (d, $J$ = 15.9 Hz, 1H, H-8′), 5.44 (br, 1H, H-1), 5.36–5.31 (m, 1H, H-3), 4.43–4.40 (m, 1H, H-4), 4.13–4.11 (m, 1H, H-5), 3.10–2.98 (m, 2H, NH-C$\underline{H}_2$), 2.25–2.22 (m, 1H, H-2), 1.97–1.94 (m, 1H, H-2), 1.75–1.74 (m, 2H, H-6), 1.41 (s, 3H, CH$_3$-C-O), 1.40–1.37 (m, 2H, NH-CH$_2$-C$\underline{H}_2$), 1.25 (s, 3H, CH$_3$-C-O), 1.25–1.19 (m, 10H, (C$\underline{H}_2$)$_5$-CH$_3$), 0.83 (t, $J$ =7.1 Hz, 3H, (CH$_2$)$_5$-C$\underline{H}_3$); $^{13}$C-NMR (150 MHz, DMSO) $\delta$: 174.9, 165.8, 148.4, 145.5, 145.2, 125.5, 121.3, 115.7, 114.8, 114.0, 107.9, 76.8, 73.9, 73.4, 70.8, 38.4, 37.0, 34.5, 31.1, 28.9, 28.6, 28.5, 27.9, 26.1, 25.8, 22.0, 13.8; HRMS (ESI): Calcd for [M + Na]$^+$ C$_{27}$H$_{39}$NNaO$_8$: 528.2573, found 528.2578.

*1-N-propargyl-4, 5-di-O-Isopropylidene-Chlorogenic Acid Amide* (**3d**): Yellowish powder, 75% yield; $^1$H-NMR (600 MHz, DMSO) $\delta$: 8.16 (t, $J$ = 5.8 Hz, 1H, NH), 7.47 (d, $J$ =15.8 Hz, 1H, H-7'), 7.04 (d, $J$ = 2.0 Hz, 1H, H-2'), 7.00 (dd, $J$ = 8.2 Hz, 2.0 Hz, 1H, H-6'), 6.76 (d, $J$ = 8.2 Hz, 1H, H-5'), 6.24 (d, $J$ = 15.8 Hz, 1H, H-8'), 5.50 (s, 1H, H-1), 5.35–5.31 (m, 1H, H-3), 4.43–4.41 (m, 1H, H-4), 4.14–4.12 (m, 1H, H-5), 3.84 (dd, $J$ = 5.8 Hz, 2.5 Hz, 2H, NH-CH$_2$), 3.03 (t, $J$ = 2.5 Hz, 1H, CCH), 2.25–2.22 (m, 1H, H-2), 2.00–1.95 (m, 1H, H-2), 1.79–1.70 (m, 2H, H-6), 1.42 (s, 3H, CH$_3$-C-O), 1.26 (s, 3H, CH$_3$-C-O); $^{13}$C-NMR (150 MHz, DMSO) $\delta$: 175.3, 166.0, 148.5, 145.7, 145.5, 125.7, 121.5, 115.9, 114.9, 114.2, 108.2, 81.2, 76.9, 74.2, 73.6, 72.6, 70.9, 37.0, 34.3, 28.4, 28.1, 26.0; HRMS (ESI): Calcd for [M + Na]$^+$ C$_{22}$H$_{25}$NNaO$_8$: 454.1478, found 454.1480.

*1-N-benzyl-4, 5-di-O-Isopropylidene-Chlorogenic Acid Amide* (**3e**): Yellowish powder, 79% yield; $^1$H-NMR (600 MHz, DMSO) $\delta$: 8.32 (t, $J$ = 6.2 Hz, 1H, NH), 7.48 (d, $J$ =15.8 Hz, 1H, H-7'), 7.31–7.28 (m, 2H, Ph-H), 7.24–7.20 (m, 3H, Ph-H), 7.05 (d, $J$ = 2.0 Hz, 1H, H-2'), 7.01 (dd, $J$ = 8.2 Hz, 1H, H-6'), 6.77 (d, $J$ = 8.2 Hz, 1H, H-5'), 6.25 (d, $J$ = 15.9 Hz, 1H, H-8'), 5.55 (br, 1H, H-1), 5.39–5.35 (m, 1H, H-3), 4.45–4.42 (m, 1H, H-4), 4.32–4.25 (m, 2H, NH-CH$_2$), 4.15–4.13 (m, 1H, H-5), 2.29–2.25 (m, 1H, H-2), 2.04–2.01 (m, 1H, H-2), 1.84–1.76 (m, 2H, H-6), 1.43 (s, 3H, CH$_3$-C-O), 1.26 (s, 3H, CH$_3$-C-O); $^{13}$C-NMR (150 MHz, DMSO) $\delta$: 175.2, 165.8, 148.4, 145.5, 145.3, 139.5, 128.1, 126.9, 126.7, 125.5, 121.3, 115.7, 114.8, 114.0, 108.0, 76.8, 74.1, 73.4, 70.8, 42.0, 37.0, 34.5, 27.9, 25.8; HRMS (ESI): Calcd for [M + Na]$^+$ C$_{26}$H$_{29}$NNaO$_8$: 506.1791, found 506.1790.

*1-N-piperidyl-4, 5-di-O-Isopropylidene-Chlorogenic Acid Amide* (**3f**): Yellowish powder, 64% yield; $^1$H-NMR (600 MHz, DMSO) $\delta$: 9.59(s, 1H, Ph-OH), 9.15(s, 1H, Ph-OH), 7.48 (d, $J$ =15.8 Hz, 1H, H-7'), 7.04 (d, $J$ = 1.5 Hz, 1H, H-2'), 7.00 (dd, $J$ = 8.1 Hz, 1.5 Hz, 1H, H-6'), 6.77 (d, $J$ = 8.0 Hz, 1H, H-5'), 6.22 (d, $J$ = 15.9 Hz, 1H, H-8'), 5.62 (s, 1H, H-1), 5.32–5.29 (m, 1H, H-3), 4.45–4.42 (m, 1H, H-4), 4.08–4.06 (m, 1H, H-5), 3.97–3.59 (m, 2H, NH-CH$_2$), 3.55–3.43 (m, 1H, NH-CH), 3.35–3.17 (m, 1H, NH-CH), 2.33–2.30 (m, 1H, H-2), 2.01–1.98 (m, 1H, H-2), 1.90–1.86 (m, 2H, H-6), 1.58–1.50 (m, 2H, CH$_2$), 1.49–1.37 (m, 2H, CH$_3$-C-O, (CH$_2$)$_2$), 1.25 (m, 3H, CH$_3$-C-O); $^{13}$C-NMR (150 MHz, DMSO) $\delta$: 171.5, 165.8, 148.5, 145.6, 145.4, 125.5, 121.3, 115.8, 114.8, 114.0, 108.0, 76.1, 74.8, 73.2, 70.6, 37.0, 36.8, 34.5, 29.0, 27.9, 25.8, 25.5, 24.2; HRMS (ESI): Calcd for [M + Na]$^+$ C$_{24}$H$_{31}$NNaO$_8$: 484.1947, found 484.1949.

*1-N-cyclohexyl-4, 5-di-O-Isopropylidene-Chlorogenic Acid Amide* (**3g**): Yellowish powder, 71% yield; $^1$H-NMR (400 MHz, DMSO) $\delta$: 7.47 (d, $J$ =15.8 Hz, 1H, H-7'), 7.38 (d, $J$ = 8.2 Hz, 1H, NH), 7.04–6.99 (m, 2H, H-2', 6'), 6.76 (d, $J$ = 8.1 Hz, 1H, H-5'), 6.24 (d, $J$ = 15.8 Hz, 1H, H-8'), 5.44 (br, 1H, H-1), 5.37–5.30 (m, 1H, H-3), 4.42–4.41 (m, 1H, H-4), 4.14–4.11 (m, 1H, H-5), 3.51–3.49 (m, 1H, NH-CH), 2.25–2.20 (m, 1H, H-2), 1.97–1.91 (m, 1H, H-2), 1.77–1.75(m, 2H, H-6), 1.67–1.65 (m, 4H, CH(CH$_2$)$_2$), 1.41 (s, 3H, CH$_3$-C-O), 1.26–1.10 (m, 9H, CH$_3$-C-O, (CH$_2$)$_3$); $^{13}$C-NMR (100 MHz, DMSO) $\delta$: 174.1, 165.8, 148.4, 145.5, 145.3, 125.5, 121.3, 115.7, 114.8, 114.0, 108.0, 76.8, 74.0, 73.5, 70.9, 47.4, 37.0, 34.4, 32.1, 28.0, 25.9, 25.1, 24.4; HRMS (ESI): Calcd for [M + Na]$^+$ C$_{25}$H$_{33}$NNaO$_8$: 498.2104, found 498.2106.

3.2.3. General Procedure for the Synthesis of Compounds **4a–g**

A mixture of compound **3a–g** (1.7 mmol) and TFA-DCM-H$_2$O (10 mL) was stirred and reacted for 4 h at room temperature. Then the reaction solvent was concentrated and the residue was purified by silica gel column chromatography (eluent: DCM-CH$_3$OH, 8:1) to get pure compound 4a-g as a light-yellow solid.

*1-N-n-butyl-Chlorogenic Acid Amide* (**4a**): Yellowish powder, 78% yield; $^1$H-NMR (600 MHz, MeOD) $\delta$: 7.58 (d, $J$ = 15.9 Hz, 1H, H-7'), 7.05 (d, $J$ = 1.9 Hz, 1H, H-2'), 6.95 (dd, $J$ = 8.2 Hz, 1.9 Hz, 1H, H-6'), 6.78 (d, $J$ = 8.1 Hz, 1H, H-5'), 6.29 (d, $J$ = 15.9 Hz, 1H, H-8'), 5.42–5.38 (m, 1H, H-3), 4.24–4.23 (m, 1H, H-4), 3.72–3.70 (m, 1H, H-5), 3.21–3.19 (m, 2H, NH-CH$_2$), 2.12–2.07 (m, 2H, H-2), 2.02–2.00 (m, 1H, H-6), 1.95–1.91 (m, 1H, H-6), 1.52–1.47 (m, 2H, NH-CH$_2$-CH$_2$), 1.36–1.32 (m, 2H, CH$_2$-CH$_3$), 0.93 (t, $J$ = 7.4 Hz, 3H, CH$_2$-CH$_3$); $^{13}$C-NMR (150 MHz, MeOD) $\delta$: 176.6, 169.0, 149.6, 147.0, 146.8, 127.8, 122.9,

116.5, 115.4, 115.2, 77.8, 74.4, 72.7, 71.9, 40.0, 39.9, 38.7, 32.6, 21.0, 14.1; HRMS (ESI): Calcd for [M + Na]$^+$ $C_{20}H_{27}NNaO_8$: 432.1634, found 432.1636.

*1-N-Isobuty-Chlorogenic Acid Amide* (**4b**): Yellowish powder, 85% yield; $^1$H-NMR (600 MHz, MeOD) $\delta$: 7.57 (d, *J* =15.7 Hz, 1H, H-7′), 7.05 (s, 1H, H-2′), 6.99–6.89 (m, 1H, H-6′), 6.78 (d, *J* = 7.7 Hz, 1H, H-5′), 6.29 (d, *J* =15.7 Hz, 1H, H-8′), 5.46–5.37 (m, 1H, H-3), 4.30–4.18 (m, 1H, H-4), 3.78–3.66 (m, 1H, H-5), 3.09–2.95 (m, 2H, NH-CH$_2$), 2.11–1.94 (m, 1H, H-2, 6), 1.85–1.73 (m, 1H, NH-CH$_2$-CH), 0.94–0.84 (m, 6H, 2 × CH$_3$); $^{13}$C-NMR (150 MHz, MeOD) $\delta$: 176.7, 169.0, 149.5, 147.0, 146.7, 127.7, 122.9, 116.5, 115.3, 115.2, 77.8, 74.3, 72.6, 71.9, 47.6, 40.0, 38.7, 29.6, 20.3; HRMS (ESI): Calcd for [M + Na]$^+$ $C_{20}H_{27}NNaO_8$: 432.1634, found 432.1638.

*1-N-n-octyl-Chlorogenic Acid Amide* (**4c**): Yellowish powder, 76% yield; $^1$H-NMR (400 MHz, MeOD) $\delta$: 7.58 (d, *J* = 15.9 Hz, 1H, H-7′), 7.05 (d, *J* = 1.9 Hz, 1H, H-2′), 6.95 (dd, *J* = 8.2 Hz, 1.9 Hz, 1H, H-6′), 6.78 (d, *J* = 8.2 Hz, 1H, H-5′), 6.29 (d, *J* = 15.9 Hz, 1H, H-8′), 5.43–5.37 (m, 1H, H-3), 4.24–4.23 (m, 1H, H-4), 3.72–3.69 (m, 1H, H-5), 3.20–3.17 (m, 2H, NH-CH$_2$), 2.13–1.90 (m, 1H, H-2, 6), 1.52–1.49 (m, 2H, NH-CH$_2$-CH$_2$), 1.34–1.26 (m, 10H, (CH$_2$)$_5$-CH$_3$), 0.87 (t, *J* =7.1 Hz, 3H, -CH$_3$); $^{13}$C-NMR (100 MHz, MeOD) $\delta$: 176.6, 169.0, 149.6, 147.0, 146.8, 127.8, 123.0, 116.5, 115.3, 115.1, 77.7, 74.4, 72.7, 72.0, 40.2, 40.0, 38.7, 33.0, 30.5, 30.4, 27.9, 23.7, 14.4; HRMS (ESI): Calcd for [M + Na]$^+$ $C_{24}H_{35}NNaO_8$: 488.2260, found 488.2252.

*1-N-Propargyl-Chlorogenic Acid Amide* (**4d**): Yellowish powder, 69% yield; $^1$H-NMR (400 MHz, MeOD) $\delta$: 7.57 (d, *J* = 15.9 Hz, 1H, H-7′), 7.11–7.00 (m, 1H, H-2′), 6.99–6.89 (m, 1H, H-6′), 6.78 (d, *J* = 8.0 Hz, 1H, H-5′), 6.29 (d, *J* = 15.9 Hz, 1H, H-8′), 5.43–5.37 (m, 1H, H-3), 4.29–4.19 (m, 1H, H-4), 4.03–4.92 (m, 2H, NH-CH$_2$), 3.74–3.71 (m, 1H, H-5), 2.61–2.49 (m, 1H, CCH), 2.15–1.95 (m, 4H, H-2, 6); $^{13}$C-NMR (100 MHz, MeOD) $\delta$: 176.5, 169.0, 149.5, 147.0, 146.7, 127.7, 123.0, 116.5, 115.3, 115.1, 80.5, 77.8, 74.3, 72.5, 72.1, 71.8, 39.8, 38.5, 29.4; HRMS (ESI): Calcd for [M + Na]$^+$ $C_{19}H_{21}NNaO_8$: 414.1165, found 414.1667.

*1-N-benzyl-Chlorogenic Acid Amide* (**4e**): Yellowish powder, 73% yield; $^1$H-NMR (600 MHz, MeOD) $\delta$: 7.58 (d, *J* = 15.9 Hz, 1H, H-7′), 7.32–7.22 (m, 5H, Ph-H), 7.05 (s, 1H, H-2′), 6.96–6.94 (m, 1H, H-6′), 6.78 (d, *J* = 8.1 Hz, 1H, H-5′), 6.29 (d, *J* =15.8 Hz, 1H, H-8′), 5.45–5.39 (m, 1H, H-3), 4.39 (m, 2H, NH-CH$_2$), 4.26–4.25 (m, 1H, H-4), 3.73–3.69 (m, 1H, H-5), 2.17–1.97 (m, 4H, H-2, 6); $^{13}$C-NMR (150 MHz, MeOD) $\delta$: 176.8, 169.0, 149.5, 147.0, 146.8, 139.9, 129.5, 128.2, 128.2, 128.0, 122.9, 116.5, 115.4, 115.2, 77.9, 74.4, 72.7, 71.9, 43.8, 40.0, 38.8; HRMS (ESI): Calcd for [M + Na]$^+$ $C_{23}H_{25}NNaO_8$: 466.1448, found 466.1484.

*1-N-piperidyl-Chlorogenic Acid Amide* (**4f**): Yellowish powder, 61% yield; $^1$H-NMR (600 MHz, MeOD) $\delta$: 7.57 (d, *J* = 15.9 Hz, 1H, H-7′), 7.05 (d, *J* = 1.9 Hz, 1H, H-2′), 6.95 (dd, *J* = 8.2 Hz, 2.0 Hz, 1H, H-6′), 6.78 (d, *J* = 8.2 Hz, 1H, H-5′), 6.26 (d, *J* = 15.9 Hz, 1H, H-8′), 5.36–5.33 (m, 1H, H-3), 4.28–4.27 (m, 1H, H-4), 4.06–3.79 (m, 2H, NH-CH$_2$), 3.73–3.71 (m, 1H, H-5), 3.63–3.38 (m, 2H, NH-CH$_2$), 2.33–2.01 (m, 4H, H-2, 6), 1.68–1.61 (m, 2H, CH$_2$), 1.60–1.48 (m, 4H, (CH$_2$)$_2$), 1.90–1.86 (m, 2H, H-6), 1.58–1.50 (m, 2H, CH$_2$), 1.49–1.37 (m, 2H, CH$_3$-C-O, (CH$_2$)$_2$); $^{13}$C-NMR (150 MHz, MeOD) $\delta$: 173.2, 168.8, 149.6, 147.2, 146.8, 127.8, 123.0, 116.5, 115.3, 115.2, 78.6, 73.3, 72.4, 71.6, 40.0, 39.2, 38.7, 30.7, 25.6; HRMS (ESI): Calcd for [M + Na]$^+$ $C_{21}H_{27}NNaO_8$: 444.1634, found 444.1637.

*1-N-cyclohexyl-chlorogenic acid amide* (**4g**): Yellowish powder, 64% yield; $^1$H-NMR (600 MHz, MeOD) $\delta$: 7.57 (d, *J* = 15.9 Hz, 1H, H-7′), 7.05 (d, *J* = 2.0 Hz, 1H, H-2′), 6.95 (dd, *J* = 8.2 Hz, 2.0 Hz, 1H, H-6′), 6.78 (d, *J* = 8.2 Hz, 1H, H-5′), 6.29 (d, *J* =16.0 Hz, 1H, H-8′), 5.42–5.36 (m, 1H, H-3), 4.24–4.22 (m, 1H, H-4), 3.72–3.69 (m, 1H, H-5), 3.65–3.58 (m, 1H, NH-CH), 2.12–1.89 (m, 4H, H-2, 6), 1.84–1.82 (m, 2H, CH$_2$), 1.77–1.73(m, 2H, CH$_2$), 1.65–1.58 (m, 2H, CH$_2$), 1.40–1.34 (m, 2H, CH$_2$), 1.26–1.23 (m, 2H, CH$_2$); $^{13}$C-NMR (150 MHz, MeOD) $\delta$: 175.7, 169.0, 149.6, 147.0, 146.8, 127.8, 122.9, 116.5, 115.4, 115.2, 77.6, 74.4, 72.7, 72.0, 49.6, 40.0, 38.7, 33.6, 30.7, 26.5, 26.1; HRMS (ESI): Calcd for [M + Na]$^+$ $C_{22}H_{29}NNaO_8$: 458.1791, found 458.1795.

The spectrograms of the compounds **3a–g** and **4a–g** are shown in the Electronic Supplementary Material (ESM).

*3.3. Evaluation of the Biological Activity*

3.3.1. Cell Culture

HepG2 cells were originated from American Type Culture Collection (ATCC) (Manassas, VA, USA) and obtained from the Peking Union Medical College (Beijing, China). HepG2 cells were cultured in Dulbecco's modified Eagle medium (DMEM) supplemented with 10% fetal bovine serum, 1% penicillin, and streptomycin at 37 $^{\circ}$C in a 5% $CO_2$ atmosphere. When grown to 70%–80% confluence, cells were incubated in serum-free DMEM containing 100 $\mu$mol/L oleic acid and co-treated with 10 $\mu$mol/L of simvastatin or chlorogenic acid, compound **3a–g**, and **4a–g** for 24 h respectively. Cells maintained in serum-free DMEM were used as the blank control. Compounds were dissolved in dimethyl sulphoxide (DMSO) and an equal volume of it was added in the control group.

3.3.2. Cell Viability Assay

Cell viability was examined using MTS assay. HepG2 cells in 96-well culture plates were treated with 10 $\mu$mol/L compounds. The cells were incubated for 24 h and the MTS reagent was added to each well according to the instruction of CellTiter 96®Aqueous One Solution Cell Proliferation Assay (promega corperation, Beijing, China). The absorbance at 490 nm was measured using a microplate reader (ThermoFisher Ltd., Shanghai, China).

3.3.3. Oil red O staining

The cells with 70–80% confluence in 96 well plates were incubated in serum-free DMEM + OA (oleic acid) (100 $\mu$mol/L) and 10 $\mu$mol/L of chlorogenic acid, compound **3a–g**, and **4a–g**, respectively, or the positive control simvastatin (10 $\mu$mol/L), for 24 h. Cells were then fixed with 4% w/v paraformal deyde (30 min, room temperature) and stained with 0.5% filtered oil-red O solution (15 min, room temperature). The staining was evaluated by a Tecan Infinite M1000Pro Microplate Reader and spectrophotometry at 358 nm.

3.3.4. Intracellular TG Quantification

HepG2 cells with 70–80% confluence in 6 well plates were incubated in serum-free DMEM + OA (100 $\mu$mol/L) and 10 $\mu$mol/L of chlorogenic acid, compound **3a–g**, and **4a–g**, respectively, or the positive control simvastatin (10 $\mu$mol/L) for 24 h. The cells were subjected to TG quantification as introduced by the protocol of Triglyceride Quantification Kit. Each experiment was repeated in triplicate, with duplicates each.

3.3.5. Statistical Analysis

Data are expressed as mean $\pm$ SEM. One-way ANOVA was used to determine significant differences among groups, after which the modified Students t-test with the Bonferroni correction was used for comparison between individual groups. All statistical analyses were performed with SPSS 17.0 software (SPSS Inc., Chicago, IL, USA). The value $p < 0.05$ was considered statistically significant.

**4. Conclusions**

In summary, fourteen caffeoylquinic acid derivatives of chlorogenic acid were designed, synthesized, and evaluated for their lipid-lowering effects. Several of the derivatives showed potent lipid-lowering activity against oleic acid-elicited lipid accumulation in HepG2 liver cells. Particularly, derivatives **3d**, **3g**, **4c**, and **4d** exhibited more potential lipid-lowering effect than the positive simvastatin and chlorogenic acid. Preliminary SAR analysis has shown that the propargyl group of the amide derivatives had a good impact on the protective effect. Further research on mechanism of **3d**, **3g**, **4c**, and **4d** revealed that it was related to the regulation of TG levels. These above results suggest that derivatives **3d**, **3g**, **4c**, and **4d** may be a promising therapeutic candidate as potential hypolipidemic agents. Further studies to design the derivatives of chlorogenic acid coupled with naturally occurring kinds of amino acids, as well as to explore the mechanism of action of this novel class of compounds is, planned to start in our group in the near future [21–23].

**Supplementary Materials:** The following are available online.

**Author Contributions:** X.-D.X. and P.G. conducted the study. Y.T. designed the detailed experiments, performed the study, and collected and analyzed data. X.-X.C., H.S., C.-M.W., X.-P.Z., and Z.X. took part in the experiments in this study. All authors commented the study and approved the final manuscript.

**Funding:** This research was funded by Beijing Natural Science Foundation (Grant No. 7192129), the National Natural Sciences Foundation of China (Grant No. 81302656 and 81502929), and the CAMS Innovation Fund for Medical Science (CIFMS) (Grant No. 2016-I2M-1-012).

**Conflicts of Interest:** The authors declare no conflict of interest.

# References

1. Barry, V.W.; Caputo, J.L.; Kang, M. The joint association of fitness and fatness on cardiovascular disease mortality: A meta-analysis. *Prog. Cardiovasc. Dis.* **2018**, *61*, 136–141. [CrossRef] [PubMed]
2. Britton, K.A.; Massaro, J.M.; Murabito, J.M.; Kreger, B.E.; Hoffmann, U.; Fox, C.S. Body fat distribution, incident cardiovascular disease, cancer, and all-cause mortality. *J. Am. Coll. Cardiol.* **2013**, *62*, 921–925. [CrossRef] [PubMed]
3. Koliaki, C.; Liatis, S.; Kokkinos, A. Obesity and cardiovascular disease: Revisiting an old relationship. *Metabolism* **2018**, *92*, 98–107. [CrossRef] [PubMed]
4. Bellan, M.; Menegatti, M.; Ferrari, C.; Carnevale Schianca, G.P.; Pirisi, M. Ultrasound-assessed visceral fat and associations with glucose homeostasis and cardiovascular risk in clinical practice. *Nutr. Metab. Cardiovas.* **2018**, *28*, 610–617. [CrossRef] [PubMed]
5. Elagizi, A.; Kachur, S.; Lavie, C.J.; Carbone, S.; Pandey, A.; Ortega, F.B.; Milani, R.V. An overview and update on obesity and the obesity paradox in cardiovascular diseases. *Prog. Cardiovasc. Dis.* **2018**, *61*, 142–150. [CrossRef] [PubMed]
6. Cholesterol Treatment Trialists' (CTT) Collaboration. Efficacy and safety of more intensive lowering of LDL cholesterol: A meta-analysis of data from 170,000 participants in 26 randomised trials. *Lancet* **2010**, *376*, 1670–1681. [CrossRef]
7. Berry, J.D.; Dyer, A.; Cai, X.; Garside, D.B.; Ning, H.Y.; Thomas, A.; Greenland, P.; Horn, L.V.; Tracy, R.P.; Lloyd-Jones, D.M. Lifetime risks of cardiovascular disease. *N. Engl. J. Med.* **2012**, *366*, 321–329. [CrossRef]
8. Jones, A. Triglycerides and cardiovascular risk. *Heart* **2013**, *99*, 1–2. [CrossRef]
9. Pilz, S.; Scharnagl, H.; Tiran, B.; Seelhorst, U.; Wellnitz, B.; Boehm, B.O.; Schaefer, J.R.; März, W. Free fatty acids are independently associated with all-cause and cardiovascular mortality in subjects with coronary artery disease. *J. Clin. Endocrinol. Metab.* **2006**, *91*, 2542–2547. [CrossRef]
10. Zhang, X.P.; Wu, C.M.; Wu, H.F.; Sheng, L.H.; Su, Y.; Zhang, X.; Luan, H.; Sun, G.B.; Sun, X.B.; Tian, Y.; et al. Anti-hyperlipidemic effects and potential mechanisms of action of the caffeoylquinic acid-rich *Pandanus tectorius* fruit extract in hamsters fed a high fat-diet. *PLoS ONE* **2013**, *8*, e61922. [CrossRef]
11. Andriani, Y.; Ramli, N.M.; Syamsumir, D.F.; Kassim, M.N.I.; Jaafar, J.; Aziz, N.A.; Marlina, L.; Musa, N.S.; Mohamad, H. Phytochemical analysis, antioxidant, antibacterial and cytotoxicity properties of keys and cores part of *Pandanus tectorius* fruits. *Arab. J. Chem.* **2015**. [CrossRef]
12. Wu, C.M.; Zhang, X.P.; Zhang, X.; Luan, H.; Sun, G.B.; Sun, X.B.; Wang, X.L.; Guo, P.; Xu, D.D. The caffeoylquinic acid-rich *Pandanus tectorius* fruit extract increases insulin sensitivity and regulates hepatic glucose and lipid metabolism in diabetic db/db mice. *J. Nutr. Biochem.* **2014**, *25*, 412–419. [CrossRef]
13. Agunloye, O.M.; Oboh, G.; Ademiluyi, A.O.; Ademosun, A.O.; Akindahunsi, A.A.; Oyagbemi, A.A.; Omobowale, T.O.; Ajibade, T.O.; Adedapo, A.A. Cardio-protective and antioxidant properties of caffeic acid and chlorogenic acid: Mechanistic role of angiotensin converting enzyme, cholinesterase and arginase activities in cyclosporine induced hypertensive rats. *Biomed. Pharmacother.* **2019**, *109*, 450–458. [CrossRef] [PubMed]
14. Zhao, M.M.; Wang, H.Y.; Yang, B.; Tao, H. Identification of cyclodextrin inclusion complex of chlorogenic acid and its antimicrobial activity. *Food Chem.* **2010**, *120*, 1138–1142. [CrossRef]

15. Gao, R.F.; Yang, H.D.; Jing, S.F.; Liu, B.; Wei, M.; He, P.F.; Zhang, N.S. Protective effect of chlorogenic acid on lipopolysaccharide-induced inflammatory response in dairy mammary epithelial cells. *Microb. Pathog.* **2018**, *124*, 178–182. [CrossRef] [PubMed]

16. Wang, D.Y.; Zhao, X.M.; Liu, Y.L. Hypoglycemic and hypolipidemic effects of a polysaccharide from flower buds of *Lonicera japonica* in streptozotocin-induced diabetic rats. *Int. J. Biol. Macromol.* **2017**, *102*, 396–404. [CrossRef] [PubMed]

17. Bender, O.; Llorent-Martínez, E.J.; Zengin, G.; Mollica, A.; Ceylan, R.; Molina-García, L.; Córdova, M.L.F.; Atalay, A. Integration of in vitro and in silico perspectives to explain chemical characterization, biological potential and anticancer effects of *Hypericum salsugineum*: A pharmacologically active source for functional drug formulations. *PLoS ONE* **2018**, *13*, e0197815. [CrossRef] [PubMed]

18. Llorent-Martínez, E.J.; Zengin, G.; Córdova, M.L.F.; Bender, O.; Atalay, A.; Ceylan, R.; Mollica, A.; Mocan, A.; Uysal, S.; Guler, G.O.; et al. Traditionally used *Lathyrus* species: Phytochemical composition, antioxidant activity, enzyme inhibitory properties, cytotoxic effects, and in silico studies of *L. czeczottianus* and *L. nissolia*. *Front. Pharmacol.* **2017**, *8*, 83. [CrossRef]

19. Jabeur, I.; Tobaldini, F.; Martins, N.; Barros, L.; Martins, I.; Calhelha, R.C.; Henriques, M.; Silva, S.; Achour, L.; Santos-Buelga, C.; et al. Bioactive properties and functional constituents of *Hypericum androsaemum* L.: A focus on the phenolic profile. *Food Res. Int.* **2016**, *89*, 422–431. [CrossRef]

20. Truong, V.D.; Mcfeeters, R.F.; Thompson, R.T.; Dean, L.L.; Shofran, B. Phenolic acid content and composition in leaves and roots of common commercial sweetpotato (*Ipomea batatas* L.) cultivars in the United States. *J. Food Sci.* **2007**, *72*, c343–c349. [CrossRef]

21. Stefanucci, A.; Macedonio, G.; Dvorácskó, S.; Tömböly, C.; Mollica, A. Novel fubinaca/rimonabant hybrids as endocannabinoid system modulators. *Amino Acids* **2018**, *50*, 1595–1605. [CrossRef] [PubMed]

22. Monti, L.; Stefanucci, A.; Pieretti, S.; Marzoli, F.; Fidanza, L.; Mollica, A.; Mirzaie, S.; Carradori, S.; Petrocellis, L.D.; Moriello, A.S.; et al. Evaluation of the analgesic effect of 4-anilidopiperidine scaffold containing ureas and carbamates. *J. Enzyme. Inhib. Med. Chem.* **2016**, *31*, 1638–1647. [CrossRef] [PubMed]

23. Mollica, A.; Costante, R.; Akdemir, A.; Carradori, S.; Stefanucci, A.; Macedonio, G.; Ceruso, M.; Supuran, C.T. Exploring new probenecid-based carbonic anhydrase inhibitors: Synthesis, biological evaluation and docking studies. *Bioorg. Med. Chem.* **2015**, *23*, 5311–5318. [CrossRef] [PubMed]

**Sample Availability:** Samples of the compounds **3a–g** and **4a–g** are available or not from the authors.

# molecules

MDPI

*Article*

# First Total Synthesis of Varioxiranol A

Angelika Lásiková [1,*], Jana Doháňošová [2], Mária Štiblariková [1], Martin Parák [1], Ján Moncol [3] and Tibor Gracza [1]

[1]  Department of Organic Chemistry, Slovak University of Technology in Bratislava, Radlinského 9, 812 37 Bratislava, Slovakia; maria.stiblarikova@stuba.sk (M.Š.); xparakm@is.stuba.sk (M.P.); tibor.gracza@stuba.sk (T.G.)
[2]  Central Laboratories, Slovak University of Technology in Bratislava, Radlinského 9, 812 37 Bratislava, Slovakia; jana.dohanosova@stuba.sk
[3]  Department of Inorganic Chemistry, Slovak University of Technology in Bratislava, Radlinského 9, 812 37 Bratislava, Slovakia; jan.moncol@stuba.sk
*  Correspondence: angelika.lasikova@stuba.sk; Tel.: +421-2-593-25-167

Academic Editors: Pavel B. Drasar and Vladimir A. Khripach
Received: 4 February 2019; Accepted: 23 February 2019; Published: 28 February 2019

check for updates

**Abstract:** The paper describes the first total synthesis of natural varioxiranol A by chiral pool approach and confirmation of its absolute configuration by single-crystal X-ray analysis. The target varioxiranol A and its 4-epimer were obtained after 10 steps from single and available chiral source 1,2-*O*-isopropylidene-D-glyceraldehyde in an overall yield of 10% and 6%, respectively. A synthetic strategy based on the Julia–Kocieński coupling reaction between aromatic sulfone and corresponding aldose derivative makes it possible to prepare other interesting polyketide derivatives (varioxiranols B-G, varioxirane, varioxiranediols).

**Keywords:** synthesis of natural products; varioxiranol A; 4-*epi*-varioxiranol A; absolute structure; *Emericella variecolor*

## 1. Introduction

The fungus *Emericella variecolor* [1] is considered as a promising source of interesting bioactive compounds. During the past several decades, the large number of natural products of this origin had been isolated and evaluated for their diversiform biological activities. First, Dunn and Johnstone [2] isolated from a static culture of a pure strain of the fungus *Aspergillus variecolor* (imperfect state of *Emericella variecolor*) 2-methoxy-6-(3,4-dihydroxyhepta-1,5-dienyl) benzyl alcohol **1** (Figure 1) along with other metabolites-6-methoxymellein, siderin, andibenin, and andilesin A–C. The structure of **1** was established by NMR spectroscopy [3], absolute configuration (3*R*,4*S*) was later determined by total synthesis [4], and the trivial name "andytriol" was kindly suggested by prof. Johnstone and used for the first time in the manuscript dealing with the synthesis [4]. In 2002, Malmstrøm et al. [5] reported the isolation of benzyl alcohols (varitriol **3**, varioxirane **2**), prenylxanthones (shamixanthone, varixanthone, tajixanthone), and cyclopentanones from a strain of *E. variecolor* derived from a Caribbean sponge. In particular, **3** showed notably increased potency toward selected renal, CNS, and breast cancer cell lines. The authors also proposed a hypothetical biogenetic relationship between these products via enzymatic intramolecular $S_N2$ epoxide ring opening and pointed out that natural andytriol **1** could be involved in this biosynthetic pathway to **3** via epoxide **2**. Recently, seven new polyketide derivatives with benzyl alcohol structural motif, namely varioxiranols A–G, isolated by chemical examination of a sponge (*Cinachyrella* sp.)-associated *E. variecolor* fungus and tested for lipid-lowering effects against oleic acid, elicited lipid accumulation in HepG2 liver cells. Among these secondary metabolites, varioxiranol A **4** exerted inhibition activity and showed no toxicity [6].

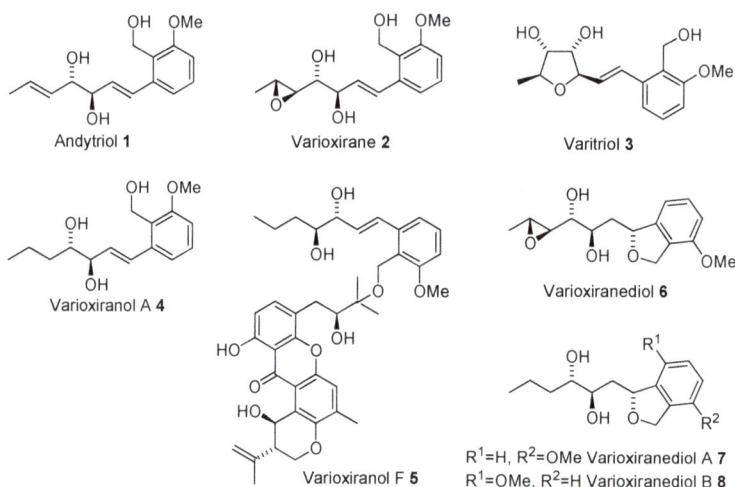

**Figure 1.** Natural compounds isolated from a strain of *Emericella variecolor* **1–8**.

Interesting activity against drug-resistant microbial pathogens was observed by varioxiranediols **6** and **7**. These metabolites of formal cyclisation of varioxirane **2** and varioxiranol A **4** were isolated from the same endophytic fungus. The structure and absolute configuration of **6** [7] and varioxiranediols A **7** and B **8** [8] were confirmed by the X-ray analysis supporting the structural relationship of the isolated natural compounds.

In the course of our long-term program directed towards the synthesis of natural compounds and secondary metabolites isolated from *E. variecolor*, we have developed the synthesis of varitriol **3** [9–13], andytriol **1** [4], and varioxirane **2** [4] and examined their antitumor activity. Herein, we describe the first total synthesis of natural varioxiranol A **4** and its 4-epimer **9** that should also be general route for the synthesis of all other varioxiranols, varioxirane, and varioxiranediols.

## 2. Results and Discussion

The synthetic strategy takes advantage of our previous synthesis of andytriol **1**. Target compounds could be readily obtained by coupling of the known sulfone **11** [4] bearing benzyl alcohol moiety with corresponding aldehydes via Julia–Kocieński olefination (Scheme 1). The aldehydic partner for the olefination, dihydroxyhexanal **10**, having the configuration of natural enantiomer **4** and its 4-epimer **9** at C-2, could be accessible from D-glyceraldehyde by introduction of the propyl group at C-1 followed by the oxidation of carbon at the other end (C-3).

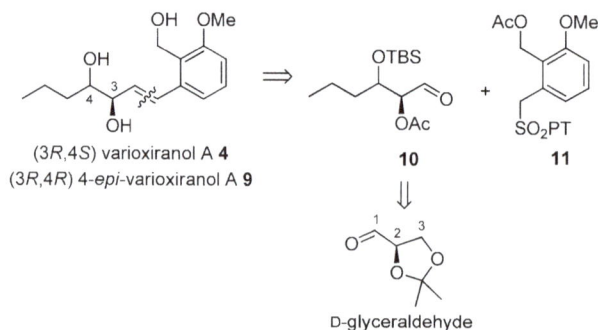

**Scheme 1.** Retrosynthetic analysis of **4** and **9**.

The synthesis of varioxiranols A **4** commenced from isopropylidene-D-glyceraldehyde, a commercially available starting material, or readily obtainable from D-mannitol via the route using standard carbohydrate chemistry [14]. The required six-carbon chain of the key fragment **10** [15] was obtained by the Grignard addition of propylmagnesium chloride to isopropylidene-D-glyceraldehyde in THF/Et$_2$O [15] (Scheme 2). A diastereomeric mixture of partially protected L-*erythro*/D-*threo* hexenetriols **12** was isolated in the ratio of 67:33 with 71% yield. To prepare both epimers of varioxiranol A **4**, we continued synthesis with the mixture of diastereomers **12**. The aldehydic partner for the Julia–Kocieński coupling, hexenose derivative **10** was prepared using a selective protection–deprotection sequence. Firstly, free hydroxyl group of **12** was protected as *tert*-butyldimethylsilylether **13** with good yield (92%). Acidic hydrolysis of **13** with trifluoroacetic acid in dichloromethane afforded vicinal diol **14** (93%) which was selectively tritylated on terminal hydroxyl group to give **15**. Subsequent acetylation of **15** provided fully protected triol **16** (87%). Selective deprotection of primary hydroxy group was then achieved by smooth tritylether hydrolysis using formic acid in ether furnishing alcohol **17** [16], however, as an inseparable mixture along with the product of the acetyl group migration **20** taking place even during MPLC. Swern oxidation of primary hydroxyl group under conventional reaction conditions provided desired aldehyde **10**, which was used in the next step without further purification. The crude aldehyde was subjected to Julia–Kocieński coupling with 2-methoxy-6-[(1-phenyl-1*H*-tetrazol-5-ylsulfonyl)methyl]benzyl acetate **11**, which was prepared according to the literature [4]. Thus, potassium hexamethyldisilazane was added to a nearly equimolar mixture of sulfone **11** and aldehyde **10** in dimethoxyethane at −60 °C and stirred for 40 min at room temperature affording coupling products **18**/**19** in 41% yield (in ratio 70:30) and with excellent *E*-selectivity. The resulting alkenes **18**/**19** could be separated by MPLC and preparative TLC. The final steps, removal of all protecting groups, were run in parallel with the pure diastereomers **18** and **19**. The first, basic hydrolysis of acetyl groups with K$_2$CO$_3$ in MeOH, was followed by treatment of the crude mixture with TBAF in THF and finally, flash chromatography purification furnished the target compounds (+)-**4** and **9** in 92% and 57% yield, respectively, over the last two steps. The $^1$H and $^{13}$C NMR, HRMS spectra, and the specific rotation {$[\alpha]_D^{20}$ +2.5 (*c* 0.207, MeOH), lit. [2], $[\alpha]_D^{20}$ +5.8 (*c* 0.53, MeOH)} of synthetic varioxiranol A **4** were in good agreement with the reported data for the natural product.

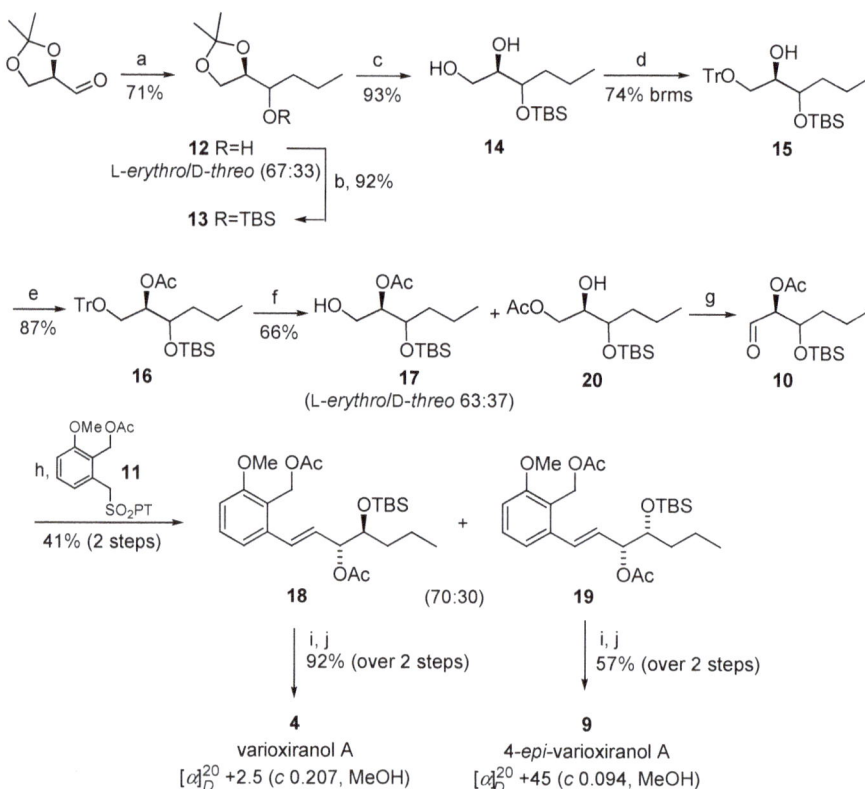

**Scheme 2.** Synthesis of natural varioxiranol A **4** and 4-*epi*-varioxiranol A **9**. Reagents and conditions: (**a**) prop-1-ylmagnesium chloride, THF, Et$_2$O, r.t., 1 h; (**b**) TBSCl, imidazole, CH$_2$Cl$_2$, 0 °C to r.t., 23 h; (**c**) TFA (50%), CH$_2$Cl$_2$, r.t., 1 h; (**d**) TrCl, Et$_3$N, DMAP, CH$_2$Cl$_2$, 0 °C to r.t., 15 h; (**e**) Ac$_2$O, DMAP, CH$_2$Cl$_2$, r.t., 30 min; (**f**) HCOOH/Et$_2$O (1/1), r.t., 50 min; (**g**) DMSO, (COCl)$_2$, Et$_3$N, CH$_2$Cl$_2$, −78 °C to r.t., 2.5 h; (**h**) KHMDS, dimethoxyethane, sulfone **11**, CH$_2$Cl$_2$, −60 °C to r.t., 40 min; (**i**) K$_2$CO$_3$, MeOH, r.t. 2.5 h; (**j**) TBAF × 3H$_2$O, THF, 0 °C to r.t., 4.5 h.

Definitive confirmation of absolute configuration of the target compound has been provided by single-crystal X-ray analysis. An X-ray study of both epimers confirmed 3*R*,4*S* (*anti*) configuration of the natural varioxiranol A **4** and 3*R*,4*R* (*syn*) configuration of its 4-epimer **9** (Figure 2). Interestingly, the crystal lattice of the 4-epimer **9** is composed of *S-cis* and *S-trans* conformers (two crystallographic independent molecules in cell; see Supplementary Materials).

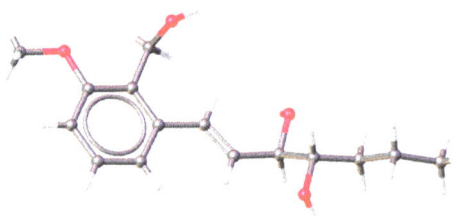

varioxiranol A **4**

**Figure 2.** *Cont.*

4-*epi*-varioxiranol A **9** (*S-trans* conformer)

4-*epi*-varioxiranol A **9** (*S-cis* conformer)

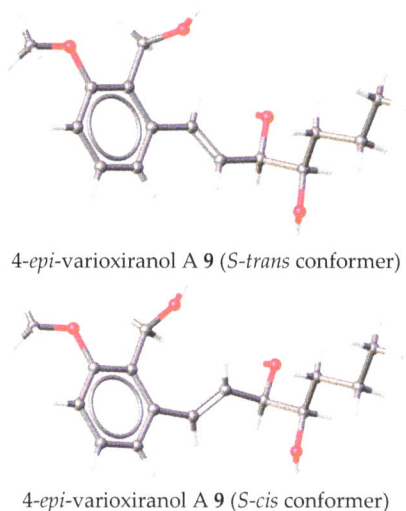

**Figure 2.** A ball-and-stick view of crystal structures **4** and **9**.

## 3. Experimental Section

### 3.1. General Methods

Commercial reagents were used without further purification. All solvents were distilled before use. Hexanes refer to the fraction boiling at 60–65 °C. Flash column liquid chromatography (FLC) was performed on silica gel Kieselgel 60 (40–63 μm, 230–400 mesh, Merck, Darmstadt, Germany) and analytical thin-layer chromatography (TLC) was performed on aluminium plates precoated with either 0.2 mm (DC-Alufolien, Merck) or 0.25 mm silica gel 60 $F_{254}$ (ALUGRAM® SIL G/$UV_{254}$, Macherey-Nagel, Fisher Scientific, Loughborough, UK). The compounds were visualized by UV fluorescence and by dipping the plates in an aqueous $H_2SO_4$ solution of cerium sulphate/ammonium molybdate followed by charring with a heat gun. Melting points were obtained using a Boecius apparatus (Büchi®melting point apparatus Model B-545, BÜCHI Labortechnik AG, Flawil, Switzerland) and are uncorrected. Optical rotations were measured with a JASCO P-2000 polarimeter (JASCO, Easton, MD, USA) and are given in units of $10^{-1}$ deg.$cm^2$.$g^{-1}$. FTIR spectra were obtained on a Nicolet 5700 spectrometer (Thermo Electron, Thermo Fisher Scientific, Waltham, MA, USA) equipped with a Smart Orbit (diamond crystal ATR) accessory, using the reflectance technique (4000–400 $cm^{-1}$).[1]H and [13]C NMR spectra were recorded on either 300 (75) MHz or 600 (150) MHz Varian spectrometer (Varian Inc., Palo Alto, CA, USA). Chemical shifts (δ) are quoted in ppm and are referenced to tetramethylsilane (TMS, δ = 0 ppm) as internal standard for [1]H NMR and to $CDCl_3$ peak (δ = 77.16 ppm in case of [13]C NMR). High-resolution mass spectra (HRMS) were recorded on an OrbitrapVelos mass spectrometer (Thermo Scientific, Waltham, MA, USA; Bremen, Germany) with a heated electrospray ionisation (HESI) source. The mass spectrometer was operated with full scan (50–2000 amu) in positive or negative FT mode (at a resolution of 100,000). The analyte was dissolved in methanol and infused via syringe pump at a rate of 5 mL/min. The heated capillary was maintained at 275 °C with a source heater temperature of 50 °C and the sheath, auxiliary, and sweep gases were at 10, 5, and 0 units, respectively. Source voltage was set to 3.5 kV.

Data collection and cell refinement of **4** and **9** were made on a Stoe StadiVari diffractometer (Stoe & Cie GmbH, Darmstadt, Germany) using a Pilatus3R 300K HPAD detector and the microfocus source Xenocs Genix3D Cu HF (λ = 1.54186 Å). The structures were solved using SHELXT [17] and refined by the full-matrix least-squares procedure with SHELXL (ver. 2018/3) (for **4**) [18] or CRYSTALS (ver. 14.61) (**9**) [19]. The structures were drawn using the OLEX2 package [20]. The absolute configurations of both compounds were determined. The Flack parameter x = −0.08(5) for **9** was calculated by Parsons method [21]. The absolute structure of very small crystal of **4** is impossible to determine based on the Flack parameter, however, using Hooft parameter [22] with (Gaussian) statistics led to the conclusive value of y = 0.07(11). The deposition numbers CCDC 1892452 (**4**) and CCDC 1892453 (**9**) contain the supplementary crystallographic data for this paper. These data can be obtained free of charge from http://www.ccdc.cam.ac.uk/conts/retrieving.html (or from the CCDC, 12 Union Road, Cambridge CB2 1EZ, UK.

*3.2. (2R,3S)-1,2-O-Isopropylidene-hexane-1,2,3-triol (L-erythro-12) and (2R,3R)-1,2-O-isopropylidene-hexane-1,2,3-triol (D-threo-12)*

A solution of propylmagnesium chloride in diethyl ether (2.0 M solution, 5.8 mL, 11.6 mmol) was added dropwise to a stirred solution of freshly distilled 1,2-O-isopropylidene-D-glyceraldehyde (1 g, 7.68 mmol) in dry THF (62 mL) at room temperature. Following the addition, the reaction mixture was stirred for 1 h (TLC control). The reaction was quenched by pouring into a sat. aqueous $NH_4Cl$ (62 mL), the aqueous layer was extracted with diethyl ether (3 × 35 mL), and the combined organic layers were dried and concentrated. The residue was purified by MPLC (gradient AcOEt/hexanes 0/100 to 30/70) to give **12** (954 mg, 71%, L-erythro-**12**/D-threo-**12** 67:33) as colorless liquid. $^1H$ NMR (600 MHz, $CDCl_3$) δ (L-erythro-**12**) 0.95 (t, H-6, *J* = 7.1 Hz, 3H), 1.33–1.44 (m, H-4, H-5a, 2 × $CH_3$, 9H), 1.51–1.59 (m, H-5b, 1H), 1.96 (d, OH, *J* = 2.9 Hz, 1H), 3.78–3.82 (m, H-3, 1H), 3.91 (dd, H-1a, *J* = 7.3, 8.0 Hz, 1H), 3.97 (dd, H-1b, *J* = 6.5, 8.0 Hz, 1H), 4.04 (ddd, H-2, *J* = 4.0, 6.5, 7.2 Hz, 1H); δ (D-threo-**12**) 0.94 (t, H-6, *J* = 7.2 Hz, 3H), 1.30–1.36 (m, H-4a, 1H), 1.37 (s, $CH_3$, 3H), 1.37–1.44 (m, H-5a, 1H), 1.44 (s, $CH_3$, 3H), 1.44–1.48 (m, H-4b, 1H), 1.53–1.59 (m, H-5b, 1H), 2.15 (d, OH, *J* = 5.2 Hz, 1H), 3.48-3.53 (m, H-3, 1H), 3.73 (dd, H-1a, *J* = 6.3, 7.7 Hz, 1H), 3.96–4.00 (m, H-2, 1H), 4.02 (dd, H-1b, *J* = 6.6, 7.8 Hz, 1H); $^{13}C$ NMR (150 MHz, $CDCl_3$) δ (L-erythro-**12**) 14.2 (C-6), 19.1 (C-5), 25.5 ($CH_3$), 26.6 ($CH_3$), 34.8 (C-4), 64.6 (C-1), 70.5 (C-3), 78.8 (C-2), 109.0 ($C_q$); δ (D-threo-**12**) 14.2 (C-6), 18.9 (C-5), 25.5 ($CH_3$), 26.8 ($CH_3$), 36.0 (C-4), 66.3 (C-1), 72.2 (C-3), 79.3 (C-2), 109.5 ($C_q$); HRMS (ESI) calcd for $C_9H_{18}O_3Na^+$ [M + Na]$^+$: 197.1148, found: 197.1148.

*3.3. (2R,3S)-1,2-O-Isopropylidene-3-O-tert-butyldimethylsilyl-hexane-1,2,3-triol (L-erythro-13) and (2R,3R)-1,2-O-isopropylidene-3-O-tert-butyldimethylsilyl-hexane-1,2,3-triol (D-threo-13)*

Imidazole (1.09 g, 16.0 mmol) was added to a solution of diastereomeric mixture **12** (928 mg, 5.33 mmol) in dry $CH_2Cl_2$ (11 mL) at room temperature. The mixture was subsequently cooled to 0 °C and tert-butyldimethylsilyl chloride (1.61 g, 10.7 mmol) was added. The reaction mixture was then stirred for 23 h at room temperature. After the dilution with $CH_2Cl_2$ (140 mL), the reaction mixture was washed with water (2 × 140 mL), the water phase was extracted with $CH_2Cl_2$ (3 × 90 mL), and combined organic layers were dried and concentrated. The residue was purified by MPLC (isocratic AcOEt/hexanes 2/98) to afford **13** (1.42 g, 92%, L-erythro-**13**/D-threo-**13** 67:33) as colorless liquid. $^1H$ NMR (600 MHz, $CDCl_3$) δ (L-erythro-**13**) 0.06 (s, $CH_3$, 3H), 0.07 (s, $CH_3$, 3H), 0.88 (s, tBu, 9H), 0.91 (t, H-6, *J* = 7.3 Hz, 3H), 1.34 (s, $CH_3$, 3H), 1.35–1.47 (m, H-4a, H-5, $CH_3$, 6H), 1.49–1.52 (m, H-4b, 1H), 3.73–3.76 (m, H-3, 1H), 3.80–3.84 (m, H-1a, 1H), 3.96–4.00 (m, H-1b, H-2c, 2H); δ (D-threo-**13**) 0.07 (s, $CH_3$, 3H), 0.08 (s, $CH_3$, 3H), 0.89 (s, tBu, 9H), 0.91 (t, H-6, *J* = 7.2 Hz, 3H), 1.30–1.39 (H-4, H-5a, $CH_3$, 6H), 1.41 (s, $CH_3$, 3H), 1.45-1.50 (m, H-5b, 1H), 3.69 (ddd, H-3, *J* = 4.2, 6.0, 7.7 Hz, 1H), 3.71 (dd, H-1a, *J* = 7.4, 8.2 Hz, 1H), 3.94 (dd, H-1b, *J* = 6.6, 8.2 Hz, 1H), 4.04 (ddd, H-2, *J* = 6.0, 6.6, 7.3 Hz, 1H); $^{13}C$ NMR (150 MHz, $CDCl_3$) δ (L-erythro-**13**) −4.2 ($CH_3$), −4.1 ($CH_3$), 14.5 (C-6), 17.8 (C-5), 18.2 (tBu), 25.6 ($CH_3$), 26.0 (tBu), 26.8 ($CH_3$), 37.0 (C-4), 66.6 (C-1), 72.4 (C-3), 78.4 (C-2), 109.0 ($C_q$); δ (D-threo-**13**) −4.5 ($CH_3$), −4.1 ($CH_3$), 14.4 (C-6), 18.4 (tBu), 19.0 (C-5), 25.4 ($CH_3$), 26.1 (tBu), 26.6 ($CH_3$), 34.7 (C-4), 65.7 (C-1), 73.2 (C-3), 78.9 (C-2), 109.2 ($C_q$); HRMS (ESI) calcd for $C_{15}H_{32}O_3SiNa^+$ [M + Na]$^+$: 311.2013, found: 311.2013.

*3.4. (2R,3S)-3-O-tert-Butyldimethylsilyl-hexane-1,2,3-triol (L-erythro-14) and*
*(2R,3R)-3-O-tert-butyldimethylsilyl-hexane-1,2,3-triol (D-threo-14)*

Trifluoroacetic acid (50%, 3.3 mL) was added dropwise to a vigorously stirred solution of compound **13** (1.39 g, 4.81 mmol) in $CH_2Cl_2$ (120 mL) at room temperature. The reaction mixture was stirred for 55 min, then diluted with $CH_2Cl_2$ (220 mL), and washed with sat. aqueous $NaHCO_3$ (60 mL) and water (2 × 130 mL). The organic phase was dried and concentrated to give crude **14** (1.11 g, 93%, L-erythro-**14**/D-threo-**14** 67:33) as colorless oil, which was used in the next step without further purification. $^1H$ NMR (600 MHz, $CDCl_3$) δ (L-erythro-**14**) 0.09 (s, $CH_3$, 3H), 0.11 (s, $CH_3$, 3H), 0.89 (s, tBu, 9H), 0.91 (t, H-6, *J* = 7.3 Hz, 3H), 1.26–1.44 (m, H-4a, H-5, 3H), 1.52–1.58 (m, H-4b, 1H), 2.20 (bs, 2x OH, 2H), 3.60 (ddd, H-2, *J* = 3.5, 3.7, 5.5 Hz, 1H), 3.66 (dd, H-1a, *J* = 3.5, 11.5 Hz, 1H), 3.79 (dd, H-1b, *J* = 5.5, 11.5 Hz, 1H), 3.83 (ddd, H-3, *J* = 3.7, 5.6, 6.6 Hz, 1H); δ (D-threo-**14**) 0.08 (s, $CH_3$, 3H), 0.09 (s, $CH_3$, 3H), 0.90 (s, tBu, 9H), 0.92 (t, H-6, *J* = 7.3 Hz, 3H), 1.28–1.44 (m, H-4a, H-5, 3H), 1.61–1.68 (m, H-4b, 1H), 2.20 (bs, 2 × OH, 2H), 3.56-3.60 (m, H-1, H-2, 3H), 3.68 (ddd, H-3, *J* = 2.8, 4.6, 6.9 Hz, 1H); $^{13}C$ NMR (150 MHz, $CDCl_3$) δ (L-erythro-**14**) −4.5 ($CH_3$), −4.4 ($CH_3$), 14.4 (C-6), 18.2 (tBu), 18.7 (C-5), 26.0 (tBu), 35.8 (C-4), 63.4 (C-1), 73.2 (C-2), 75.3 (C-3); δ (D-threo-**14**) −4.6 ($CH_3$), −4.0 ($CH_3$), 14.4 (C-6), 18.2 (tBu), 18.4 (C-5), 26.0 (tBu), 36.2 (C-4), 64.7 (C-1), 72.4 (C-3), 73.1 (C-2); HRMS (ESI) calcd for $C_{12}H_{28}O_3SiNa^+$ $[M + Na]^+$: 271.1700, found: 271.1700.

*3.5. (2R,3S)-3-O-tert-Butyldimethylsilyl-1-O-trityl-hexane-1,2,3-triol (L-erythro-15) and*
*(2R,3R)-3-O-tert-butyldimethylsilyl-1-O-trityl-hexane-1,2,3-triol (D-threo-15)*

A solution of trityl chloride (1.01 g, 3.64 mmol) in dry $CH_2Cl_2$ (2.4 mL) was cooled to 0 °C, and triethylamine (0.95 mL, 6.84 mmol) and DMAP (46 mg, 0.37 mmol) were added. Subsequently, a solution of compound **14** (772 mg, 3.11 mmol) in $CH_2Cl_2$ (2.4 mL) was added dropwise. After warming to room temperature, the reaction mixture was stirred for 15 h, then quenched with sat. aqueous $NH_4Cl$ (25 mL), the aqueous layer was extracted with $CH_2Cl_2$ (3 × 10 mL), and the combined organic layers were dried and concentrated. The residue was purified by MPLC (gradient AcOEt/hexanes 0/100 to 5/95) to afford **15** (1.01 g, 74% brsm, L-erythro-**15**/D-threo-**15** 64:36) as colorless oil. $^1H$ NMR (600 MHz, $CDCl_3$) δ (L-erythro-**15**) −0.07 (s, $CH_3$, 3H), 0.01 (s, $CH_3$, 3H), 0.79 (s, tBu, 9H), 0.85 (t, H-6, *J* = 7.2 Hz, 3H), 1.21–1.28 (m, H-4a, H-5a, 2H), 1.36–1.47 (m, H-4b, H-5b, 2H), 2.36 (d, OH, *J* = 2.8 Hz, 1H), 3.17 (dd, H-1a, *J* = 7.7, 9.4 Hz, 1H), 3.23 (dd, H-1b, *J* = 4.5, 9.4 Hz, 1H), 3.69 (ddd, H-3, *J* = 4.1, 4.5, 6.9 Hz, 1H), 3.79-3.83 (m, H-2, 1H), 7.22–7.25 (m, Tr-$H_p$, 3H), 7.28–7.32 (m, Tr-$H_m$, 6H), 7.42–7.45 (m, Tr-$H_o$, 6H); δ (D-threo-**15**) -0.13 (s, $CH_3$, 3H), 0.02 (s, $CH_3$, 3H), 0.78 (s, tBu, 9H), 0.90 (t, H-6, *J* = 7.2 Hz, 3H), 1.30-1.35 (m, H-4a, H-5, 3H), 1.58–1.63 (m, H-4b, 1H), 2.37 (d, OH, *J* = 7.5 Hz, 1H), 3.03 (dd, H-1a, *J* = 6.1, 9.4 Hz, 1H), 3.22 (dd, H-1b, *J* = 6.3, 9.4 Hz, 1H), 3.65–3.69 (m, H-2, 1H), 3.79-3.82 (m, H-3, 1H), 7.22–7.25 (m, Tr-$H_p$, 3H), 7.27–7.31 (m, Tr-$H_m$, 6H), 7.42–7.45 (m, Tr-$H_o$, 6H); $^{13}C$ NMR (150 MHz, $CDCl_3$) δ (L-erythro-**15**) −4.4 ($CH_3$), −4.3 ($CH_3$), 14.4 (C-6), 18.2 (tBu), 18.3 (C-5), 26.0 (tBu), 34.5 (C-4), 65.1 (C-1), 73.3 (C-3), 73.5 (C-2), 86.9 (Trt-$C_q$), 127.2 (3 × Tr-$C_p$), 128.0 (6 × Tr-$C_m$), 128.8 (6 × Tr-$C_o$), 144.1 (3 × Tr-$C_{ipso}$); δ (D-threo-**15**) −4.7 ($CH_3$), −4.1 ($CH_3$), 14.4 (C-6), 18.2 (tBu), 18.7 (C-5), 26.0 (tBu), 36.1 (C-4), 65.3 (C-1), 71.8 (C-2), 71.9 (C-3), 86.8 (Tr-$C_q$), 127.1 (3 × Tr-$C_p$), 127.9 (6 × Tr-$C_m$), 128.8 (6 × Tr-$C_o$), 144.2 (3 × Tr-$C_{ipso}$); HRMS (ESI) calcd for $C_{31}H_{42}O_3SiNa^+$ $[M + Na]^+$: 513.2795, found: 513.2795.

*3.6. (2R,3S)-2-O-Acetyl-3-O-tert-butyldimethylsilyl-1-O-trityl-hexane-1,2,3-triol (L-erythro-16) and*
*(2R,3R)-2-O-acetyl-3-O-tert-butyldimethylsilyl-1-O-trityl-hexane-1,2,3-triol (D-threo-16)*

DMAP (771 mg, 6.32 mmol) and $Ac_2O$ (0.60 mL, 6.32 mmol) were added to a soluion of triol **15** (1.03 g, 2.11 mmol) in dry $CH_2Cl_2$ (30 mL) at room temperature. The reaction mixture was stirred for 30 min and quenched with sat. aqueous $NaHCO_3$ (30 mL). The water phase was extracted with $CH_2Cl_2$ (3 × 40 mL), combined organic layers were dried and concentrated. The residue was purified by MPLC (gradient AcOEt/hexanes 0/100 to 5/95) to give **16** (975 mg, 87%, L-erythro-**16**/D-threo-**16** 65:35) as colorless oil. $^1H$ NMR (600 MHz, $CDCl_3$) δ (L-erythro-**16**) -0.10 (s, $CH_3$, 3H), -0.04 (s, $CH_3$,

3H), 0.74 (s, tBu, 9H), 0.85 (t, H-6, *J* = 7.1 Hz, 3H), 1.22–1.39 (m, H-4, H-5, 4H), 2.10 (s, C(O)CH$_3$, 3H), 3.25–3.29 (m, H-1, 2H), 3.78 (ddd, H-3, *J* = 3.5, 5.0, 6.7 Hz, 1H), 5.15 (ddd, H-2, *J* = 3.5, 4.7, 6.8 Hz, 1H), 7.20–7.24 (m, Tr-H$_p$, 3H), 7.27–7.30 (m, Tr-H$_m$, 6H), 7.40–7.43 (m, Tr-H$_o$, 6H); δ (D-threo-**16**) 0.01 (s, CH$_3$, 3H), 0.02 (s, CH$_3$, 3H), 0.79 (s, tBu, 9H), 0.80 (t, H-6, *J* = 7.1 Hz, 3H), 1.16–1.30 (m, H-4a, H-5, 3H), 1.33–1.38 (m, H-4b, 1H), 2.15 (s, C(O)CH$_3$, 3H), 3.17 (dd, H-1a, *J* = 6.8, 10.2 Hz, 1H), 3.25 (dd, H-1b, *J* = 2.7, 10.2 Hz, 1H), 3.86 (ddd, H-3, *J* = 4.3, 5.5, 6.8 Hz, 1H), 5.07 (ddd, H-2, *J* = 2.7, 5.5, 6.8 Hz, 1H), 7.20–7.24 (m, Tr-H$_p$, 3H), 7.27–7.30 (m, Tr-H$_m$, 6H), 7.40–7.43 (m, Tr-H$_o$, 6H); $^{13}$C NMR (150 MHz, CDCl$_3$) δ (L-erythro-**16**) −4.6 (CH$_3$), −4.4 (CH$_3$), 14.3 (C-6), 18.1 (tBu), 18.7 (C-5), 21.4 (C(O)CH$_3$), 25.9 (tBu), 36.1 (C-4), 62.3 (C-1), 72.5 (C-3), 76.0 (C-2), 86.8 (Tr-C$_q$), 127.1 (3 × Tr-C$_p$), 127.9 (6 × Tr-C$_m$), 128.8 (6 × Tr-C$_o$), 144.1 (3 × Tr-C$_{ipso}$), 170.5 (C(O)CH$_3$); δ (D-threo-**16**) −4.4 (CH$_3$), −4.4 (CH$_3$), 14.3 (C-6), 18.1 (tBu), 18.5 (C-5), 21.5 (C(O)CH$_3$), 25.9 (tBu), 35.1 (C-4), 62.6 (C-1), 70.9 (C-3), 75.8 (C-2), 86.5 (Tr-C$_q$), 127.1 (3 × Tr-C$_p$), 127.9 (6 × Tr-C$_m$), 128.8 (6 × Tr-C$_o$), 144.1 (3 × Tr-C$_{ipso}$), 170.7 (C(O)CH$_3$); HRMS (ESI) calcd for C$_{33}$H$_{44}$O$_4$SiNa$^+$ [M + Na]$^+$: 555.2901, found: 555.2901.

### 3.7. (2R,3S)-2-O-Acetyl-3-O-tert-butyldimethylsilyl-hexane-1,2,3-triol (L-erythro-17) and (2R,3R)-2-O-acetyl-3-O-tert-butyldimethylsilyl-hexane-1,2,3-triol (D-threo-17)

Formic acid (12.2 mL) was added to a solution of protected triol **16** (959 mg, 1.80 mmol) in diethyl ether (12.2 mL) at 0 °C. The reaction mixture was stirred for 50 min at room temperature, diluted with diethyl ether (30 mL), and cooled to 0 °C. Sat. aqueous NaHCO$_3$ (equimolar to formic acid, 323 mmol) was added with vigorous stirring to neutralize the reaction mixture (accompanied by the separation of two layers). The water phase was extracted with diethyl ether (3 × 50 mL) and the combined organic layers were dried and concentrated. The residue was purified by MPLC (gradient AcOEt/hexanes 0/100 to 10/90) to afford 346 mg (66%) of yellowish oil as an inseparable mixture of **17** (L-erythro-**17**/D-threo-**17** 63:37) and **20** (L-erythro-**20**/D-threo-**20** 58:42, the product of the acetyl group migration of **17** taking place even during MPLC) in ratio 82:18. $^1$H NMR (600 MHz, CDCl$_3$) δ (L-erythro-**17**) 0.07 (s, CH$_3$, 3H), 0.11 (s, CH$_3$, 3H), 0.90 (s, tBu, 9H), 0.93 (t, H-6, *J* = 7.3 Hz, 3H), 1.31–1.36 (m, H-5a, 1H), 1.48–1.55 (m, H-4, H-5b, 3H), 2.12 (s, C(O)CH$_3$, 3H), 2.68 (dd, OH, *J* = 3.7, 7.7 Hz, 1H), 3.81–3.85 (m, H-1a, 1H), 3.90–3.95 (m, H-1b, H-3, 2H), 4.77 (dt, H-2, *J* = 3.1, 4.9 Hz, 1H); δ (D-threo-**17**) 0.08 (s, CH$_3$, 3H), 0.12 (s, CH$_3$, 3H), 0.90 (s, tBu, 9H), 0.91 (t, H-6, *J* = 7.2 Hz, 3H), 1.24–1.33 (m, H-5a, 1H), 1.35–1.45 (m, H-4a, H-5b, 2H), 1.46–1.54 (m, H-4b, 1H), 2.11 (s, C(O)CH$_3$, 3H), 2.13–2.16 (m, OH, 1H), 3.69–3.73 (m, H-1a, 1H), 3.84–3.90 (m, H-1b, H-3, 2H), 4.87 (ddd, H-2, *J* = 4.1, 4.8, 6.5 Hz, 1H); δ (L-erythro-**20**) 0.08 (s, CH$_3$, 3H), 0.09 (s, CH$_3$, 3H), 0.90 (s, tBu, 9H), 0.92 (t, H-6, *J* = 7.1 Hz, 3H), 1.29–1.34 (m, H-4a, 1H), 1.38–1.44 (m, H-5, 2H), 1.55–1.61 (m, H-4b, 1H), 2.10 (s, C(O)CH$_3$, 3H), 2.32 (d, OH, *J* = 4.7 Hz, 1H), 3.75-3.78 (m, H-3, 1H), 3.78-3.82 (m, H-2, 1H), 4.06 (dd, H-1a, *J* = 7.7, 11.6 Hz, 1H), 4.23 (dd, H-1b, *J* = 3.0, 11.6 Hz, 1H); δ (D-threo-**20**) 0.08 (s, CH$_3$, 3H), 0.10 (s, CH$_3$, 3H), 0.90 (s, tBu, 9H), 0.92 (t, H-6, *J* = 7.1 Hz, 3H), 1.31-1.44 (m, H-4a, H-5, 3H), 1.64–1.71 (m, H-4a, 1H), 2.09 (s, C(O)CH$_3$, 3H), 2.40 (d, OH, *J* = 7.8 Hz, 1H), 3.68-3.71 (m, H-3, 1H), 3.72-3.75 (m, H-2, 1H), 4.05 (dd, H-1a, *J* = 5.1, 11.3 Hz, 1H), 4.08 (dd, H-1b, *J* = 7.0, 11.3 Hz, 1H); $^{13}$C NMR (150 MHz, CDCl$_3$) δ (L-erythro-**17**) −4.5, −4.5 (2 × CH$_3$), 14.3 (C-6), 18.2 (tBu), 18.8 (C-5), 21.4 (C(O)CH$_3$), 25.9 (tBu), 36.8 (C-4), 61.9 (C-1), 73.8 (C-3), 76.6 (C-2), 171.2 (C(O)CH$_3$); δ (D-threo-**17**) = −4.4 (CH$_3$), −4.4 (CH$_3$), 14.4 (C-6), 18.1 (tBu), 18.9 (C-5), 21.3 (C(O)CH$_3$), 25.9 (tBu), 34.8 (C-4), 62.0 (C-1), 71.5 (C-3), 76.7 (C-2), 171.2 (C(O)CH$_3$); δ (L-erythro-**20**) −4.5 (CH$_3$), −4.3 (CH$_3$), 14.4 (C-6), 18.2 (tBu), 18.4 (C-5), 21.1 (C(O)CH$_3$), 26.0 (tBu), 34.9 (C-4), 65.8 (C-1), 72.6 (C-2), 73.2 (C-3), 171.5 (C(O)CH$_3$); δ (D-threo-**20**) −4.7 (CH$_3$),-4.1 (CH$_3$), 14.3 (C-6), 18.2 (tBu), 18.6 (C-5), 21.1 (C(O)CH$_3$), 26.0 (tBu), 35.9 (C-4), 66.2 (C-1), 70.8 (C-2), 71.9 (C-3), 171.2 (C(O)CH$_3$); HRMS (ESI) calcd for C$_{14}$H$_{30}$O$_4$SiNa$^+$ [M + Na]$^+$: 313.1806, found: 313.1806.

### 3.8. (2R,3S)-2-O-Acetyl-4,5,6-trideoxy-3-O-tert-butyldimethylsilyl- L-erythro-hexose (L-erythro-10) and (2R,3R)-2-O-acetyl-4,5,6-trideoxy-3-O-tert-butyldimethylsilyl- D-threo-hexose (D-threo-10)

Oxalyl chloride (2.0 M in CH$_2$Cl$_2$, 0.85 mL, 1.70 mmol) was added dropwise to a solution of dimethyl sulfoxide (0.18 mL, 2.56 mmol) in dry CH$_2$Cl$_2$ (3.9 mL) at -78 °C. After 30 min of stirring at −78 °C, a solution of alcohol **17** (330 mg, 82:18 mixture with **20**, 0.93 mmol of **17**) in dry $CH_2Cl_2$ (1.9 mL) was added dropwise. The reaction mixture was stirred at −78 °C for 30 min and $Et_3N$ (0.63 mL, 4.54 mmol) was added. After 1 h of stirring at −78 °C, the reaction mixture was allowed to reach room temperature slowly (additional 1 h). Subsequently, the reaction mixture was concentrated, dry diethyl ether was added to the residue, and the mixture was filtered through a short silicagel column. The filtrate was concentrated to give the crude aldehyde **10** which was immediately used in the following step without further purification.

*3.9. (3R,4S)-3-O-Acetyl-1-(2-acetoxymethyl-3-methoxyphenyl)-4-O-tert-butyldimethylsilyl-hept-1-ene-3,4-diol (18) and*
*(3R,4R)-3-O-acetyl-1-(2-acetoxymethyl-3-methoxyphenyl)-4-O-tert-butyldimethylsilyl-hept-1-ene-3,4-diol (19)*

Sulfone **11** (457 mg, 1.14 mmol) in dry DME (10.5 mL) was slowly added to a solution of the crude aldehyde **10** (theor. 0.93 mmol) in dry DME (10.5 mL) and the mixture was cooled to −60 °C. Subsequently, KHMDS (0.5 M in toluene, 3.97 mL, 1.99 mmol) was added dropwise keeping −60 °C and the reaction mixture was allowed to reach room temperature. The reaction mixture was stirred for 40 min, quenched with sat. aqueous $NH_4Cl$ (20 mL), and diluted with AcOEt (20 mL). The aqueous layer was extracted with AcOEt (3 × 20 mL) and the combined organic layers were dried and concentrated. The residue (**18/19** 70:30) was repeatedly purified by MPLC (gradient AcOEt/hexanes 0/100 to 5/95) and preparative TLC to afford **18** (79 mg, 18%), **19** (22 mg, 5%), and the mixture of **18** and **19** (78 mg, 18%) as colorless oils over two steps (41% overall yield).

[1]H NMR (600 MHz, $CDCl_3$) δ (**18**) 0.06 (s, $CH_3$, 3H), 0.10 (s, $CH_3$, 3H), 0.90 (t, H-7, $J$ = 7.0 Hz, 3H), 0.91 (s, tBu, 9H), 1.30–1.36 (m, H-6a, 1H), 1.39–1.48 (m, H-5, H-6b, 3H), 2.06 (s, $C(O)CH_3$, 3H), 2.09 (s, $C(O)CH_3$, 3H), 3.84 (s, $OCH_3$, 3H), 3.85–3.88 (m, H-4, 1H), 5.25 (d, $CH_2OAc$, $J$ = 11.8 Hz, 1H), 5.27 (d, $CH_2OAc$, $J$ = 11.8 Hz, 1H), 5.33 (ddd, H-3, $J$ = 1.0, 2.8, 7.6 Hz, 1H), 6.18 (dd, H-2, $J$ = 7.6, 15.9 Hz, 1H), 6.84 (d, H-4′, $J$ = 8.2 Hz, 1H), 6.85 (d, H-1, $J$ = 15.9 Hz, 1H), 7.11 (d, H-6′, $J$ = 7.8 Hz, 1H), 7.30 (t, H-5′, $J$ = 8.1 Hz, 1H); [13]C NMR (150 MHz, $CDCl_3$) δ (**18**) −4.4 ($CH_3$), −4.2 ($CH_3$), 14.3 (C-7), 18.4 (tBu), 18.9 (C-6), 21.1 ($C(O)CH_3$), 21.5 ($C(O)CH_3$), 26.0 (tBu), 36.3 (C-5), 56.0 ($OCH_3$), 57.7 ($CH_2OAc$), 73.8 (C-4), 77.9 (C-3), 110.3 (C-4′), 118.9 (C-6′), 121.5 (C-2′), 127.5 (C-2), 130.0 (C-5′), 131.2 (C-1), 139.0 (C-1′), 158.6 (C-3′), 170.2 ($C(O)CH_3$), 171.2 ($C(O)CH_3$); HRMS (ESI) calcd for $C_{25}H_{40}O_6SiNa^+$ [M + Na]$^+$: 487.2486, found: 487.2486; $[\alpha]_D^{20}$ −48.8 (c 1.131 MeOH).

[1]H NMR (600 MHz, $CDCl_3$) δ (**19**) 0.08 (s, $CH_3$, 3H), 0.11 (s, $CH_3$, 3H), 0.89 (t, H-7, $J$ = 7.0 Hz, 3H), 0.90 (s, tBu, 9H), 1.30–1.37 (m, H-6a, 1H), 1.38–1.50 (m, H-5, H-6b, 3H), 2.06 (s, $C(O)CH_3$, 3H), 2.11 (s, $C(O)CH_3$, 3H), 3.80 (ddd, H-4, $J$ = 3.6, 6.0, 7.2 Hz, 1H), 3.84 (s, $OCH_3$, 3H), 5.26 (s, $CH_2OAc$, 2H), 5.38 (ddd, H-3, $J$ = 1.4, 6.1, 6.1 Hz, 1H), 6.11 (dd, H-2, $J$ = 6.2, 15.9 Hz, 1H), 6.83 (d, H-4′, $J$ = 8.3 Hz, 1H), 6.84 (dd, H-1, $J$ = 1.3, 15.9 Hz, 1H), 7.09 (d, H-6′, $J$ = 7.8 Hz, 1H), 7.29 (t, H-5′, $J$ = 8.1 Hz, 1H); [13]C NMR (150 MHz, $CDCl_3$) δ (**19**) −4.4 ($CH_3$), −4.2 ($CH_3$), 14.5 (C-7), 18.2 (tBu), 18.5 (C-6), 21.1 ($C(O)CH_3$), 21.4 ($C(O)CH_3$), 26.0 (tBu), 35.3 (C-5), 56.0 ($OCH_3$), 57.7 ($CH_2OAc$), 73.0 (C-4), 76.8 (C-3), 110.2 (C-4′), 118.8 (C-6′), 121.4 (C-2′), 128.1 (C-2), 129.5 (C-1), 130.0 (C-5′), 139.0 (C-1′), 158.6 (C-3′), 170.2 ($C(O)CH_3$), 171.2 ($C(O)CH_3$); HRMS (ESI) calcd for $C_{25}H_{40}O_6SiNa^+$ [M + Na]$^+$: 487.2486, found: 487.2486; $[\alpha]_D^{20}$ +5.5 (c 0.431 MeOH).

*3.10. Varioxiranol A (4)*

Compound **18** (21.1 mg, 0.045 mmol) was dissolved in methanol (2 mL) and $K_2CO_3$ (12.6 mg, 0.091 mol) was added. The reaction mixture was stirred at room temperature for 2.5 h, diluted with AcOEt (4 mL) and with water (4 mL). The water phase was extracted with AcOEt (4 × 2 mL) and combined organic layers were dried and concentrated. The residue was diluted in THF (0.5 mL), the solution was cooled to 0 °C, and TBAF × $3H_2O$ in THF (1.0 M solution, 46 µL, 0.046 mmol) was added. The reaction mixture was stirred at room temperature for 4.5 h and quenched with sat. aqueous $NH_4Cl$ (5 mL), the aqueous layer was extracted with $CH_2Cl_2$ (3 × 5 mL), and the combined organic layers were dried and concentrated. The residue was purified by FLC (isocratic acetone/$CH_2Cl_2$ 20/80 then 50/50) to afford varioxiranol A (**4**, 11.1 mg, 92% over two steps) as colorless crystalline solid that was subsequently recrystallized from AcOEt-hexanes. [1]H NMR (600 MHz, CDCl$_3$) δ (**4**) 0.94 (t, H-7, *J* = 7.2 Hz, 3H), 1.36-1.43 (m, H-6a, 1H), 1.43–1.48 (m, H-5, 2H), 1.51–1.57 (m, H-6b, 1H), 2.09 (d, OH, *J* = 4.3 Hz, 1H), 2.20 (t, OH, *J* = 5.8 Hz, 1H), 2.22 (d, OH, *J* = 4.6 Hz, 1H), 3.78–3.82 (m, H-4, 1H), 3.87 (s, OCH$_3$, 3H), 4.26–4.30 (m, H-3, 1H), 4.76–4.83 (m, CH$_2$OH, 2H), 6.19 (dd, H-2, *J* = 6.9, 15.8 Hz, 1H), 6.84 (d, H-4′, *J* = 8.2 Hz, 1H), 7.02 (d, H-1, *J* = 15.8 Hz, 1H), 7.10 (d, H-6′, *J* = 7.9 Hz, 1H), 7.25 (t, H-5′, *J* = 8.2 Hz, 1H); [13]C NMR (150 MHz, acetone-d$_6$) δ (**4**) 14.5 (C-7), 19.8 (C-6), 35.7 (C-5), 55.5 (CH$_2$OH), 56.1 (OCH$_3$), 75.0 (C-4), 76.7 (C-3), 110.4 (C-4′), 119.4 (C-6′), 127.9 (C-2′), 129.2 (C-1), 129.3 (C-5′), 133.7 (C-2), 139.6 (C-1′), 158.9 (C-3′); HRMS (ESI) calcd for C$_{15}$H$_{22}$O$_4$Na$^+$ [M+Na]$^+$: 289.1410, found: 289.1410; [α]$_D^{20}$ +2.5 (c 0.207 MeOH); mp 120–121 °C.

### 3.11. 4-epi-Varioxiranol A (9)

Compound **19** (20.8 mg, 0.045 mmol) was dissolved in methanol (2 mL) and K$_2$CO$_3$ (12.4 mg, 0.090 mmol) was added. The reaction mixture was stirred at room temperature for 2.5 h and diluted with AcOEt (4 mL) and with water (4 mL). The water phase was extracted with AcOEt (4 × 2 mL) and combined organic layers were dried and concentrated. The residue was diluted in THF (0.5 mL), the solution was cooled to 0 °C, and TBAF × 3H$_2$O in THF (1.0 M solution, 45 μL, 0.045 mmol) was added. The reaction mixture was stirred at room temperature for 4.5 h and quenched with sat. aqueous NH$_4$Cl (5 mL). The aqueous layer was extracted with CH$_2$Cl$_2$ (3 × 5 mL) and the combined organic layers were dried and concentrated. The residue was purified by FLC (isocratic acetone/CH$_2$Cl$_2$ 30/70) to afford 4-epi-varioxiranol A (**9**, 6.8 mg, 57% over two steps) that was subsequently recrystallized from CH$_2$Cl$_2$-hexanes yielding colorless crystalline solid. [1]H NMR (600 MHz, CDCl$_3$) δ (**9**) 0.94 (t, H-7, *J* = 7.1 Hz, 3H), 1.39–1.45 (m, H-6a, 1H), 1.46–1.56 (m, H-5, H-6b, 3H), 2.20 (t, OH, *J* = 6.3 Hz, 1H), 2.28 (d, OH, *J* = 4.3 Hz, 1H), 2.40 (d, OH, *J* = 4.3 Hz, 1H), 3.56–3.61 (m, H-4, 1H), 3.87 (s, OCH$_3$, 3H), 4.10–4.14 (m, H-3, 1H), 4.78 (dd, CH$_2$OH, *J* = 6.3, 12.1 Hz, 1H), 4.81 (dd, CH$_2$OH, *J* = 6.3, 12.1 Hz, 1H), 6.12 (dd, H-2, *J* = 6.8, 15.8 Hz, 1H), 6.84 (d, H-4′, *J* = 8.2 Hz, 1H), 7.04 (d, H-1, *J* = 15.8 Hz, 1H), 7.08 (d, H-6′, *J* = 7.8 Hz, 1H), 7.25 (t, H-5′, *J* = 8.0 Hz, 1H); [13]C NMR (150 MHz, CDCl$_3$) δ (**9**) 14.2 (C-7), 19.0 (C-6), 35.4 (C-5), 55.8 (OCH$_3$), 56.6 (CH$_2$OH), 74.5 (C-4), 76.3 (C-3), 110.0 (C-4′), 119.4 (C-6′), 126.4 (C-2′), 129.1 (C-5′), 129.8 (C-1), 132.6 (C-2), 137.7 (C-1′), 158.2 (C-3′); HRMS (ESI) calcd for C$_{15}$H$_{22}$O$_4$Na$^+$ [M + Na]$^+$: 289.1410, found: 289.1410; [α]$_D^{20}$ +45.0 (c 0.094 MeOH); mp 112–113 °C.

**Supplementary Materials:** The following are available online, S1: Copies of [1]H NMR and [13]C NMR spectra for all new compounds.

**Author Contributions:** Conceptualization, A.L. and T.G.; synthesis, M.Š. and M.P.; data curation and NMR analysis, J.D.; X-ray analysis, J.M.; writing—original draft preparation, A.L. and T.G.

**Funding:** This research received no external funding.

**Acknowledgments:** This work was supported by Slovak Grant Agencies (VEGA No. 1/0552/18, APVV-14-0147 and ASFU, Bratislava, ITMS projects No. 26240120025 and 26240220084).

**Conflicts of Interest:** The authors declare no conflict of interest.

### References

1. Zadar, P.; Frisvad, J.C.; Gunde-Cimerman, N.; Varga, J.; Samson, R.A. Four new species of *Emericella* from the Mediterranean region of Europe. *Mycologia* **2008**, *100*, 779–795. [CrossRef]
2. Dunn, A.W.; Johnstone, R.A.W. Fungal metabolites. Part 7. Structures of C$_{25}$ compounds from *Aspergillus variecolor*. *J. Chem. Soc. Perkin Trans. 1* **1979**, 2113–2117. [CrossRef]
3. Dunn, A.W.; Johnstone, R.A.W. Fungal metabolites. Part 8. Isolation of 2-methoxy-6-(3,4-dihydroxyhepta-1,5-dienyl)benzyl alcohol. *J. Chem. Soc. Perkin Trans. 1* **1979**, 2122–2123. [CrossRef]
4. Markovič, M.; Lopatka, P.; Koóš, P.; Gracza, T. First total synthesis of natural andytriol and a biomimetic approach to varioxiranes. *Tetrahedron* **2015**, *71*, 8407–8415. [CrossRef]

5. Malmstrøm, J.; Christophersen, C.; Barrero, A.F.; Oltra, J.E.; Justicia, J.; Rosales, A. Bioactive Metabolites from a Marine-Derived Strain of the Fungus *Emericella variecolor. J. Nat. Prod.* **2002**, *65*, 364–367. [CrossRef]

6. Wu, Q.; Wu, C.; Long, H.; Chen, R.; Liu, D.; Proksch, P.; Guo, P.; Lin, W. Varioxiranols A–G and 19-*O*-Methyl-22-methoxypre-shamixanthone, PKS and Hybrid PKS-Derived Metabolites from a Sponge-Associated *Emericella variecolor* Fungus. *J. Nat. Prod.* **2015**, *78*, 2461–2470. [CrossRef] [PubMed]

7. Liangsakul, J.; Pornpakakul, S.; Sangvichien, E.; Muangsin, N.; Sihanonth, P. Emervaridione and varioxiranediol, two new metabolites from the endophytic fungus, *Emericella variecolor. Tetrahedron Lett.* **2011**, *52*, 6427–6430. [CrossRef]

8. He, Y.; Hu, Z.; Li, Q.; Huang, J.; Li, X.-N.; Zhu, H.; Liu, J.; Wang, J.; Xue, Y.; Zhang, Y. Bioassay-Guided Isolation of Antibacterial Metabolites from *Emericella* sp. TJ29. *J. Nat. Prod.* **2017**, *80*, 2399–2405. [CrossRef] [PubMed]

9. Palík, M.; Karlubíková, O.; Lásiková, A.; Kožíšek, J.; Gracza, T. Total Synthesis of (+)-Varitriol. *Eur. J. Org. Chem.* **2009**, 709–715. [CrossRef]

10. Palík, M.; Karlubíková, O.; Lackovičová, D.; Lásiková, A.; Gracza, T. Formal synthesis of (+)-varitriol. Application of Pd(II)/Cu(II)-catalysed bicyclisation of unsaturated polyols. *Tetrahedron* **2010**, *66*, 5244–5249. [CrossRef]

11. Karlubíková, O.; Palík, M.; Lásiková, A.; Gracza, T. An Efficient Total Synthesis of (+)-Varitriol from D-Ribonolactone. *Synthesis* **2010**, 3449–3452. [CrossRef]

12. Caletková, O.; Lásiková, A.; Hajdúch, M.; Džubák, P.; Gracza, T. Synthesis and antitumour activity of varitriol and its analogues. *ARKIVOC* **2012**, 365–383. [CrossRef]

13. Antošová, A.; Šípošová, K.; Bednáriková, Z.; Lásiková, A.; Dohánošová, J.; Gracza, T.; Gažová, Z. Natural tetrahydrofuran derivatives reduce insulin amyloid aggregation. *Eur. Biophys. J.* **2013**, *42*, S62. [CrossRef]

14. Schmid, C.R.; Bryant, J.D. D-(*R*)-Glyceraldehyde acetonide. *Org. Synth.* **1995**, *72*, 6–8. [CrossRef]

15. de Napoli, L.; Messere, A.; Palomba, D.; Piccialli, V.; Evidente, A.; Gennaro Piccialli, G. Studies toward the Synthesis of Pinolidoxin, a Phytotoxic Nonenolide from the Fungus *Ascochyta pinodes*. Determination of the Configuration at the C-7, C-8, and C-9 Chiral Centers and Stereoselective Synthesis of the $C_6$–$C_{18}$ Fragment. *J. Org. Chem.* **2000**, *65*, 3432–3442. [CrossRef] [PubMed]

16. Heo, J.N.; Micalizio, G.C.; Roush, W.R. Enantio- and Diastereoselective Synthesis of Cyclic β-Hydroxy Allylsilanes via Sequential Aldehyde γ-Silylallylboration and Ring-Closing Metathesis Reactions. *Org. Lett.* **2003**, *5*, 1693–1696. [CrossRef] [PubMed]

17. Sheldrick, G.M. *SHELXT*—Integrated space-group and crystal-structure determination. *Acta Crystallogr.* **2015**, *A71*, 3–8. [CrossRef] [PubMed]

18. Sheldrick, G.M. Crystal structure refinement with *SHELXL. Acta Crystallogr.* **2015**, *C71*, 3–8. [CrossRef]

19. Betteridge, P.W.; Carruthers, J.R.; Copper, R.I.; Prout, K.; Watkin, D.J. *CRYSTALS* version 12: Software for guided crystal structure analysis. *J. Appl. Crystallogr.* **2003**, *36*, 1487. [CrossRef]

20. Dolomanov, O.; Bourhis, L.J.; Gildea, R.I.; Howard, J.A.K.; Puschmann, H. *OLEX2*: A complete structure solution, refinement and analysis program. *J. Appl. Crystallogr.* **2009**, *42*, 339–341. [CrossRef]

21. Parsons, S.; Flack, H.D.; Wagner, T. Use of intensity quotients and differences in absolute structure refinement. *Acta Crystallogr.* **2013**, *B69*, 249–259. [CrossRef] [PubMed]

22. Hooft, R.W.W.; Straver, L.H.; Spek, A.L. Determination of absolute structure using Bayesian statistics on Bijvoet differences. *J. Appl. Crystallogr.* **2008**, *41*, 96–103. [CrossRef] [PubMed]

**Sample Availability:** Samples of the compounds are available from the authors.

*Article*

# Preparation of Polysaccharides from *Ramulus mori*, and Their Antioxidant, Anti-Inflammatory and Antibacterial Activities

**Wansha Yu, Hu Chen, Zhonghuai Xiang and Ningjia He \***

State Key Laboratory of Silkworm Genome Biology, Southwest University, Beibei, Chongqing 400715, China; yuwansha123456@163.com (W.Y.); c924992@sina.com (H.C.); xbxzh@swu.edu.cn (Z.X.)
\* Correspondence: hejia@swu.edu.cn; Tel.: +86-23-6825-0797; Fax: +86-23-6825-1128

Academic Editors: Pavel B. Drasar and Vladimir A. Khripach
Received: 15 January 2019; Accepted: 24 February 2019; Published: 28 February 2019

**Abstract:** The extraction of *Ramulus mori* polysaccharides (RMPs) was optimized using response surface methodology (RSM). The optimal process conditions, which gave the highest yield of RMPs (6.25%) were 80 °C, 50 min, and a solid–liquid ratio of 1:40 (g/mL), with the extraction performed twice. The RMPs contained seven monosaccharides, namely, mannose, rhamnose; glucuronic acid, glucose, xylose, galactose, and arabinose, in a 1.36:2.68:0.46:328.17:1.53:21.80:6.16 molar ratio. The glass transition and melting temperatures of RMPs were 83 and 473 °C, respectively. RMPs were α-polysaccharides and had surfaces that resembled a porous sponge, as observed by scanning electron microscopy. RMPs inhibited the proliferation of *Escherichia coli, Staphylococcus aureus,* and *Pseudomonas aeruginosa* and showed antioxidant activity (assessed by three different methods), although it was generally weaker than that of vitamin C. RMPs showed anti-inflammatory activity in a concentration-dependent manner. This study provides a basis for exploring the potential uses of RMPs.

**Keywords:** *Ramulus mori*; polysaccharides; bioactivity

## 1. Introduction

Mulberry (*Morus alba* L.) is a perennial woody plant with many ecological effects, including improving air quality, protecting of water resources, soil integration, and improving of microclimates [1]. In recent years, mulberry has been used to restore vegetation in the Three Gorges Reservoir Region, control rocky desertification, and reconstruct ecological landscapes in China [2,3]. Mulberry leaves are used to rear the domesticated silkworm, with constant pruning critical for the accumulation of leaves to feed more silkworms. During annual cutting, tons of mulberry branches are discarded and burned, representing a significant waste of resources and causing substantial environmental pollution. Mulberry branch, *Ramulus mori,* is a traditional Chinese medicine. Modern studies have shown that *R. mori* contains flavonoids, polyphenols, alkaloids, polysaccharides, and other active ingredients [4–6] that can have curative effects, including diabetes-alleviating and liver-protecting effects [7,8].

Response surface methodology (RSM), is a collection of statistical and mathematical techniques that are effective for developing, improving, and optimizing processes [9] and products [10]. RSM has been widely used to optimize process variables for the extraction of polysaccharides [11], flavonoids [12], alkaloids [13], and saponins [14] from various materials.

Anti-diabetic effects mediated by *R. mori* polysaccharides (RMPs) have long been the focus of studies on these natural products [7,8]. However, the structural characterization, and antibacterial, antioxidant, and anti-inflammatory activities of RMPs have yet to be reported. A basic understanding of the structure and biological activity of RMPs is essential for future applications of these polysaccharides.

This study aimed to obtain an optimal extraction method for RMPs and explore the potential value of RMPs in the pharmaceutical and food industries.

## 2. Results and Discussion

### 2.1. Single-Factor Experiments

Preliminary studies were conducted to investigate the influence of the solid–liquid ratio on the RMP extraction yield when the other extraction parameters were fixed as follows: Extraction temperature, 70 °C; extraction time, 30 min; number of extractions, one. As shown in Figure 1a, the RMP yield increased from 3.96% to 5.00% with an increase in the solid–liquid ratio from 1:10 to 1:40. When the solid–liquid ratio was further increased, the RMP yield decreased to 4.26%. The polysaccharides were rapidly dissolved, which resulted in an increased yield. However, the viscosity of the solvent increased with an increasing solid–liquid ratio, resulting in an increase in the diffusion distance toward the internal tissues [15]. Therefore, the RMP yield increased slowly with increasing solid–liquid ratio, but with a downward trend at the highest solid–liquid ratio (Figure 1a). To achieve increased RMP production using less solvent and in a shorter time, 1:40 was selected as the optimum solid–liquid ratio.

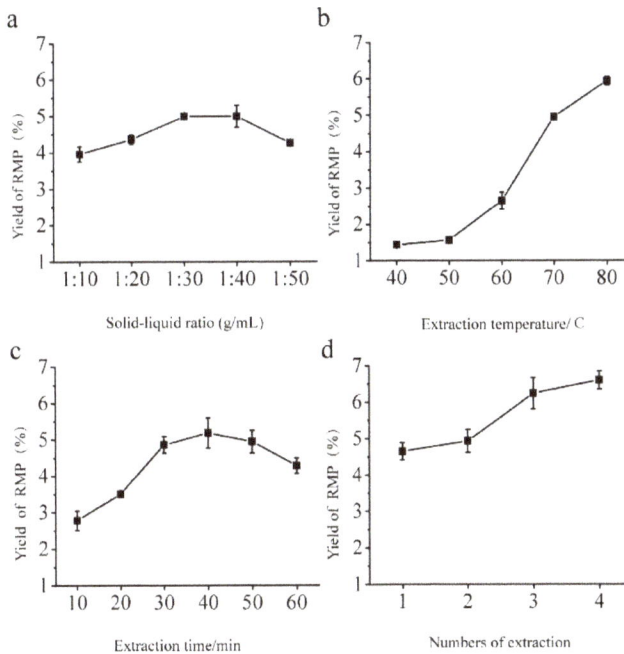

**Figure 1.** Relationships between (**a**) solid–liquid ratio, (**b**) extraction temperature, (**c**) extraction time, and (**d**) number of extractions and *R. mori* polysaccharide (RMP) yield in single-factor experiments.

Figure 1b shows the effect of extraction temperature on the RMP yield. Different extraction temperatures were used, with the other extraction parameters fixed as follows: Solid–liquid ratio, 1:40; extraction time, 30 min; number of extractions, one. The RMP yield increased from 1.44% to 5.94% when the extraction temperature was increased from 40 to 80 °C, perhaps due to the higher solubility and diffusivity of the polysaccharides in water at higher temperatures [16]. Accordingly, 80 °C was selected as the optimum extraction temperature.

The effect of different extraction times (10, 20, 30, 40, 50, and 60 min) on the RMP yield was investigated, with the other extraction parameters fixed as follows: Solid–liquid ratio, 1:40; extraction temperature, 80 °C; number of extractions, one. As shown in Figure 1c, the RMP yield increased with increasing extraction time from 10 to 40 min, with RMP production approaching a maximum at 40 min (5.18%) and decreasing thereafter. This decrease indicated that longer extraction times led to thermal instability and degradation of the RMPs [17]. Consequently, 40 min was selected as the optimal RMP extraction times.

The effect of a different number of extractions (1–4) on the RMP yield was investigated, with the other extraction parameters fixed as follows: Solid–liquid ratio, 1:40; extraction temperature, 80 °C; and extraction time, 40 min (Figure 1d). The yield increased as the number of extractions increased from one to four. Therefore, three extractions was selected as the optimum number for subsequent experiments.

## 2.2. Model Fitting and Statistical Analysis

According to these single-factor experiments, the ultrasound assisted extraction variables were established (Table 1). The actual and predictive values of responses (RMP yields) under different conditions are shown in Table 1. The predicted response value ($Y_{RMP}$) was based on the following second order polynomial equation:

$$Y_{RMP} = 4.38 + 1.79X_1 + 0.24X_2 + 0.031X_3 + 0.51X_4 + 0.38X_1X_2 + 0.35X_1X_3 - 0.56X_1X_4 + 0.14X_2X_3 - 0.23X_2X_4 + 0.23X_3X_4 - 0.47 X_1^2 - 0.33 X_2^2 - 0.50 X_3^2 + 0.073X_4^2 \tag{1}$$

where $Y_{RMP}$ is the RMP yield and $X_1$, $X_2$, $X_3$, and $X_4$ are the coded variables for extraction temperature, extraction time, solid–liquid ratio, and number of extractions, respectively. From the response surface method, analysis of variance (ANOVA) for the screening test model in Table 2 gave a determination coefficient ($R^2$) of 0.9997. The coefficient of variation was low, at only 0.83%, indicating that this model (Equation (1)) had a high and reliable degree of precision. As shown in Table 2, the large $F$-value (3356.14) and low $p$-value (<0.0001 **) indicated that this model was accurate.

**Table 1.** Experimental design with predicted and experimental extraction yield of RMPs.

| Independent Variables | Symbol | Range and Level | | |
|---|---|---|---|---|
| | | −1 | 0 | +1 |
| Extraction temperature (°C) | $X_1$ | 60 | 70 | 80 |
| Extraction time (min) | $X_2$ | 30 | 40 | 50 |
| Solid-liquid ratio (g/mL) | $X_3$ | 1:30 | 1:40 | 1:50 |
| Numbers of extraction | $X_4$ | 2 | 3 | 4 |

| Run | Coded Variable Levels | | | | Extraction Yield (%) | |
|---|---|---|---|---|---|---|
| | $X_1$ | $X_2$ | $X_3$ | $X_4$ | Experimental | Predicted |
| 1 | 60 | 30 | 1:40 | 3 | 1.91 | 1.93 |
| 2 | 80 | 30 | 1:40 | 3 | 4.74 | 4.75 |
| 3 | 60 | 50 | 1:40 | 3 | 1.63 | 1.65 |
| 4 | 80 | 50 | 1:40 | 3 | 5.97 | 5.99 |
| 5 | 70 | 40 | 1:30 | 2 | 3.63 | 3.63 |
| 6 | 70 | 40 | 1:50 | 2 | 3.22 | 3.24 |
| 7 | 70 | 40 | 1:30 | 4 | 4.20 | 4.21 |
| 8 | 70 | 40 | 1:50 | 4 | 4.69 | 4.72 |
| 9 | 60 | 40 | 1:40 | 2 | 1.12 | 1.12 |
| 10 | 80 | 40 | 1:40 | 2 | 5.81 | 5.82 |
| 11 | 60 | 40 | 1:40 | 4 | 3.27 | 3.26 |
| 12 | 80 | 40 | 1:40 | 4 | 5.73 | 5.73 |
| 13 | 70 | 30 | 1:30 | 3 | 3.41 | 3.42 |
| 14 | 70 | 50 | 1:30 | 3 | 3.59 | 3.61 |

**Table 1.** *Cont.*

| Run | Coded Variable Levels | | | | Extraction Yield (%) | |
|-----|----|----|------|----|-------------|-----------|
| | $X_1$ | $X_2$ | $X_3$ | $X_4$ | Experimental | Predicted |
| 15 | 70 | 30 | 1:50 | 3 | 3.22 | 3.20 |
| 16 | 70 | 50 | 1:50 | 3 | 3.97 | 3.96 |
| 17 | 60 | 40 | 1:30 | 3 | 1.95 | 1.93 |
| 18 | 80 | 40 | 1:30 | 3 | 4.85 | 4.82 |
| 19 | 60 | 40 | 1:50 | 3 | 1.31 | 1.30 |
| 20 | 80 | 40 | 1:50 | 3 | 5.59 | 5.58 |
| 21 | 70 | 30 | 1:40 | 2 | 3.15 | 3.14 |
| 22 | 70 | 50 | 1:40 | 2 | 4.10 | 4.08 |
| 23 | 70 | 30 | 1:40 | 4 | 4.64 | 4.63 |
| 24 | 70 | 50 | 1:40 | 4 | 4.67 | 4.65 |
| 25 | 70 | 40 | 1:40 | 3 | 4.42 | 4.38 |
| 26 | 70 | 40 | 1:40 | 3 | 4.39 | 4.38 |
| 27 | 70 | 40 | 1:40 | 3 | 4.37 | 4.38 |
| 28 | 70 | 40 | 1:40 | 3 | 4.41 | 4.38 |
| 29 | 70 | 40 | 1:40 | 3 | 4.31 | 4.38 |

**Table 2.** Analysis of variance for the response surface regression model.

| Source | Sum of Squares | df | Mean Square | F | P |
|--------|----------------|----|-------------|------|---|
| Model | 48.41 | 14 | 3.46 | 3356.14 | <0.0001 ** |
| $X_1$ | 38.52 | 1 | 38.52 | 37385.9 | <0.0001 ** |
| $X_2$ | 0.68 | 1 | 0.68 | 661.55 | <0.0001 ** |
| $X_3$ | 0.011 | 1 | 0.011 | 11.07 | 0.0050 ** |
| $X_4$ | 3.17 | 1 | 3.17 | 3078.94 | <0.0001 ** |
| $X_1X_2$ | 0.57 | 1 | 0.57 | 553.23 | <0.0001 ** |
| $X_1X_3$ | 0.48 | 1 | 0.48 | 462.07 | <0.0001 ** |
| $X_1X_4$ | 1.24 | 1 | 1.24 | 1206.6 | <0.0001 ** |
| $X_2X_3$ | 0.081 | 1 | 0.081 | 78.83 | <0.0001 ** |
| $X_2X_4$ | 0.21 | 1 | 0.21 | 205.37 | <0.0001 ** |
| $X_3X_4$ | 0.2 | 1 | 0.2 | 196.53 | <0.0001 ** |
| $X_1{}^2$ | 1.44 | 1 | 1.44 | 1393.12 | <0.0001 ** |
| $X_2{}^2$ | 0.71 | 1 | 0.71 | 687.3 | <0.0001 ** |
| $X_3{}^2$ | 1.63 | 1 | 1.63 | 1584.35 | <0.0001 ** |
| $X_4{}^2$ | 0.035 | 1 | 0.035 | 33.86 | <0.0001 ** |
| Residual | 0.014 | 14 | 0.001 | | |
| Lack of fit | 0.0068 | 10 | 0.0068 | 0.36 | 0.9138 |
| Pure error | 0.0076 | 4 | 0.0019 | | |
| Cor. total | 48.43 | 28 | | | |

$R^2 = 0.9997$; $R^2{}_{adj} = 0.9994$; $R^2{}_{pred} = 0.9989$; $R_{SN} = 210.828$; CV = 0.83%

** $P < 0.01$.

## 2.3. Optimization of RMP Extraction

The 3D response surface plots are shown in Figure 2a–f. The optimal values of the tested variables provided the highest RMP yield of 6.37%. The predicted variable parameters were as follows: Extraction temperature, 80 °C; extraction time, 50 min; solid–liquid ratio, 1:42.86 (g/mL); number of extractions, two. However, the solid–liquid ratio was modified to 1:40 (g/mL) in consideration of the actual production process. Under these conditions, the experimental RMP yield was 6.25 ± 0.38% ($n = 3$), which was close to the predicted value ($P > 0.05$). This result validated the response model and the existence of an ideal optimum point.

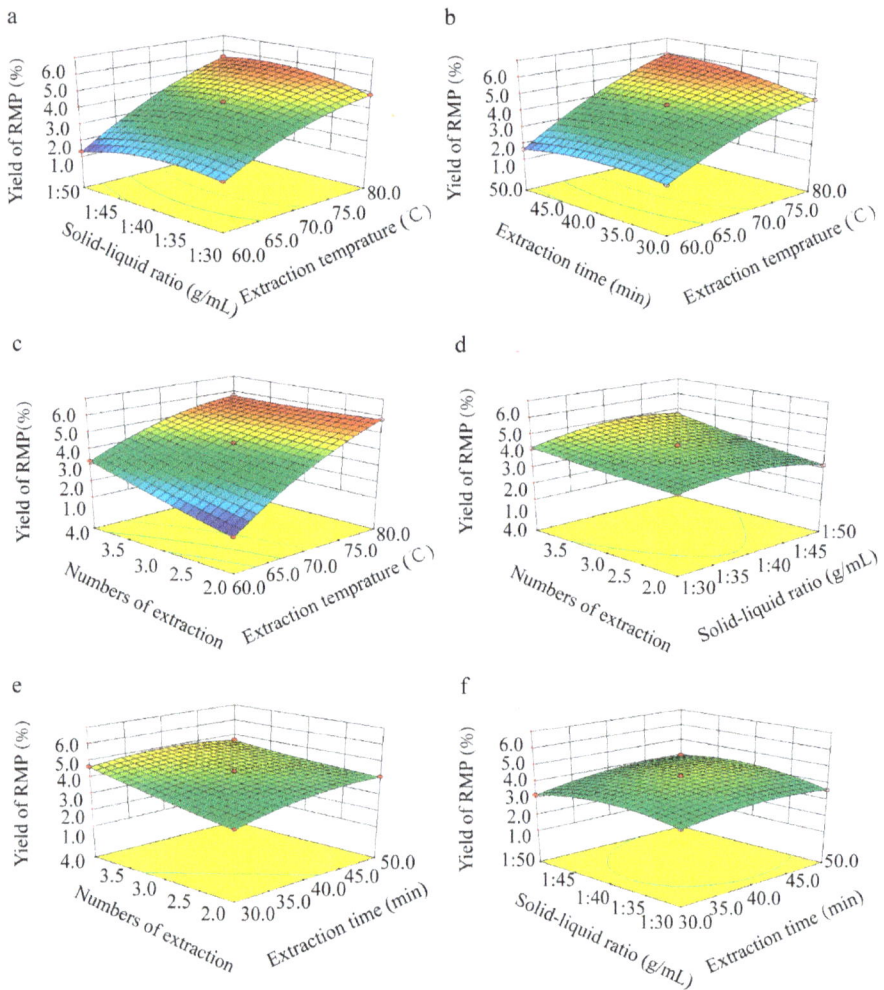

**Figure 2.** 3D response surface plots showing the interaction effects on the RMP extraction yield: (**a**) Solid–liquid ratio and extraction temperature; (**b**) extraction time and extraction temperature; (**c**) number of extractions and extraction temperature; (**d**) number of extractions and solid–liquid ratio; (**e**) number of extractions and extraction time; and (**f**) and solid–liquid ratio and extraction time.

## 2.4. Identification of Monosaccharides

The monosaccharide composition of the RMPs was analyzed by ultra-performance liquid chromatography coupled with a tunable ultraviolet detector (UPLC-TUV; Figure 3a). The RMPs consisted of mannose, rhamnose, glucuronic acid, glucose, xylose, galactose, and arabinose at a molar ratio of 1.36:2.68:0.46:328.17:1.53:21.80:6.16 (Table 3).

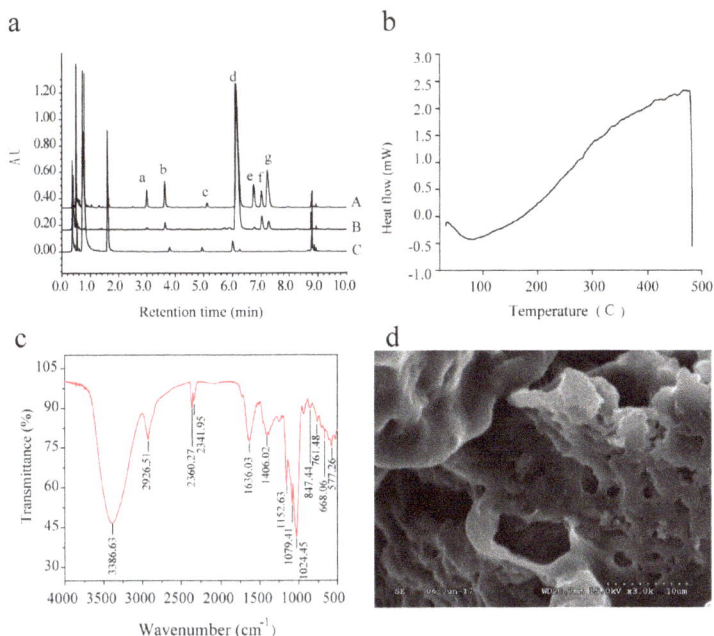

**Figure 3.** (**a**) Chromatograms of monosaccharide standards and samples (A: monosaccharide standards; B: hydrolyzed RMPs; C: unhydrolyzed RMPs; a: mannose; b: rhamnose; c: glucuronic acid; d: glucose; e; xylose; f: galactose; and g: arabinose); (**b**) DSC thermogram of RMPs; (**c**) FT-IR spectra of RMPs; (**d**) SEM photograph of RMPs (3000×).

**Table 3.** Chromatography information and contents of monosaccharide.

| Monosaccharide | Regression Equations | $R^2$ | Sample Hydrolyzed (µg/mL) | Sample Unhydrolyzed (µg/mL) | RMPs (µg/mL) |
|---|---|---|---|---|---|
| a-Mannose | $Y = 1.53e + 0.04X - 7.38$ | 0.9998 | 2.79 | 0.34 | 2.45 |
| b-Rhamnose | $Y = 1.29e + 0.04X + 1.62$ | 0.9999 | 12.07 | 7.67 | 4.40 |
| c-Glucuronic Acid | $Y = 1.13e + 0.04X + 1.25$ | 0.9996 | 1.15 | 0.25 | 0.90 |
| d- Glucose | $Y = 1.31e + 0.04X + 1.44$ | 0.9998 | 593.59 | 2.88 | 590.71 |
| e-Xylose | $Y = 3.08e + 0.04X - 6.61$ | 0.9992 | 2.69 | 0.40 | 2.29 |
| f-Galactose | $Y = 1.09e + 0.04X - 7.97$ | 0.9979 | 40.33 | 1.05 | 39.28 |
| g-Arabinose | $Y = 3.30e + 0.04X - 2.12$ | 0.9986 | 9.52 | ND | 9.52 |

Means and standard deviations are based on three replicates. ND: Not detected.

## 2.5. Differential Scanning Calorimetry (DSC) Analysis

Figure 3b shows the DSC diagram of the RMPs. During the heating process, the solid structure of the RMPs was altered at 83 °C. An exothermic peak was observed at 470 °C, which indicated that the RMPs were in a molten state, and then oxidized and decomposed. Previous studies have suggested that the thermal behavior of polysaccharides is influenced by their chemical composition, physical form, mannose content, and molecular weight [18,19].

## 2.6. FT-IR Spectroscopy Analysis

The IR spectrum of the RMPs (Figure 3c) exhibited bands at 3386.63, 2926.51, and 1406.02 cm$^{-1}$ indicating –OH stretching, C–H stretching, and C–H bending vibrations, respectively, which are characteristic absorption bands of carbohydrates [20]. The band at 847.44 cm$^{-1}$ was characteristic of an α-polysaccharide [21]. Furthermore, absorption peaks at 1024.45, 1079.41, and 1152.63 cm$^{-1}$ suggested the presence of C–O and C–C bands in RMPs [22]. The peaks at 2360.27 cm$^{-1}$ and 2341.95 cm$^{-1}$ were attributed to $CO_2$ and $H_2O$ trapped in the sample, respectively [23].

### 2.7. Morphological Analysis

The surface of the RMPs resembled a rugged sponge (Figure 3d), perhaps due to freeze-drying and water evaporation. Zhu found that the antitumor activity of polysaccharides isolated from *Cordyceps gunnii* differed depending on the extraction method [24]. PPS$_{MAE}$ (polysaccharides from microwave-assisted extraction) had the strongest antitumor activity, perhaps due to the small and thin lamellar structure of PPS$_{MAE}$, such that the tumor cells could be fully exposed to the polysaccharides. Therefore, observing the surface structure of the polysaccharides could provide a scientific basis for the biological activity of RMPs.

### 2.8. Antibacterial Activity of RMPs

Figure 4a–c illustrate the antimicrobial activity of RMPs against *E. coli*, *S. aureus*, and *P. aeruginosa*. The antibacterial system consisted of three different bacteria ($10^5$ CFU/mL) incubated with different RPM concentrations, and the absorbance was measured at 600 nm. The results indicated that Gram-negative bacteria *P. aeruginosa* had the highest sensitivity ($P < 0.01$) to RMPs at the studied concentrations (Figure 4c). The highest bacterial resistance against RMPs was observed for *S. aureus* (Figure 4b). It has been suggested that polysaccharides might change the cell wall and membrane permeability of bacteria, or act as a barrier that inhibits bacterial growth by blocking nutrient import [25]. As shown in Figure 4a, the results for *E. coli*, in which the absorbance increased with increasing RMP concentration, did not support these theories. This might be attributed to the ability of *E. coli* to hydrolyze the RMPs and use the produced monosaccharides as a nutritional source [26].

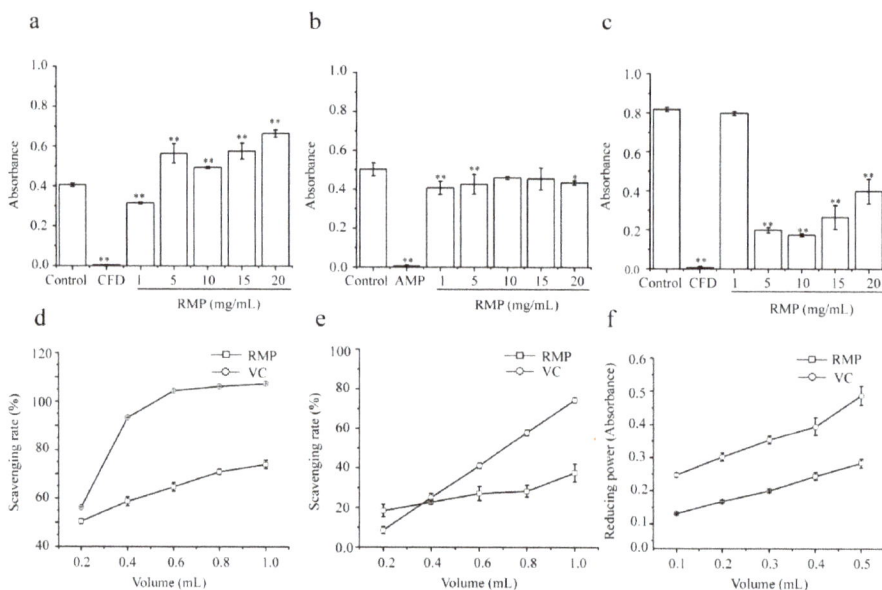

**Figure 4.** Antibacterial and antioxidant activity of RMPs. Effects of RMPs on the proliferation of (**a**) *E. coli*, (**b**) *S. aureus*, and (**c**) *P. aeruginosa*. (**d**) Hydroxyl radical scavenging activity; (**e**) superoxide-radical scavenging activity; and (**f**) reducing the power of RMPs. Values shown are means ± SD obtained from three measurements. Abbreviations: CFD, Cefobid; AMP, ampicillin; and VC, vitamin C. ** $P < 0.01$ and * $P < 0.05$.

## 2.9. Antioxidant Activity of RMPs

Antioxidant mechanisms include the suppression of hydroxyl radical generations and the scavenging of generated hydroxyl radicals [27]. As shown in Figure 4e, the scavenging activity of the RMPs solution (1.0 mL) toward hydroxyl radicals reached 73.97%, which was 33% lower than that of VC. The superoxide radical is the most active reactive oxygen species and is involved in many physiological and pathological processes [28]. As shown in Figure 4f, the scavenging effects of RMPs increased as the volume was increased from 0.2 to 1.0 mL. When the volume was 1.0 mL, the scavenging rates of RMPs and VC were 37.61% and 74.50%, respectively. In the reducing-power assay, RMPs reduced $Fe^{3+}$ to $Fe^{2+}$, which was monitored by measuring the formation of Perl's Prussian blue at 700 nm [29]. Although the reducing power of RMPs at volumes of 0.1–0.5 mL was lower than that of VC, it still reached 0.28 at a volume of 0.5 mL (Figure 4g). According to the results of the above three analyses, RMPs possessed antioxidant activity but were less active than VC. We speculated that the antioxidant activity of RMPs might be related to the high glucose content, although further study is needed to validate this claim.

## 2.10. Effect of RMPs on NO Production Inhibition

Inflammation is a complex process associated with the immune response. When pathogens invade the human body, endotoxins or cytokines induce macrophages and other cells to express an inducible NO synthase that, through NO generation, plays an important role in the cytotoxicity of activated macrophages and the immunoinflammatory response [30]. Significant inhibition of NO production in a concentration-dependent manner was observed at an RMP concentration of 0.5–10.0 mg/mL (Figure 5a, Table 4). The effect on NO production was even greater than that of the positive control group at an RMP concentration of 10.0 mg/mL. These results indicated that RMPs exhibited antibacterial action and would relieve the inflammatory response caused by infection.

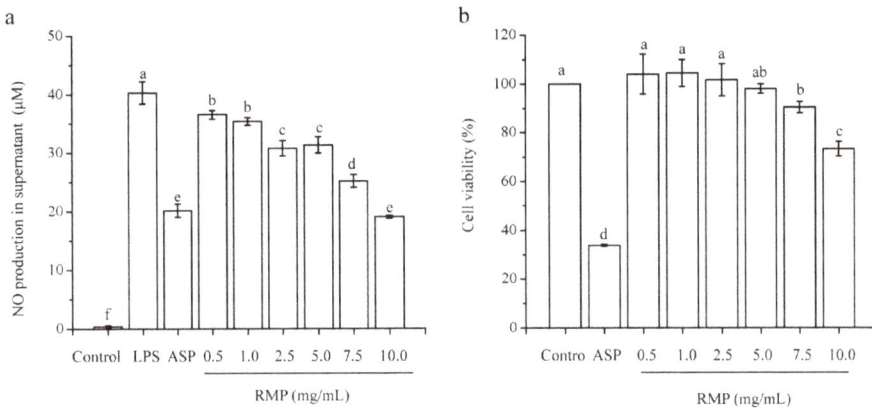

**Figure 5.** Anti-inflammatory activity of RMPs. (**a**) Effect of RMPs on inhibiting NO production and (**b**) the cell viability of RAW 264.7 cells. Abbreviations: LPS, lipopolysaccharides; ASP, aspirin. Different letters (a-f) indicate significant difference between groups ($P < 0.05$) and same letters indicate $P > 0.05$.

**Table 4.** Anti-inflammatory activity of RMPs.

| Group | NO Production (μM) | Cell Viability (%) |
|---|---|---|
| Control | 0.40 ± 0.17 [f] | 100.00 ± 0.00 [a] |
| LPS | 40.36 ± 1.92 [a] | — |
| ASP 1.0 mg/mL | 20.14 ± 1.12 [e] | 33.80 ± 0.32 [d] |
| RMP 0.5 mg/mL | 36.56 ± 0.77 [b] | 104.02 ± 8.10 [a] |
| RMP 1.0 mg/mL | 35.38 ± 0.65 [b] | 104.42 ± 5.48 [a] |
| RMP 2.5 mg/mL | 30.79 ± 1.33 [c] | 101.68 ± 6.54 [a] |
| RMP 5.0 mg/mL | 31.36 ± 1.37 [c] | 98.03 ± 1.88 [a,b] |
| RMP 7.5 mg/mL | 25.23 ± 1.09 [d] | 90.52 ± 2.28 [b] |
| RMP 10.0 mg/mL | 19.03 ± 0.22 [e] | 73.29 ± 2.91 [c] |

Same letter means $P > 0.05$; different letters mean $P < 0.05$. Abbreviations: LPS, lipopolysaccharides; ASP, aspirin.

*2.11. Effect of RMPs on the Cell Viability of RAW 264.7 Cells*

RAW 264.7 cells were treated with RMPs at concentrations of 0.5, 1.0, 2.5, 5.0, 7.5, and 10.0 mg/mL with 1.0 μg/mL LPS. The results showed that treatment with RMPs at concentrations of 0.5, 1.0, 2.5, and 5.0 mg/mL had no obvious toxic effect on cell growth compared with the control group. The cell viability of aspirin (1.0 mg/mL) was 33.8%, which was the lowest among all the groups tested. In general, RMPs did not exhibit any toxic effects in the concentration range of 0.5–2.5 mg/mL.

## 3. Materials and Methods

*3.1. Materials*

*R. mori* was obtained from the mulberry breeding center at Southwest University, Chongqing, China. *E. coli, P. aeruginosa*, and *S. aureus* were obtained from the Laboratory of Silkworm Pathophysiology and Application of Microbial Research of Southwest University. RAW 264.7 cells were provided by Procell Co., Ltd (Wuhan, China). Aspirin, vitamin C, lipopolysaccharides, and DMSO were purchased from Sigma-Aldrich (St. Louis, MO, USA). Streptomycin, penicillin, fetal bovine serum, trypsin, and Dulbecco's modified Eagle medium (DMEM) were purchased from Gibco (Grand Island, NY, USA).

*3.2. Extraction of RMPs*

The dry biomass powder (10 g) was extracted with water (400 mL). The mixture was treated with ultrasound for 50 min in a water bath at 80 °C. After being centrifuged at 10,000× *g* for 5 min, the supernatant was collected. The RMP in pellets was extracted again using the same method. The supernatant gathered from two extractions was combined. Four times the volume of 95% ethanol was slowly added to the supernatant and the mixture was stored at 4 °C overnight for sedimentation. The precipitate was collected by centrifugation at 10,000× *g* for 5 min. The coarse RMPs were then washed and deproteinized using the Sevag method. The total carbohydrate content was measured by the phenol–$H_2SO_4$ assay for RMPs [31], using glucose as a standard ($R^2 = 0.9969$).

*3.3. Experimental Design and Statistical Analysis*

The Box–Behnken design (BBD, design expert software, version 8.0.5) was applied to determine the experimental conditions, which combined four independent variables at three levels, namely, extraction temperature ($X_1$: 60, 70, and 80 °C ), extraction time ($X_2$: 30, 40, and 50 min), solid–liquid ratio ($X_3$: 1:30, 1:40, and 1:50), and number of extractions ($X_4$: 2, 3, and 4). Data were analyzed using a quadratic polynomial model that expressed the response as a function of the independent variables as follows:

$$Y = A_0 + \sum_{i=1}^{4} A_i X_i + \sum_{i=1}^{4} A_{ii} X_{ii}^2 + \sum_{i=1}^{3} \sum_{j=i+1}^{4} A_{ij} X_i X_j, \tag{2}$$

where $Y$ is the value of the studied response predicted by the model, $A_0$ is a constant coefficient, $A_i$ is the linear coefficient for each independent variable, $A_{ii}$ is the interaction coefficient, and $X_i$ and $X_j$ are the actual values of the independent variables.

*3.4. UPLC Analysis of The Monosaccharide Composition of RMPs*

According to a literature method [32], RMPs were hydrolyzed into monosaccharides and 1-phenyl-3-methyl-5-pyrazolone (PMP)-labeled monosaccharides were derived for use in the UPLC system. Chromatographic separation was conducted on a Waters Acquity UPLC I-Class system, including a tunable UV detector and an ACQUITY UPLC BEH $C_{18}$ column (1 mm × 100 mm, 1.7 μm, Waters, Milford, MA, USA). The column temperature was set at 40 °C. Gradient elution was conducted by varying the proportion of each mobile phase at a flow rate of 0.17 mL/min. Mobile phase A consisted of 50 mM $NH_4OAc–NH_3$ in $H_2O$ (pH 9.5) and mobile phase B was acetonitrile. The gradient elution comprised a linear increase from 8% to 15% B over 5.5 min and was then held at 15% B for 2.5 min. The wavelength was 250 nm and the injection volume was 1 μL. Glucose, glucuronic acid, mannose, rhamnose, xylose, galactose, and arabinose with purity greater than 98% were purchased from ChromaBio (Chengdu, China) and used to prepare the standard solution. The correlation coefficients ($R^2$) and linearity ranges of the seven monosaccharides were as follows: Glucose ($R^2$ = 0.9998, 0–1600 μg/mL), glucuronic acid ($R^2$ = 0.9996, 0–10 μg/mL), mannose ($R^2$ = 0.9998, 0–40 μg/mL), rhamnose ($R^2$ = 0.9999, 0–80 μg/mL), xylose ($R^2$ = 0.9992, 0–80 μg/mL), galactose ($R^2$ = 0.9979, 0–60 μg/mL), and arabinose ($R^2$ = 0.9986, 0–100 μg/mL).

*3.5. FT-IR Analysis of RMPs*

RMPs were identified by Fourier transform infrared spectroscopy (Thermo Scientific, MA, USA) in the frequency range of 4000–500 $cm^{-1}$ using the KBr pressed-disk method. The dried RMPs were mixed with KBr powder and pressed into 1-mm pellets for measurement. Three replicate spectra were obtained.

*3.6. Antibacterial Experiments In Vitro*

Three different bacteria ($10^5$ CFU/mL) were inoculated into sterile LB liquid medium, containing RMPs at concentrations of 1, 5, 10, 15, and 20 mg/mL. Ampicillin (AMP) and Cefobid (CFD) were used as positive control groups. The absorbance of the cell concentrations was measured at 600 nm to assess the antibacterial activity of RMPs after incubation for 24 h at 37 °C [26].

*3.7. Antioxidant Activity of RMPs*

Three different methods were used to analyze the antioxidant activity of RMPs, namely, a hydroxyl radical scavenging assay, a superoxide-radical scavenging assay, and a reducing-power assay. The hydroxyl radical scavenging activity, superoxide radical scavenging activity, and reducing-power of the RMPs were determined using a previously reported method with slight modification [33–35]. Vitamin C ($V_C$) was diluted in deionized water and used as a positive control in the above three experiments.

*3.8. Determination of Anti-Inflammatory Activity*

The cell viability was analyzed using a CKK-8 assay in vitro. RAW 264.7 cells were seeded into 96-well culture plates ($10^5$ cells/well) and incubated at 37 °C with 5% $CO_2$ for 24 h. Cells were exposed to the culture medium containing RMPs at concentrations of 0.5, 1.0, 2.5, 5.0, 7.5, and 10 mg/mL. After incubation for 24 h, 10 μL of CKK-8 solution was added to each well. The absorbance was then detected at 450 nm after incubation at 37 °C for 1 h. RAW 264.7 cells ($10^5$ cells/mL) were plated in 96-well plates and subsequently treated with lipopolysaccharides (LPS, $10^3$ ng/mL) in the presence of different RMPs concentrations (0.5, 1.0, 2.5, 5.0, 7.5, and 10 mg/mL) for 24 h. Aspirin (ASP) was used as a positive control. The supernatant of each culture (50 μL) was mixed with Griess reagent (100 μL) and then the amount of NO production was determined.

## 4. Conclusions

RSM was applied to optimize the RMP extraction conditions. The optimum conditions for maximum biomass in RMP production were a solid–liquid ratio of 1:40 (g/mL), 80 °C, 50 min, with extraction performed twice. The RMPs contained seven monosaccharides, namely, mannose, rhamnose, glucuronic acid, glucose, xylose, galactose, and arabinose with molar a ratio of 1.36:2.68:0.46:328.17:1.53:21.80:6.16. The RMPs were α-polysaccharides with characteristic absorption bands of carbohydrates and a loose porous sponge-like surface. The RMPs showed significant antibacterial, antioxidant, and anti-inflammatory activities. This study provides a basis for exploring the potential uses of RMPs.

**Author Contributions:** W.Y. and N.H. conceived and designed the experiments. W.Y. and H.C. performed the experiments. N.H., W.Y., and Z.X. analyzed the data. W.Y. and N.H. wrote the paper.

**Funding:** This research was founded by the National Key Research and Development Program of China, grant number 2018YFD1000602, National Natural Science Foundation of China, grant number 31572323, and Project of extraction, detection and variation of alkaloids and polyphenols in Mulberry, grant number cstc2016jcyjys0002.

**Conflicts of Interest:** The authors declare no conflicts of interest.

## References

1.  Jian, Q.; He, N.; Wang, Y.; Xiang, Z. Ecological issues of mulberry and sustainable development. *J. Resour. Ecol.* **2012**, *3*, 330–339. [CrossRef]
2.  Liu, Y. Application prospect of mulberry plants to vegetation restoration in Three Gorges Reservoir area. *Sci. Seric.* **2011**, *37*, 93–97.
3.  Wu, Y.; Liang, Z.; Xing, D. Comparison of the physiological characteristics of paper mulberry (*Broussonetia papyrifera*) and mulberry (*Morus alba*) undersimulated drought stress. *Guihaia* **2011**, *31*, 92–96. [CrossRef]
4.  Shi, Y.; Wang, C.; Wang, X.; Zhang, Y.; Liu, L.; Wang, R.; Ye, J.; Hu, L.; Kong, L. Uricosuric and nephroprotective properties of *Ramulus mori* ethanol extract in hyperuricemic mice. *J. Ethnopharmacol.* **2012**, *143*, 896–904. [CrossRef] [PubMed]
5.  Guo, C.; Li, R.; Zheng, N.; Xu, L.; Liang, T.; He, Q. Anti-diabetic effect of *Ramulus mori* polysaccharides, isolated from *Morus alba* L., on STZ-diabetic mice through blocking inflammatory response and attenuating oxidative stress. *Int. Immunopharmaco.* **2013**, *16*, 93–99. [CrossRef] [PubMed]
6.  Yang, S.; Wang, B.; Xia, X.; Li, X.; Wang, R.; Sheng, L.; Li, D.; Liu, Y.; Li, Y. Simultaneous quantification of three active alkaloids from a traditional Chinese medicine *Ramulus mori* (Sangzhi) in rat plasma using liquid chromatography-tandem mass spectrometry. *J. Pharmaceut. Biomed.* **2015**, *109*, 177–183. [CrossRef] [PubMed]
7.  Guo, C.; Liang, T.; He, Q.; Wei, P.; Zheng, N.; Xu, L. Renoprotective effect of *Ramulus mori* polysaccharides on renal injury in STZ-diabetic mice. *Int. J. Biol. Macromol.* **2013**, *62*, 720–725. [CrossRef] [PubMed]
8.  Xu, L.; Yang, F.; Wang, J.; Huang, H.; Huang, Y. Anti-diabetic effect mediated by *Ramulus mori* polysaccharides. *Carbohydr. Polym.* **2015**, *117*, 63–69. [CrossRef] [PubMed]
9.  Anderson-Cook, C.M.; Borror, C.M.; Montgomery, D.C. Response surface design evaluation and comparison. *J. Stat. Plan. Infer.* **2009**, *139*, 629–641. [CrossRef]
10. Granato, D.; Ribeiro, J.C.B.; Castro, I.A.; Masson, M.L. Sensory evaluation and physicochemical optimisation of soy-based desserts using response surface methodology. *Food Chem.* **2010**, *121*, 899–906. [CrossRef]
11. Li, F.; Gao, J.; Xue, F.; Yu, X.; Shao, T. Extraction optimization, purification and physicochemical properties of polysaccharides from *Gynura medica. Molecules* **2016**, *21*, 397. [CrossRef]
12. Yang, R.; Geng, L.; Lu, H.; Fan, X. Ultrasound-synergized electrostatic field extraction of total flavonoids from *Hemerocallis citrina baroni. Ultrason. Sonochem.* **2017**, *34*, 571–579. [CrossRef] [PubMed]
13. Xiong, W.; Chen, X.; Lv, G.; Hu, D.; Zhao, J.; Li, S. Optimization of microwave-assisted extraction of bioactive alkaloids from lotus plumule using response surface methodology. *J. Pharmaceut. Biomed.* **2016**, *6*, 382–388. [CrossRef] [PubMed]

14. Hu, T.; Guo, Y.; Zhou, Q.; Zhong, X.; Zhu, L.; Piao, J.; Chen, J.; Jiang, J. Optimization of ultrasonic-assisted extraction of total saponins from *Eclipta prostrasta* L. using response surface methodology. *J. Food. Sci.* **2012**, *77*, 975–982. [CrossRef] [PubMed]

15. Ying, Z.; Han, X.; Li, J. Ultrasound-assisted extraction of polysaccharides from mulberry leaves. *Food Chem.* **2011**, *127*, 1273–1279. [CrossRef] [PubMed]

16. Du, H.; Chen, J.; Tian, S.; Gu, H.; Li, N.; Sun, Y.; Ru, J.; Wang, J. Extraction optimization, preliminary characterization and immunological activities in vitro of polysaccharides from *Elaeagnus angustifolia* L. Pulp. *Carbohydr. Polym.* **2016**, *151*, 348–357. [CrossRef] [PubMed]

17. Qu, Y.; Li, C.; Zhang, C.; Zeng, R.; Fu, C. Optimization of infrared-assisted extraction of *Bletilla striata* polysaccharides based on response surface methodology and their antioxidant activities. *Carbohydr. Polym.* **2016**, *148*, 345–353. [CrossRef] [PubMed]

18. Cerqueira, M.A.; Souza, B.W.S.; Simões, J.; Teixeira, J.A.; Domingues, M.R.M.; Coimbra, M.A.; Vicente, A.A. Structural and thermal characterization of galactomannans from non-conventional sources. *Carbohydr. Polym.* **2011**, *83*, 179–185. [CrossRef]

19. Zhang, S.; Zhong, G.; Liu, B.; Wang, B. Physicochemical and functional properties of fern rhizome (*Pteridium aquilinum*) starch. *Starch-Starke* **2011**, *63*, 468–474. [CrossRef]

20. Mecozzi, M.; Pietrantonio, E.; Pietroletti, M. The roles of carbohydrates, proteins and lipids in the process of aggregation of natural marine organic matter investigated by means of 2D correlation spectroscopy applied to infrared spectra. *Spectrochim. Acta Part A* **2009**, *71*, 1877–1884. [CrossRef] [PubMed]

21. Kačuráková, M.; Wilson, R.H. Developments in mid-infrared FT-IR spectroscopy of selected carbohydrates. *Carbohydr. Polym.* **2001**, *44*, 291–303. [CrossRef]

22. Yan, J.; Wang, Y.; Ma, H.; Wang, Z.; Pei, J. Structural characteristics and antioxidant activity in vivo of a polysaccharide isolated from *Phellinus linteus* mycelia. *J. Taiwan Inst. Chem. E.* **2016**, *65*, 110–117.

23. Mayer, S.G.; Boyd, J.E.; Heser, J.D. A high-pressure attenuated total reflectance cell for collecting infrared spectra of carbon dioxide mixtures. *Vib. Spectrosc.* **2010**, *53*, 311–313. [CrossRef]

24. Zhu, Z.; Dong, F.; Liu, X.; Lv, Q.; Yang, Y.; Liu, F.; Chen, L.; Wang, T.; Wang, Z.; Zhang, Y. Effects of extraction methods on the yield, chemical structure and anti-tumor activity of polysaccharides from *Cordyceps gunnii* mycelia. *Carbohydr. Polym.* **2016**, *140*, 461–471. [CrossRef] [PubMed]

25. Han, Q.; Wu, Z.; Huang, B.; Sun, L.; Ding, C.; Yuan, S.; Zhang, Z.; Chen, Y.; Hu, C.; Zhou, L.; Liu, J.; Huang, Y.; Liao, J.; Yuan, M. Extraction, antioxidant and antibacterial activities of *Broussonetia papyrifera* fruits polysaccharides. *Int. J. Biol. Macromol.* **2016**, *92*, 116–124. [CrossRef] [PubMed]

26. Mazarei, F.; Jooyandeh, H.; Noshad, M.; Hojjati, M. Polysaccharide of caper (*Capparis spinosa* L.) Leaf: Extraction optimization, antioxidant potential and antimicrobial activity. *Int. J. Biol. Macromol.* **2017**, *95*, 224–231. [CrossRef] [PubMed]

27. Guo, X.; Shang, X.; Zhou, X.; Zhao, B.; Zhang, J. Ultrasound-assisted extraction of polysaccharides from *Rhododendro aganniphum*: Antioxidant activity and rheological properties. *Ultrason. Sonochem.* **2017**, *38*, 246–255. [CrossRef] [PubMed]

28. Liu, Y.; Liu, X.; Liu, Y.; Liu, G.; Ding, L.; Lu, X. Construction of a highly sensitive non-enzymatic sensor for superoxide anion radical detection from living cells. *Biosens. Bioelectron.* **2017**, *90*, 39–45. [CrossRef] [PubMed]

29. Raza, A.; Li, F.; Xu, X.; Tang, J. Optimization of ultrasonic-assisted extraction of antioxidant polysaccharides from the stem of *Trapa quadrispinosa* using response surface methodology. *Int. J. Biol. Macromol.* **2017**, *94*, 35–344. [CrossRef] [PubMed]

30. Chen, Y.; Li, C.; Zhu, J.; Xie, W.; Hu, X.; Song, L.; Zi, J.; Yu, R. Purification and characterization of an antibacterial and anti-inflammatory polypeptide from *Arca subcrenata*. *Int. J. Biol. Macromol.* **2017**, *96*, 177–184. [CrossRef] [PubMed]

31. Masuko, T.; Minami, A.; Iwasaki, N.; Majima, T.; Nishimura, S.-I.; Lee, Y.C. Carbohydrate analysis by a phenol-sulfuric acid method in microplate format. *Anal. Biochem.* **2005**, *339*, 69–72. [CrossRef] [PubMed]

32. Wen, Z.; Xiang, X.; Jin, H.; Guo, X.; Liu, L.; Huang, Y.; OuYang, X.; Qu, Y. Composition and anti-inflammatory effect of polysaccharides from *Sargassum horneri* in RAW264.7 macrophages. *Int. J. Biol. Macromol.* **2016**, *88*, 403–413. [CrossRef] [PubMed]

33. Hong, J.; Hu, J.; Liu, J.; Zhou, Z.; Zhao, A. In vitro antioxidant and antimicrobial activities of flavonoids from *Panax notoginseng* flowers. *Nat. Prod. Res.* **2014**, *28*, 1260–1266. [CrossRef] [PubMed]

34. Ballesteros, L.F.; Teixeira, J.A.; Mussatto, S.I. Extraction of polysaccharides by autohydrolysis of spent coffee grounds and evaluation of their antioxidant activity. *Carbohydr. Polym.* **2017**, *157*, 258–266. [CrossRef] [PubMed]

35. Kumar, S.; Yadav, M.; Yadav, A.; Yadav, J.P. Impact of spatial and climatic conditions on phytochemical diversity and in vitro antioxidant activity of Indian *Aloe vera* (L.) Burm.f. *South Afr. J. Bot.* **2017**, *111*, 50–59. [CrossRef]

**Sample Availability:** Samples of the compounds are not available from the authors.

*molecules*

MDPI

*Article*

# The Effect of Enzymolysis on Performance of Soy Protein-Based Adhesive

Yantao Xu [1,2], Yecheng Xu [1,2], Yufei Han [1,2], Mingsong Chen [1,2], Wei Zhang [1,2], Qiang Gao [1,2,*] and Jianzhang Li [1,2,*]

[1] Key Laboratory of Wood Material Science and Utilization, Beijing Forestry University, Beijing 100083, China; xuyantao@bjfu.edu.cn (Y.X.); xuyecheng@bjfu.edu.cn (Y.X.); hanyufei96@163.com (Y.H.); chen_bjfu@163.com (M.C.); zhangwei@bjfu.edu.cn (W.Z.)

[2] Beijing Key Laboratory of Wood Science and Engineering, Ministry of Education, College of Materials Science and Technology, Beijing Forestry University, Beijing 100083, China

* Correspondence: gaoqiang@bjfu.edu.cn (Q.G.); lijzh@bjfu.edu.cn (J.L.); Tel.: +86-010-6233-6912 (Q.G.)

Academic Editors: Pavel B. Drasar and Vladimir A. Khripach
Received: 6 October 2018; Accepted: 23 October 2018; Published: 24 October 2018

check for
updates

**Abstract:** In this study, bromelain was used to break soy protein molecules into polypeptide chains, and triglycidylamine (TGA) was added to develop a bio-adhesive. The viscosity, residual rate, functional groups, thermal behavior, and fracture surface of different adhesives were measured. A three-ply plywood was fabricated and evaluated. The results showed that using 0.1 wt% bromelain improved the soy protein isolate (SPI) content of the adhesive from 12 wt% to 18 wt%, with viscosity remaining constant, but reduced the residual rate by 9.6% and the wet shear strength of the resultant plywood by 69.8%. After the addition of 9 wt% TGA, the residual rate of the SPI/bromelain/TGA adhesive improved by 13.7%, and the wet shear strength of the resultant plywood increased by 681.3% relative to that of the SPI/bromelain adhesive. The wet shear strength was 30.2% higher than that of the SPI/TGA adhesive, which was attributed to the breakage of protein molecules into polypeptide chains. This occurrence led to (1) the formation of more interlocks with the wood surface during the curing process of the adhesive and (2) the exposure and reaction of more hydrophilic groups with TGA to produce a denser cross-linked network in the adhesive. This denser network exhibited enhanced thermal stability and created a ductile fracture surface after the enzymatic hydrolysis process.

**Keywords:** soy protein isolate; bromelain; triglycidylamine; viscosity; water resistance; adhesive

## 1. Introduction

Biomass adhesives, such as tannin, lignin, carbohydrate, unsaturated oil, and protein-based adhesives, have been widely studied as alternatives to formaldehyde-based adhesives to eliminate formaldehyde hazard in wood panels [1]. Among these biomass adhesives, the soy protein adhesive is a rich, formaldehyde-free, low-cost raw material and exhibits considerable potential for development [2]. However, poor water resistance limits the application of soy protein adhesives [3]. Most studies have focused on using chemical modification to improve the performance of soy protein-based adhesives [4], such as denaturing agent modification [5], graft modification [6], biomimetic modification [7], latex modification [8], and synthetic resin modification [9]. Polyacrylamide and epoxide have been proven to be effective as cross-linkers for soy protein-based adhesives, with the resultant plywood meeting the requirements for interior plywood [10,11]. However, these modified adhesives have a low solid content and high viscosity, resulting in a panel that is difficult to apply and has poor production stability.

Wood is a porous material. The bond strength of a wood panel mainly comes from mechanical interlocking after the curing process of the adhesive [12]. The high molecule weight of the soy protein-based adhesive impedes its penetration into the wood, thus barely forming an interlock, resulting in a low bond strength and wood failure. In addition, soy protein is the aggregation of high-molecular-weight polypeptide chains with the complex quaternary structure, which implies numerous active groups in the interior of the protein, resulting in a low reactivity of protein. From another perspective, a high molecular weight of protein leads to a high viscosity and low solid content of the resultant adhesive, which means a lot water is introduced into the wood panel during the fabrication process, leading to poor production stability of the resultant panel.

In recent years, enzyme technology has gradually developed. Driving this new thrust are three major new goals, that is, maximizing the exploitation of renewable resources as sources of raw materials for the production of multifunctional polymers, development of an environmentally friendly process, and development of biodegradable products [13]. Enzyme technology has been widely used. For example, the use of enzymes for the selective hydrolysis/treatment of polymers and materials [14], the mild surface functionalization of polymers such as polyethylene terephthalate (PET) and polylactic acid (PLA), and the subsequent coupling of molecules and grafting of molecules on wood after enzymatic pre-treatment has achieved certain results [15,16]. The enzyme technology is becoming increasinly more mature. Enzyme modified soy protein is also a feasible method.

In the current study, a protein endonuclease–bromelain was used to break down protein molecules into polypeptide chains to reduce the viscosity and improve the solid content of the soy protein isolate (SPI) adhesive. The active groups on soy protein molecule chains were also exposed during this process. These polypeptide chains then reacted with a laboratory-made cross-linker triglycidylamine (TGA) to develop a soy protein-based adhesive. The effects of the low molecular weight of protein on the performance of the resultant adhesive, including the viscosity, residual rate and the functional groups, thermostability, and fracture surface, were characterized. Three-ply plywood samples were fabricated using the resultant adhesives, and their wet shear strengths were evaluated.

## 2. Materials and Methods

### 2.1. Materials

SPI with 95% protein content was obtained from Yuwang Ecological Food Industry Co, Ltd. (Jinan, China) [17]. Poplar veneer (200 × 200 × 1.5 mm, 8% moisture content) from Hebei Province of China was provided. Bromelain (BR, 300 u/mg, CAS # 37189-34-7) was purchased from Shanghai Yuanye Group (Shanghai, China). Triethylamine, a laboratory-made epoxy cross-linker, was also used. The reaction pathway of TGA is systhsized following our previous research [18] and illustrated in Figure 1. Epichlorohydrin and aqueous ammonia with a mole ratio of 5:1 was placed into a three-necked flask equipped with a condenser and a stirrer. The mixture was stirred continuously at a rate of 800 rpm. Ammonium triflate was used to catalyze the reaction at 23 °C for 48 h, and then at 35 °C for 3 h. The residual epichlorohydrin and ammonium hydroxide were removed by a vacuum distillation, and the result was a colorless syrup consisting mostly of tris(3-chloro-2-hydroxypropyl) amine. An excess of sodium hydroxide solution (50%) was added for the epoxy-ring closure reaction at 20 °C for 2 h. Because the reaction was highly exothermal, an external ice-water cooling circulator was required to hold the temperature. The precipitate of sodium chloride was filtered off, and the residue was vacuum distilled to obtain pure viscous TGA.

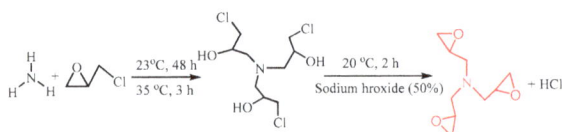

**Figure 1.** The synthesis procedure of cross-linker triglycidylamine (TGA).

## 2.2. Preparation of Soy Protein Adhesive

Protein and water were mixed to develop an SPI adhesive (Table 1). Bromelain was added to the SPI adhesive and then stirred in a water bath at 50 °C for 20 min, allowing full digestion. The mixture was then placed in a water bath at 90 °C and then stirred for 10 min to deactivate bromelain. TGA was ultimately added into the mixture to develop the final adhesives.

**Table 1.** Various adhesive formulations. SPI—Soy protein isolate; TGA—Triglycidylamine.

| Sample | SPI (g) | Distilled Water (g) | Bromelain (g) | TGA (g) |
|--------|---------|---------------------|---------------|---------|
| 0 | 12 | 88 | 0 | 0 |
| 1 | 18 | 82 | 0.1 | 0 |
| 2 | 18 | 82 | 0.1 | 3 |
| 3 | 18 | 82 | 0.1 | 6 |
| 4 | 18 | 82 | 0.1 | 9 |
| 5 | 18 | 82 | 0.1 | 12 |
| 6 | 18 | 82 | 0 | 9 |

## 2.3. Preparation and Evaluation of Plywood

Three layers of poplar plywood were prepared in this study. The adhesive was evenly coated on both sides of the core veneer with glue spreading of 200 g/m$^2$. The coated plywood was placed between two uncoated veneers, perpendicular to the grain of the adjacent veneer. The laminated plywood was hot-pressed at 120 °C and 1.0 MPa for 6 min, and two sheets of plywood were produced using the same adhesive formulation. The shear strength of the plywood was determined in accordance with the Chinese National Standard GB/T 17657 (2013) [19]. The prepared plywood was allowed to remain at room temperature for at least 24 h. Twelve specimens measuring 100 mm × 25 mm (glue area, 25 mm × 25 mm) were uniformly cut from the center and the edges of the two sheets of plywood.

## 2.4. Viscosity

The viscosity of the soybean adhesives was measured using a Brook field DV-II viscometer, employing the rotor with a spinning rate of 100 rpm. An average of three replicate measurements was reported as the viscosity of each sample.

## 2.5. Residual Rate Test

The adhesive sample was placed in an oven at 120 ± 2 °C until a constant weight was obtained and then ground into 100 mesh powder (0.15 mm) using a ceramic mortar. To determine mass loss, the cured adhesive was wrapped with a qualitative filter paper and then placed in a glass with distilled water [20]. After blistering for 6 h in an oven at 60 ± 2 °C, the sample was dried (120 ± 2 °C, 3 h) and weighed. The mass loss was determined by calculating the difference in weight before and after hydrolysis.

## 2.6. Wet Shear Strength Measurement

In accordance with the Chinese National Standard (GB/T 17657-2013), the wet shear strength of the second-grade plywood (interior use plywood) was determined. Twelve plywood specimens (25 mm × 100 mm) were cut from two pieces of plywood, immersed in water at 63 °C for 3 h, dried at room temperature for 10 min, and subjected to tensile testing. Wet shear strength was calculated using Equation (1). The standard deviation of the data was calculated.

$$\text{Shear strength} = \frac{\text{Force (N)}}{\text{Gluing area (m}^2)} \tag{1}$$

*2.7. Fourier Transform Infrared (FTIR) Spectroscopy*

The different adhesive samples prepared were cured in an oven at $120 \pm 2\,^{\circ}$C until a constant weight was obtained. Then, we ground the adhesive into a powder. FTIR spectra of the different cured adhesives were recorded using a Nicolet 7600 spectrometer (Nicolet Instrument Corporation, Madison, WI, USA) from 500 to 4000 cm$^{-1}$ with a 4 cm$^{-1}$ resolution using 32 scans.

*2.8. Thermogravimetric (TG)*

The different adhesives were cured in an oven at $120 \pm 2\,^{\circ}$C until a constant weight was obtained, and we then ground the adhesive into a powder. The thermal stabilities of the cured adhesive samples were tested using a TGA instrument (TA Q50, Waters Company, Milford, MA, USA). Approximately 5 mg powdered samples were weighed in a platinum cup and scanned from 30 to 600 $^{\circ}$C at a heating rate of 10 $^{\circ}$C min$^{-1}$ in a nitrogen environment while recording the weight change.

*2.9. Scanning Electron Microscopy (SEM)*

The fracture surface micrographs of cured adhesives were measured using a JSM-6500F field emission scanning electron microscope (FE-SEM) (JEOL USA Inc., Peabody, MA, USA). Prior to testing, the fracture surface was placed on an aluminum stub and a 10 nm gold film was coated on using an ion sputter (HITACHI MCIOOO, Tokyo, Japan).

## 3. Results and Discussion

*3.1. Viscosity*

Extremely high viscosity rendered the coating process difficult and ineffective; meanwhile, extremely low viscosity led to the over penetration of adhesive into the wood surface in the manufacture of plywood [21]. A wood adhesive requires a suitable viscosity to ensure strong contact with wood. In addition, it should exhibit adequate penetration and mechanical interlocking with the substrates [22].

Table 2 shows that the native SPI adhesive (adhesive 0) contains 12 wt% of SPI; the viscosity is 61,000 cP, which presents no flowability in the adhesive. As a protein endonuclease, bromelain broke soy protein molecules into polypeptide chains, reducing the molecular weight of the soy protein and exhibiting low viscosity. When the SPI content was increased to 18 wt%, the development of a uniform pure soy protein adhesive was impeded. However, when using bromelain in the adhesive formulation, the viscosity of adhesive (1) reached 62,880 cP, which was similar to that that of adhesive (0). This similarity indicated that enzymatic hydrolysis effectively increased the solid content of the soy protein adhesive. Further addition of TGA into the adhesive treated with enzymatic hydrolysis led to a gradual decrease in viscosity from 62,880 cP to 285 cP. With the addition of 3 wt% TGA into the adhesive formulation, the viscosity of adhesive (2) decreased by 48.64% relative to that of adhesive (1). Further addition of TGA to 12 wt% caused a reduction in the viscosity of adhesive (5) by 99% to 285 cP. This decrease was attributed to the low molecular weight of TGA, which reduced the friction in the decomposed soy protein macromolecules, consequently decreasing the viscosity of the adhesive. However, the viscosity of adhesive (5) was too low, such that the adhesive could easily over penetrate the wood surface during gluing, preventing the formation of an adhesive layer. As a control, adhesive (6) contained 18 wt% SPI and 9 wt% TGA, and its viscosity was 2200 cP, which was 211% higher than that of adhesive (4). This result also indicated that enzymatic hydrolysis effectively decreased the viscosity of the adhesive.

**Table 2.** Initial viscosity of different adhesive samples: 0 (12 wt% SPI), 1 (18 wt% SPI/bromelain), 2 (18 wt% SPI/bromelain/3 wt% TGA), 3 (18 wt% SPI/bromelain/6wt% TGA), 4 (18 wt% SPI/bromelain/9 wt% TGA), 5 (18 wt% SPI/bromelain /12 wt% TGA), and 6 (18 wt% SPI/9 wt% TGA).

| Sample | 0 | 1 | 2 | 3 | 4 | 5 | 6 |
|---|---|---|---|---|---|---|---|
| Viscosity (cP) | 61,000 ± 2896 | 62,880 ± 3263 | 32,293 ± 1892 | 2439 ± 433 | 707 ± 82 | 285 ± 57 | 2200 ± 387 |

## 3.2. Residual Rate Test

When the flour of the cured adhesive was immersed in water, the amount of insoluble mass determined the cross-link density and water resistance of the adhesive [23]. The bond strength of the native SPI adhesive was mainly based on the hydrogen bond between active groups. Owing to the hydrophilicity of most active groups, SPI adhesive (0) exhibited poor water resistance.

Figure 2 showed that the residual rate of SPI adhesive (0) was 92.17%. After enzymatic hydrolysis, the residual rate of adhesive (1) decreases to 83.29%, indicating a reduction in the water resistance of the adhesive. This effect was attributed to the following reasons: First, bromelain reduced the molecular weight of soy protein by breaking down protein into molecular chains. This process produced small soy protein molecules, which were easier to dissolve in water. Second, enzymatic hydrolysis exposed the hydrophilic groups of the protein, such as $-NH_2$, $-COOH$, which further reduced the water resistance of the SPI adhesives. When 3 wt% TGA was added to the adhesive formulation, the residue rate of the adhesive increased by 4.75% relative to that of adhesive (1). With a further increase in TGA addition to 9 wt%, the residual rate increased by 13.70% to the maximum value, which was higher than that of adhesive (6). This effect was attributed to the cross-linking of the epoxy groups of TGA with the exposed active groups of soy protein chains, as well as the formation of a more compact cross-linked network structure, resulting in improved water resistance of the adhesive. With an increase in the number of epoxy groups, more cross-linking reactions led to the formation of a more compact cross-linked network structure. However, when TGA reached 12 wt%, the residue ratio decreased by 4% to 90.90%, indicating excessive TGA dosage. Soluble TGA was eluted from the filter paper, resulting in the reduction of the residue ratio.

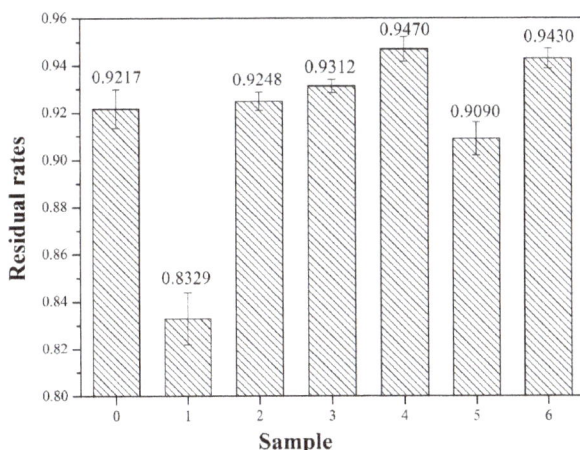

**Figure 2.** Residual rates of different adhesive samples: 0 (12 wt% soy protein isolate (SPI)), 1 (18 wt% SPI/bromelain), 2 (18 wt% SPI/bromelain/3 wt% TGA), 3 (18 wt% SPI/bromelain/6 wt% TGA), 4 (18 wt% SPI/bromelain/9 wt% TGA), 5 (18 wt% SPI/bromelain/12 wt% TGA), and 6 (18 wt% SPI/ 9 wt% TGA).

## 3.3. Wet Shear Strength Measurement

Generally speaking, the wet shear strength of interior plywood required over 0.7 MPa according to the Chinese national standard. The wet shear strength of the plywood bonded by commercial adhesives are ranged from 0.7 to 1.2 MPa. The wet shear strength of the different adhesive samples is shown in Figure 3. The bond strength of the native SPI adhesive (adhesive 0) primarily resulted from the intermolecular hydrogen bond of soy protein, which was easily broken by moisture [24]. Thus, the wet shear strength of the plywood bonded with adhesive (0) was 0.53 MPa. After enzymatic hydrolysis, the wet shear strength of the plywood bonded with adhesive (1) decreased by 69.8% to 0.16 MPa. The 1, 2 level structures of protein are important for the bond strength formation of the protein adhesive. When the 1, 2 level structure was broken by bromelain, the bond strength was markedly reduced. When 3 wt% TGA was added to the adhesive formulation, the wet shear strength of the plywood bonded with adhesive (2) increased by 381% relative to that of adhesive (1). With a further increase in TGA addition to 9 wt%, the wet shear strength of the plywood reached 1.25 MPa, which increased by six times compared with that of adhesive (1). When TGA was added to the SPI-based adhesives, cross-linking occurred between the epoxy group and the reactive group ($-NH_2$, $-COOH$), followed by the replacement of the weak hydrogen bond with a stable chemical bond. Simultaneously, a compact cross-linked network structure and a rigid curing system were formed, which improved the wet shear strength of the resultant plywood. As a control, the wet shear strength of the plywood bonded with adhesive (6) was 0.96 MPa, which was 30.2% lower than that of adhesive (4). This result could be attributed to the following reasons: First, the SPI molecule was degraded to polypeptide chains by bromelain, which exposed more active groups and produced more reactive sites to increase the reactivity of the adhesive, resulting in a denser cross-linked network structure formation and an increase in the water resistance of the adhesive. Second, with enzymatic hydrolysis, the viscosity of the adhesive was markedly reduced, and the permeability of the adhesive was improved. These changes led to enhanced mechanical interlocking with the wood formed, further improving the wet shear strength of the adhesive. Third, enzymatic hydrolysis also improved the solid content of the adhesive, which helped to improve the adhesive bond performance. A schematic of the adhesive reaction is presented in Figure 4. With the addition of 12 wt% TGA, the wet shear strength of the plywood was reduced to 0.91 MPa. This reduction was attributed to the considerably low viscosity of the adhesive, which led to the overpenetration of the wood surface and a reduction in the bond strength of the plywood.

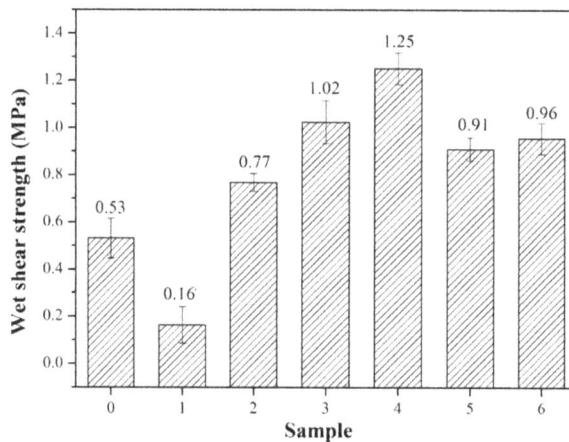

**Figure 3.** Wet shear strength of the different adhesive samples: 0 (12 wt% SPI), 1 (18 wt% SPI/bromelain), 2 (18 wt% SPI/bromelain/3 wt% TGA), 3 (18 wt% SPI/bromelain/6 wt% TGA), 4 (18 wt% SPI/bromelain/9 wt% TGA), 5 (18 wt% SPI/bromelain/12 wt% TGA), and 6 (18 wt% SPI/ 9 wt% TGA).

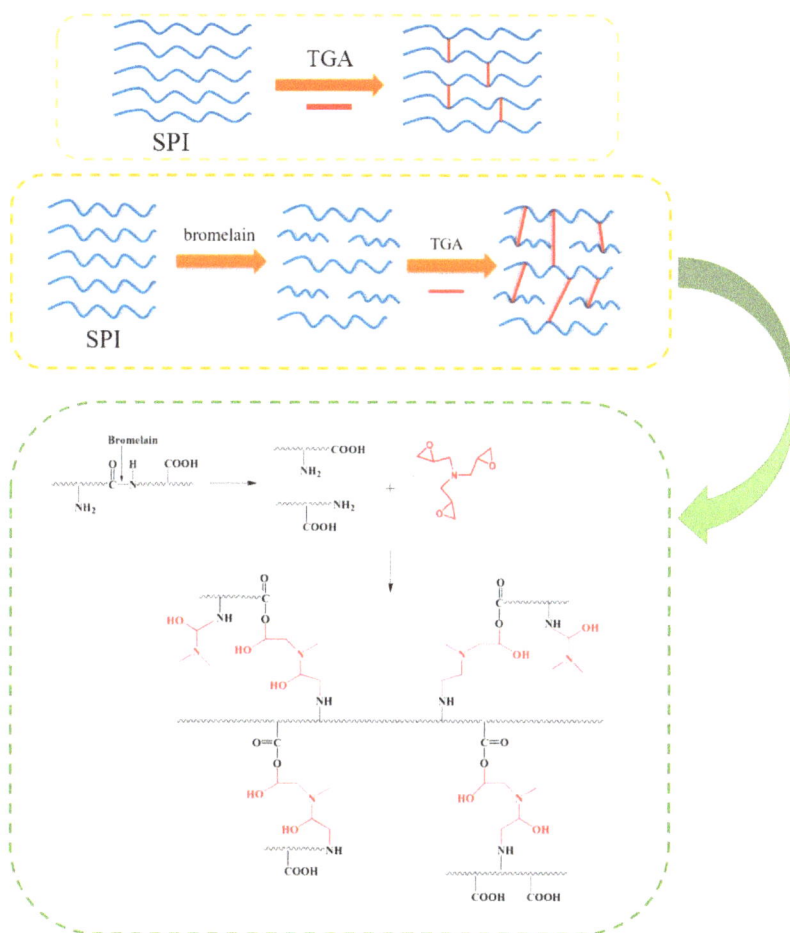

**Figure 4.** Cross-linking network of the soy protein adhesive.

*3.4. Fourier Transform Infrared (FTIR) Spectroscopy Analysis*

Figure 5 presents the Fourier transform infrared spectra of the adhesives. The corresponding bending vibrations of free and bound N–H and O–H groups were approximately located at 3303 cm$^{-1}$, which formed hydrogen bonds with the carbonyl group of the peptide linkage in soy protein [25]. The peak observed at about 2930 cm$^{-1}$ was attributed to the symmetric and asymmetric stretching vibrations of the –CH$_2$ group in the different adhesives [26]. In all adhesives, three characteristic bands of amides, namely, C=O stretching (amide I); N–H bending (amide II); N–H in-plane vibrations and the C–N stretching vibration (amide III), were observed at 1661, 1515, and 1238 cm$^{-1}$, respectively [27]. The peaks at 1441 and 1384 cm$^{-1}$ were the –CH$_2$ deformation vibrations of the methyl group and the COO– stretching vibration, respectively [28].

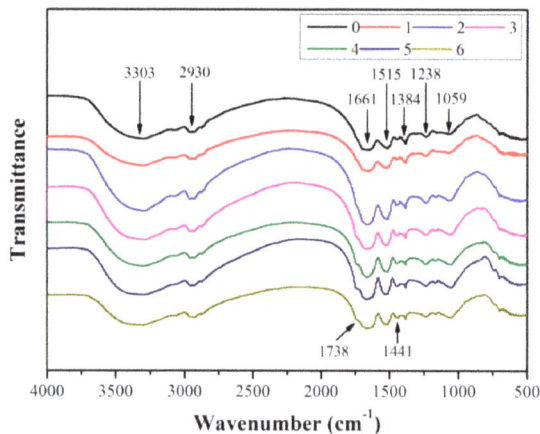

**Figure 5.** Fourier-transform infrared spectra of the different adhesive samples: 0 (12 wt% SPI), 1 (18 wt% SPI/bromelain), 2 (18 wt% SPI/bromelain/3 wt% TGA), 3 (18 wt% SPI/bromelain/6 wt% TGA), 4 (18 wt% SPI/bromelain/9 wt% TGA), 5 (18 wt% SPI/ bromelain/12 wt% TGA), and 6 (18 wt% SPI/ 9 wt% TGA).

In the mixed adhesives, the C–O bending absorption peak at 1059 cm$^{-1}$ increased gradually with an increase in TGA addition. This result indicated that the TGA was well distributed in the adhesive system. No new peaks of the epoxy groups were found around 910 cm$^{-1}$ after adding TGA into the adhesive formulation, indicating that the epoxy groups reacted with the active groups. With an increase in TGA addition, the peak of COO– at 1384 cm$^{-1}$ gradually decreased, and a new peak of the carbonyl group gradually appeared at 1738 cm$^{-1}$, which resulted from the esterification between the epoxy group and the carbonyl group of soy protein molecules. This finding was consistent with previous studies [29]. In addition, the soy protein adhesive contained numerous amino groups (–NH$_2$) because the activation energy of the epoxy group with the amino group reaction was lower than that of the epoxy group with the carbonyl group reaction. TGA reacted faster with the amino groups in the soy protein molecules, indicating the occurrence of cross-linking. The peak of the C–O group at 1059 cm$^{-1}$ in adhesive (4) was lower than that of adhesive (6), which might have resulted from the cross-linking of more TGA with reactive groups and the formation of a more compact cross-linked network structure. The cross-linking reaction between the TGA and the active groups led to the conversion of weak hydrogen bonds in the soy protein to rigid chemical bonds. This conversion reduced the number of hydrophilic groups and increased the cross-link density, thereby improving the water resistance of the soy protein adhesive.

### 3.5. Thermogravimetric (TG) Analysis

Figure 6 shows the thermogravimetric and derivative thermal gravimetric curves of various adhesives. The thermal degradation of the adhesive can be divided into three stages. The slight weight loss before the temperature reached 130 °C was attributed to the evaporation of the residual moisture in the adhesive samples. The first stage was the post-reaction stage in the 130–200 °C temperature range. This was the result of the further curing reaction between SPI and the cross-linking agent, which produced vapor and gases, resulting in mass loss [30]. The second stage was the initial degradation stage in the 200–270 °C temperature range, which was attributed to the degradation of small molecules and the breakdown of some unstable chemical bonds. The third stage was the degradation phase of the framework structure in the 270–370 °C temperature range, which was caused by the degradation of the cross-linked network structure [31]. After the third degradation stage, further heating caused the breakdown of the C–C, C–N, and C–O linkages and the decomposition of soy protein backbone peptide bonds, which produced gases such as CO, $CO_2$, $NH_3$, and $H_2S$ [32,33].

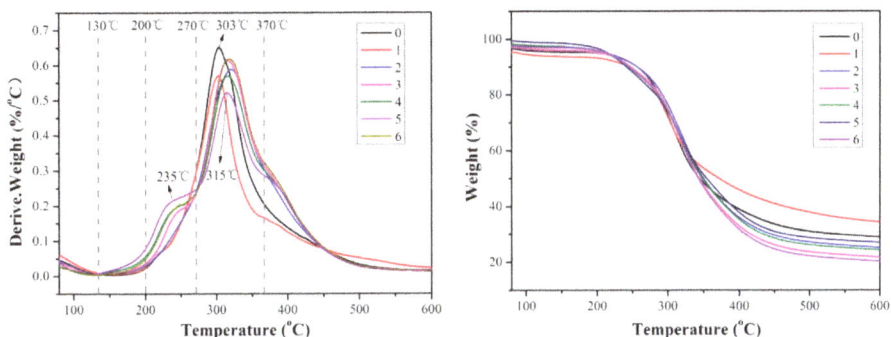

**Figure 6.** Thermogravimetric (TG, **left**) and derivative thermogravimetric (DTG, **right**) curves of the different adhesive samples: 0 (12 wt% SPI), 1 (18 wt% SPI/bromelain), 2 (18 wt% SPI/bromelain/ 3 wt% TGA), 3 (18 wt% SPI/bromelain/6 wt% TGA), 4 (18 wt% SPI/bromelain/9 wt% TGA), 5 (18 wt% SPI/bromelain/12 wt% TGA), and 6 (18 wt% SPI/9 wt% TGA).

Compared with the adhesives without TGA, the adhesives with TGA showed a new peak around 235 °C in the second stage. This new peak was improved with an increase in TGA addition, indicating that TGA reacted with active groups and that the structure of the adhesive changed. In the third stage, the peak of the adhesives with TGA shifted to a higher temperature, indicating an improvement in the thermal stability of the adhesives by forming a new structure. In the second and third stages, the degradation rate of adhesive (1) was lower than that of adhesive (0), indicating that the reduction in molecular weight by enzymatic hydrolysis improved the thermal stability of the adhesive. The peak degradation rate of adhesive (4) was lower than that of adhesive (6) in the third stage, also suggesting that enzymatic hydrolysis improved the thermal stability of the resultant adhesive. The cross-linking reaction between TGA and the enzymatic hydrolysis of soy protein molecules formed a more stable cross-linked network structure than that of the native soy protein adhesive.

### 3.6. Scanning Electron Microscopy (SEM) Analysis

Fracture surface micrographs of various types of cured adhesives are shown in Figure 7. SPI adhesive (0) showed a loose surface with sparse rifts. These sparse rifts could be the channels for subsequent water intrusion, reducing the water resistance of the adhesive [34]. After enzymatic hydrolysis, broken SPI molecular chains exposed more hydrophilic groups and increased the hydrophilic characteristic of adhesive (1). This occurrence resulted in a more disordered fracture surface and larger rifts, leading to a reduction in the water resistance of adhesive (1). However, with an increase in TGA addition, the cracks of enzymatic hydrolysis adhesives in the fracture surface became uniform, and the rifts disappeared gradually. These effects indicated that TGA cross-linked with soy protein molecules to increase the cross-link density, which improved the water resistance of the adhesive. Compared with adhesive (0), adhesive (6) had a smoother fracture surface and fewer cracks, indicating that TGA addition increased the brittleness of the adhesive. Enzymatic hydrolysis created a ductile fracture surface of adhesive (4) relative to adhesive (6), indicating an increase in the toughness of the adhesive, which contributed to the bond performance of the adhesive.

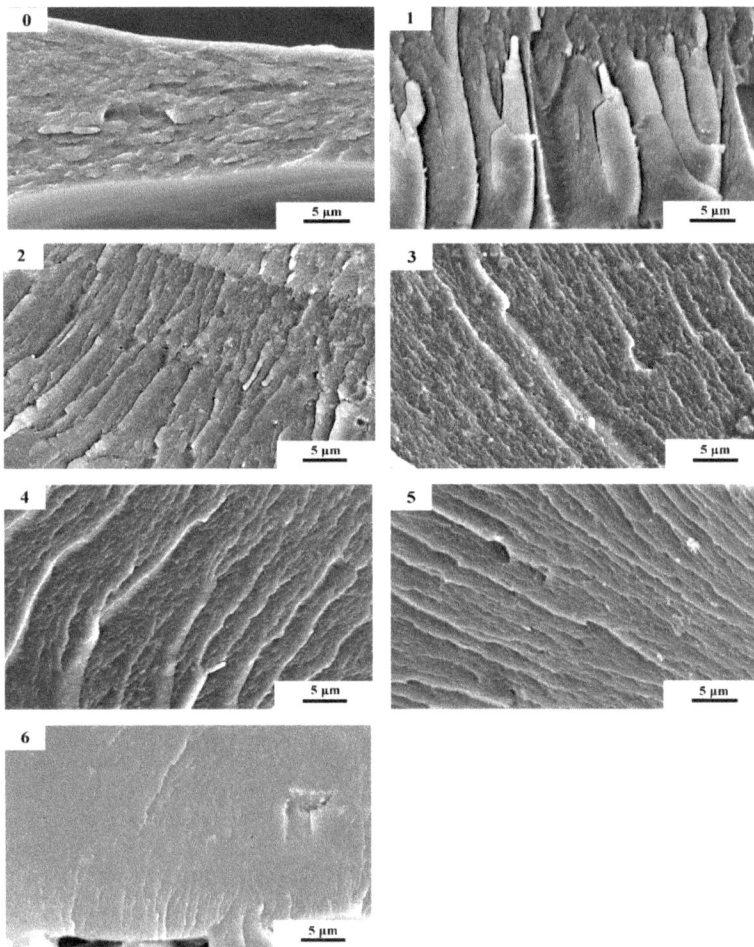

**Figure 7.** Fracture surface micrographs of the different cured adhesive samples: 0 (12 wt% SPI), 1 (18 wt% SPI/bromelain), 2 (18 wt% SPI/bromelain/3 wt% TGA), 3 (18 wt% SPI/bromelain/6 wt% TGA), 4 (18 wt% SPI/bromelain/9 wt% TGA), 5 (18 wt% SPI/bromelain/12 wt% TGA), and 6 (18 wt% SPI/9 wt% TGA).

## 4. Conclusions

From the study on the modification of soy protein adhesives by enzymatic hydrolysis, the following conclusions were drawn:

(1)  Using 0.1 wt% bromelain effectively reduced the viscosity of the SPI adhesive by 67.9% and improved the SPI content of the adhesive from 12 wt% to 18 wt%, while maintaining a similar viscosity. After the enzymatic hydrolysis process, the residual rate of the SPI/bromelain adhesive markedly decreased by 9.6%, and the wet shear strength of the resultant plywood was reduced to 70.4%. These reductions were attributed to the breakdown of the soy protein molecules into polypeptide chains and the exposure of more hydrophilic groups.

(2)  With the addition of 9 wt% TGA, the residual rate of the SPI/bromelain/TGA adhesive improved by 13.7%, and the wet shear strength of the resultant plywood increased by 681.3% to 1.25 MPa, relative to that of the SPI/bromelain adhesive. This wet shear strength was 30.2% higher than

that of the SPI/TGA adhesive. This improvement was attributed to the breakdown of soy protein molecules into polypeptide chains. This occurrence led to (1) the formation of more interlocks with the wood surface during the curing process of the adhesive and (2) the exposure of more hydrophilic groups and increase in the reactivity of protein with TGA, leading to a denser cross-linked network produced in the adhesive.

(3) The formed cross-linked structure exhibited a higher thermal stability after enzymatic hydrolysis, indicating an improvement in the cross-link density of the adhesive. This structure also created a ductile fracture surface of the adhesive, indicating an improvement in the toughness of the adhesive.

**Author Contributions:** Y.X. (Yantao Xu) and Q.G. conceived and designed the experiments; Y.X. (Yecheng Xu) and Y.X. (Yantao Xu) performed the experiments; Y.X. (Yantao Xu) and Y.H. analyzed the data; M.C. and Y.X. (Yantao Xu) contributed reagents/materials/analysis tools; W.Z., Q.G., and J.L. supervised and directed the project; all authors reviewed the manuscript.

**Funding:** The National Key Research and Development Program of China (2016YFD0600705).

**Conflicts of Interest:** The authors declare no conflict of interest.

## References

1. Nordqvist, P.; Nordgren, N.; Khabbaz, F.; Malmström, E. Plant proteins as wood adhesives: Bonding performance at the macro- and nanoscale. *Ind. Crop. Prod.* **2013**, *44*, 246–252. [CrossRef]

2. Vnučec, D.; Kutnar, A.; Goršek, A. Soy-based adhesives for wood-bonding—A review. *J. Adhes. Sci. Technol.* **2016**, *31*, 910–931. [CrossRef]

3. Liu, H.; Li, C.; Sun, X.S. Improved water resistance in undecylenic acid (ua)-modified soy protein isolate (spi)-based adhesives. *Ind. Crop. Prod.* **2015**, *74*, 577–584. [CrossRef]

4. Eslah, F.; Jonoobi, M.; Faezipour, M.; Ashori, A. Chemical modification of soybean flour-based adhesives using acetylated cellulose nanocrystals. *Polym. Compos.* **2017**, *39*, 3618–3625. [CrossRef]

5. Li, J.; Luo, J.; Li, X.; Yi, Z.; Gao, Q.; Li, J. Soybean meal-based wood adhesive enhanced by ethylene glycol diglycidyl ether and diethylenetriamine. *Ind. Crop. Prod.* **2015**, *74*, 613–618. [CrossRef]

6. Zhao, Y.; Xu, H.; Mu, B.; Xu, L.; Yang, Y. Biodegradable soy protein films with controllable water solubility and enhanced mechanical properties via graft polymerization. *Polym. Degrad. Stab.* **2016**, *133*, 75–84. [CrossRef]

7. Kang, H.; Song, X.; Wang, Z.; Zhang, W.; Zhang, S.; Li, J. High-performance and fully renewable soy protein isolate-based film from microcrystalline cellulose via bio-inspired poly(dopamine) surface modification. *ACS Sustain. Chem. Eng.* **2016**, *4*, 4354–4360. [CrossRef]

8. Jong, L. Reinforcement effect of soy protein nanoparticles in amine-modified natural rubber latex. *Ind. Crop. Prod.* **2017**, *105*, 53–62. [CrossRef]

9. Eslah, F.; Jonoobi, M.; Faezipour, M.; Afsharpour, M.; Enayati, A.A. Preparation and development of a chemically modified bio-adhesive derived from soybean flour protein. *Int. J. Adhes. Adhes.* **2016**, *71*, 48–54. [CrossRef]

10. Xu, F.; Dong, Y.; Zhang, W.; Zhang, S.; Li, L.; Li, J. Preparation of cross-linked soy protein isolate-based environmentally-friendly films enhanced by ptge and pam. *Ind. Crop. Prod.* **2015**, *67*, 373–380. [CrossRef]

11. Luo, J.; Li, X.; Zhang, H.; Gao, Q.; Li, J. Properties of a soybean meal-based plywood adhesive modified by a commercial epoxy resin. *Int. J. Adhes. Adhes.* **2016**, *71*, 99–104. [CrossRef]

12. Vnučec, D.; Mikuljan, M.; Kutnar, A.; Šernek, M.; Goršek, A. Influence of process parameters on the bonding performance of wood adhesive based on thermally modified soy proteins. *Eur. J. Wood Wood Prod.* **2016**, *74*, 553–561. [CrossRef]

13. Nyanhongo, G.; Kudanga, T.; Prasetyo, E.N.; Guebitz, G.M. Mechanistic insights into laccase-mediated functionalisation of lignocellulose material. *Biotechnol. Gen. Eng.* **2010**, *27*, 305–330. [CrossRef]

14. Slagman, S.; Zuilhof, H.; Franssen, M.C.R. Laccase-Mediated Grafting on Biopolymers and Synthetic Polymers: A Critical Review. *ChemBioChem* **2018**, *19*, 288–311. [CrossRef] [PubMed]

15. Ribitsch, D.; Herrero Acero, E.; Greimel, K.; Dellacher, A.; Zitzenbacher, S.; Marold, A.; Rodriguez, R. A New Esterase from Thermobifida halotolerans Hydrolyses Polyethylene Terephthalate (PET) and Polylactic Acid (PLA). *Polymers* **2012**, *4*, 617–629. [CrossRef]

16. Areskogh, D.; Henriksson, G. Immobilisation of laccase for polymerisation of commercial lignosulphonates. *Process Biochem.* **2011**, *46*, 1071–1075. [CrossRef]

17. Lee, H.; Yildiz, G.; dos Santos, L.C.; Jiang, S.; Andrade, J.E.; Engeseth, N.J.; Feng, H. Soy protein nano-aggregates with improved functional properties prepared by sequential ph treatment and ultrasonication. *Food Hydrocoll.* **2016**, *55*, 200–209. [CrossRef]

18. Luo, J.; Luo, J.; Li, X.; Li, K.; Gao, Q.; Li, J. Toughening improvement to a soybean meal-based bioadhesive using an interpenetrating acrylic emulsion network. *J. Mater. Sci.* **2016**, *51*, 9330–9341. [CrossRef]

19. GB/T17657-2013. Standardization Administration of the People's Republic of China. Available online: https://www.chinesestandard.net/PDF.aspx/GBT17657-2013 (accessed on 24 October 2018).

20. Wang, X.; He, Z.; Zeng, M.; Qin, F.; Adhikari, B.; Chen, J. Effects of the size and content of protein aggregates on the rheological and structural properties of soy protein isolate emulsion gels induced by caso4. *Food Chem.* **2017**, *221*, 130–138. [CrossRef] [PubMed]

21. Cheng, E.Z.; Sun, X. Effects of wood-surface roughness, adhesive viscosity and processing pressure on adhesion strength of protein adhesive. *J. Adhes. Sci. Technol.* **2006**, *20*, 997–1017. [CrossRef]

22. Gao, Q.; Qin, Z.Y.; Li, C.C.; Zhang, S.F.; Li, J.Z. Preparation of wood adhesives based on soybean meal modified with pegda as a crosslinker and viscosity reducer. *Bioresources* **2013**, *8*, 5380–5391. [CrossRef]

23. Luo, J.; Luo, J.; Zhang, J.; Bai, Y.; Gao, Q.; Li, J.; Li, L. A new flexible soy-based adhesive enhanced with neopentyl glycol diglycidyl ether: Properties and application. *Polymers* **2016**, *8*, 346. [CrossRef]

24. Luo, J.; Luo, J.L.; Yuan, C.; Zhang, W.; Li, J.Z.; Gao, Q.; Chen, H. An eco-friendly wood adhesive from soy protein and lignin: Performance properties. *RSC Adv.* **2015**, *5*, 100849–100855. [CrossRef]

25. Liu, C.; Zhang, Y.; Li, X.; Luo, J.; Gao, Q.; Li, J. "Green" bio-thermoset resins derived from soy protein isolate and condensed tannins. *Ind. Crop. Prod.* **2017**, *108*, 363–370. [CrossRef]

26. Liu, C.; Zhang, Y.; Li, X.N.; Luo, J.; Gao, Q.; Li, J.Z. A high-performance bio-adhesive derived from soy protein isolate and condensed tannins. *RSC Adv.* **2017**, *7*, 21226–21233. [CrossRef]

27. Li, X.; Li, J.; Luo, J.; Li, K.; Gao, Q.; Li, J. A novel eco-friendly blood meal-based bio-adhesive: Preparation and performance. *J. Polym. Environ.* **2017**, *26*, 607–615. [CrossRef]

28. Luo, J.; Li, L.; Luo, J.; Li, X.; Li, K.; Gao, Q. A high solid content bioadhesive derived from soybean meal and egg white: Preparation and properties. *J. Polym. Environ.* **2016**, *25*, 948–959. [CrossRef]

29. Kumar, R.; Choudhary, V.; Mishra, S.; Varma, I.K. Enzymatically-modified soy protein part 2: Adhesion behaviour. *J. Adhes. Sci. Technol.* **2004**, *18*, 261–273. [CrossRef]

30. Yuan, C.; Luo, J.; Luo, J.L.; Gao, Q.; Li, J.Z. A soybean meal-based wood adhesive improved by a diethylene glycol diglycidyl ether: Properties and performance. *RSC Adv.* **2016**, *6*, 74186–74194. [CrossRef]

31. Luo, J.; Luo, J.L.; Bai, Y.Y.; Gao, Q.; Li, J.Z. A high performance soy protein-based bio-adhesive enhanced with a melamine/epichlorohydrin prepolymer and its application on plywood. *RSC Adv.* **2016**, *6*, 67669–67676. [CrossRef]

32. Schmidt, V.; Giacomelli, C.; Soldi, V. Thermal stability of films formed by soy protein isolate–sodium dodecyl sulfate. *Polym. Degrad. Stab.* **2005**, *87*, 25–31. [CrossRef]

33. Li, J.J.; Li, X.N.; Li, J.Z.; Gao, Q. Investigating the use of peanut meal: A potential new resource for wood adhesives. *RSC Adv.* **2015**, *5*, 80136–80141. [CrossRef]

34. Zhang, Y.; Zhu, W.; Lu, Y.; Gao, Z.; Gu, J. Nano-scale blocking mechanism of mmt and its effects on the properties of polyisocyanate-modified soybean protein adhesive. *Ind. Crop. Prod.* **2014**, *57*, 35–42. [CrossRef]

**Sample Availability:** Samples of the adhesive are available from the authors.

*molecules*

MDPI

*Article*

# New Octadecanoid Enantiomers from the Whole Plants of *Plantago depressa*

Xiu-Qing Song [1,2], Kongkai Zhu [2], Jin-Hai Yu [2], Qianqian Zhang [2], Yuying Zhang [2], Fei He [2], Zhi-Qiang Cheng [2], Cheng-Shi Jiang [2] , Jie Bao [2,*] and Hua Zhang [2,*]

[1] School of Chemistry and Chemical Engineering, University of Jinan, 336 West Road of Nan Xinzhuang, Jinan 250022, China; 13156019194@163.com
[2] School of Biological Science and Technology, University of Jinan, 336 West Road of Nan Xinzhuang, Jinan 250022, China; hkhhh.k@163.com (K.Z.); yujinhai12@sina.com (J.-H.Y.); zhangqian464x@163.com (Q.Z.); yuyingzhang2008@163.com (Y.Z.); 18864838287@163.com (F.H.); czq13515312897@163.com (Z.-Q.C.); jiangchengshi-20@163.com (C.-S.J.)
* Correspondence: bio_baoj@ujn.edu.cn (J.B.); bio_zhangh@ujn.edu.cn (H.Z.); Tel.: +86-531-8973-6199 (H.Z.)

Academic Editor: Pavel B. Drasar
Received: 25 June 2018; Accepted: 11 July 2018; Published: 14 July 2018

check for updates

**Abstract:** In this study, 19 octadecanoid derivatives—four pairs of enantiomers (**1–8**), two racemic /scalemic mixtures (**9–10**), and nine biosynthetically related analogues—were obtained from the ethanolic extract of a Chinese medicinal plant, *Plantago depressa* Willd. Their structures were elucidated on the basis of detailed spectroscopic analyses, with the absolute configurations of the new compounds assigned by time-dependent density functional theory (TD-DFT)-based electronic circular dichroism (ECD) calculations. Six of them (**1**, **3–6**, and **9**) were reported for the first time, while **2**, **7**, and **8** have been previously described as derivatives and are currently obtained as natural products. Our bioassays have established that selective compounds show in vitro anti-inflammatory activity by inhibiting lipopolysaccharide-induced nitric oxide (NO) production in mouse macrophage RAW 264.7 cells.

**Keywords:** *Plantago depressa*; octadecanoid; fatty acid; natural enantiomer; anti-inflammation

## 1. Introduction

The genus *Plantago* L. (family Plantaginaceae) consists of more than 190 species that are widely distributed in temperate and tropical areas all over the world. There are 20 *Plantago* plants that grow in China, including two invasive and one cultivated species [1]. *P. depressa* Willd. is a very common species found in most Asian countries [1], and its whole plants have long been used in traditional Chinese medicine as "Cheqian Cao" for the treatment of oedema, cough, carbuncle, etc. [2]. Previous chemical investigations of this medicinal plant have revealed the presence of phenylethanoid glycosides [3–5], iridoid glucosides [6,7], alkaloids [8], and so on [6,7,9]. However, few reports have dealt with the lipid constituents from *P. depressa* until now [10]. In the present work, we carried out an intensive chemical study on the EtOAc partition generated from the ethanolic extract of the whole plants of *P. depressa*, which resulted in the isolation of a series of fatty acid derivatives—four pairs of enantiomers (**1–8**), two racemic/scalemic mixtures (**9–10**) (Figure 1), and nine related analogues (**11–19**). The structures of these compounds were fully characterized by comprehensive spectroscopic analyses, with the absolute stereochemistry of the new compounds established via calculated ECD data. The in vitro antimicrobial, anti-acetylcholinesterase, and anti-inflammatory activities of these lipid molecules were tested; only two known compounds exhibited moderate anti-inflammatory effects. Herein, we describe the separation, structural characterization, and biological evaluations of these plant lipids.

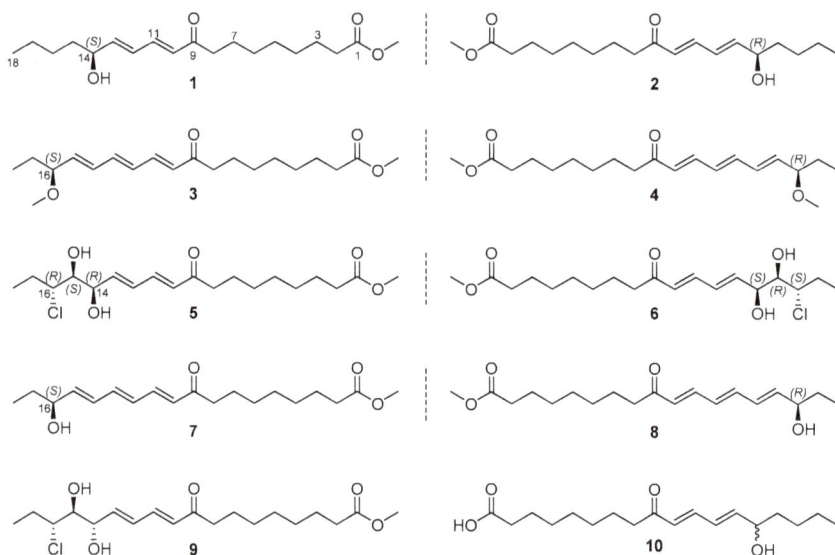

**Figure 1.** Chemical structures of **1–10** from *Plantago depressa*.

## 2. Results

Compounds **1/2**—colorless gum—were assigned the molecular formula of $C_{19}H_{32}O_4$ by positive mode high resolution electrospray ionization mass spectrometry [(+)-HR-ESIMS] analysis at $m/z$ 325.2368 ([M + H]$^+$, calcd 325.2373). The $^1$H NMR (nuclear magnetic resonance) data (Table 1) revealed the presence of a conjugated diene [6.20 (d, $J$ = 15.6 Hz, H-10), 6.25 (dd, $J$ = 15.3, 5.9 Hz, H-13), 6.41 (dd, $J$ = 15.3, 10.8 Hz, H-12), 7.27 (dd, $J$ = 15.6, 10.8 Hz, H-11)], an oxygenated methine ($\delta_H$ 4.17, m), a methoxy ($\delta_H$ 3.65, s), and a methyl [$\delta_H$ 0.92 (t, $J$ = 7.4 Hz)] group. The $^{13}$C NMR data (Table 2) showed signals of a conjugated ketone ($\delta_C$ 203.7, C-9), an ester carbonyl ($\delta_C$ 176.0, C-1), four olefinic ($\delta_C$ 128.8, 130.3, 144.3, 148.5; C-10 to C-13), an oxygenated methine ($\delta_C$ 72.6, C-14), a methoxy ($\delta_C$ 52.0), ten aliphatic methylene ($\delta_C$ 23.7, 25.5, 26.0, 28.7, 30.0, 30.1, 30.2, 37.8, 34.8, 41.0), and a methyl ($\delta_C$ 14.4, C-18) carbon. These spectral features were similar to those of compound **10** isolated from the fungus *Pleurocybella porrigens* [11] but with an additional methoxy group, suggestive of a methyl ester derivative. Detailed examination of 2D $^1$H-$^1$H COSY (correlated spectroscopy) and HMBC (heteronuclear multiple-bond correlation) data (Figure 2) confirmed the above conclusion with key HMBC correlations from H$_2$-7, H$_2$-8, H-10 and H-11 to C-9 ($\delta_C$ 203.7), and H$_2$-2, H$_2$-3 and OC$\underline{H}_3$ to C-1 ($\delta_C$ 176.0,). Therefore, compounds **1/2** were characterized as methyl (10*E*,12*E*)-14-hydroxy-9-oxo-10,12-octadecadienoate. Further spectroscopic analyses revealed that compared with the reported (−)-enantiomer of compound **10** (porrigenic acid) [12], **1/2** were neither active enough in the [$\alpha$]$_D$ measurement, nor showed decent Cotton effect in the ECD experiment; this alerted us of its racemic or scalemic nature. Subsequent chiral high performance liquid chromatography (HPLC) analysis clearly revealed a pair of enantiomers in a ratio of *ca.* 55:45 (Supplementary Information Figure S2). On reviewing the literature, the (+)-enantiomer (**1**) was identified to be a new compound, while the (−)-enantiomer (**2**) was a new natural product that had been reported as the methyl ester of porrigenic acid (**10**) during structure characterization [12]. It is worth noting that the absolute configuration of compound **2** was initially determined as *S* using allylic benzoate method [12]. However, the assignment was apparently not rigorous because this ECD method was originally developed to assign absolute stereochemistry of allylic alcohol (hydroxyl group adjacent to a double bond chromophore) [13], but the chromophore in compound **2** is an $\alpha,\beta,\gamma,\delta$-conjugated ketone.

We therefore employed the time-dependent density functional theory (TD-DFT) method to calculate the ECD spectra (Figure 3) of the two enantiomers and finally differentiated them from each other.

**Table 1.** $^1$H NMR data for **1–6** and **9** (600 MHz).

| Position | 1/2 [a] | 3/4 [a] | 5/6 [b] | 9 [b] |
|---|---|---|---|---|
| 2 | 2.32, t (7.4) | 2.31, t (7.4) | 2.29, t (7.6) | 2.30, t (7.5) |
| 3 | 1.60, m | 1.61, m | 1.61, m | 1.62, m |
| 4 | 1.34, m | 1.33, m | 1.32, m | 1.32, m |
| 5 | 1.34, m | 1.33, m | 1.32, m | 1.32, m |
| 6 | 1.34, m | 1.33, m | 1.32, m | 1.32, m |
| 7 | 1.60, m | 1.61, m | 1.61, m | 1.61, m |
| 8 | 2.61, t (7.3) | 2.61, t (7.4) | 2.55, t (7.5) | 2.55, t (7.4) |
| 10 | 6.20, d (15.6) | 6.23, d (15.5) | 6.20, d (15.6) | 6.21, d (15.5) |
| 11 | 7.27, dd (15.6, 10.8) | 7.31, dd (15.5, 11.2) | 7.15, dd (15.6, 10.8) | 7.16, dd (15.6, 11.0) |
| 12 | 6.41, dd (15.3, 10.8) | 6.43, dd (15.1, 11.2) | 6.49, dd (15.5, 10.8) | 6.50, dd (15.3, 11.0) |
| 13 | 6.25, dd (15.3, 5.9) | 6.75, dd (15.1, 10.9) | 6.21, dd (15.5, 6.0) | 6.25, dd (15.3, 6.8) |
| 14 | 4.17, m | 6.39, dd (15.2, 10.9) | 4.70, dd (6.0, 2.7) | 4.62, dd (6.8, 5.1) |
| 15 | 1.55, m | 5.81, dd (15.2, 7.8) | 3.62, dd (7.6, 2.7) | 3.86, dd (7.7, 5.1) |
| 16 | 1.34, m | 3.62, m | 3.97, ddd (9.3, 7.6, 2.9) | 3.76, ddd (9.4, 7.7, 2.8) |
| 17 | 1.37, m | 1.61, m | 1.76, m 2.07, m | 1.74, m 2.08, m |
| 18 | 0.92, t (7.4) | 0.90, t (7.4) | 1.08, t (7.3) | 1.07, t (7.2) |
| 1-OMe | 3.65, s | 3.65, s | 3.67, s | 3.67, s |
| 16-OMe | | 3.28, s | | |

[a] In CD$_3$OD; [b] in CDCl$_3$.

**Table 2.** $^{13}$C NMR data for **1–6** and **9** (150 MHz).

| Position | 1/2 [a] | 3/4 [a] | 5/6 [b] | 9 [b] |
|---|---|---|---|---|
| 1 | 176.0 | 176.0 | 174.5 | 174.5 |
| 2 | 34.8 | 34.8 | 34.2 | 34.2 |
| 3 | 26.0 | 26.0 | 25.0 | 25.0 |
| 4 | 30.0 [c] | 30.0 [d] | 29.1 [e] | 29.1 [f] |
| 5 | 30.1 [c] | 30.2 [d] | 29.2 [e] | 29.2 [f] |
| 6 | 30.2 [c] | 30.2 [d] | 29.2 [e] | 29.2 [f] |
| 7 | 25.5 | 25.5 | 24.3 | 24.3 |
| 8 | 41.0 | 41.1 | 40.8 | 40.8 |
| 9 | 203.7 | 203.5 | 201.0 | 200.9 |
| 10 | 130.3 | 130.5 | 130.6 | 130.9 |
| 11 | 144.3 | 144.5 | 141.1 | 141.0 |
| 12 | 128.8 | 131.9 | 130.3 | 131.5 |
| 13 | 148.5 | 142.3 | 141.7 | 139.4 |
| 14 | 72.6 | 133.2 | 71.0 | 72.5 |
| 15 | 37.8 | 140.2 | 76.5 | 76.6 |
| 16 | 28.7 | 84.5 | 64.6 | 64.7 |
| 17 | 23.7 | 29.2 | 26.8 | 26.6 |
| 18 | 14.4 | 9.9 | 10.7 | 10.6 |
| 1-OMe | 52.0 | 52.0 | 51.7 | 51.6 |
| 16-OMe | | 56.7 | | |

[a] In CD$_3$OD; [b] in CDCl$_3$; [c–f] Interchangeable assignments.

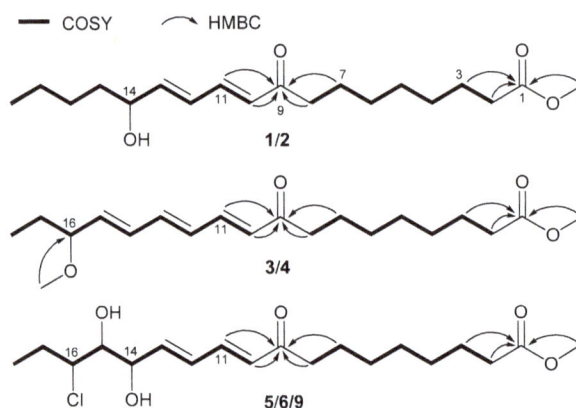

**Figure 2.** $^1$H-$^1$H COSY and selected HMBC correlations for **1–6** and **9**.

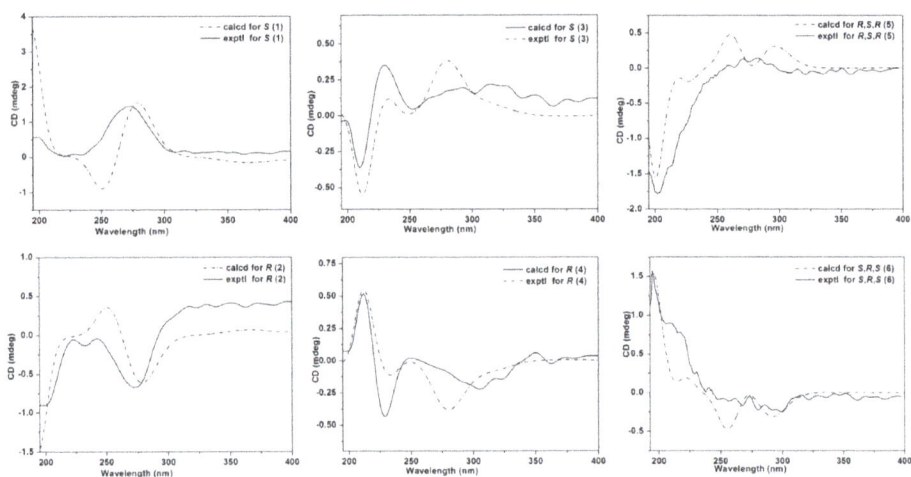

**Figure 3.** Experimental and calculated ECD spectra for **1–6**.

Compounds **3/4** had the molecular formula of $C_{20}H_{32}O_4$ as deduced from the (+)-HR-ESIMS ion peak at $m/z$ 337.2369 ([M + H]$^+$, calcd 337.2373), which was 14 mass units (CH$_2$) more than that of compounds **7/8** [14,15] indicative of a methylated analogue. Analysis of the NMR data (Tables 1 and 2) for compounds **3/4** confirmed this hypothesis, with extra signals for a methoxy group ($\delta_H$ 3.28, $\delta_C$ 56.7) and the downfield shifted C-16 resonance ($\delta_C$ 84.5) in contrast with that ($\delta_C$ 74.2) in compounds **7/8**. Further inspection of 2D NMR data (Figure 2) corroborated this structural assignment, revealing key HMBC correlations with the methoxy protons to C-16. Compounds **3/4** were thus characterized to be methyl (10*E*,12*E*,14*E*)-16-methoxy-9-oxo-10,12,14-octadecatrienoate. Similar to compounds **1/2**, the optical rotation and ECD data of compounds **3/4** suggested a scalemic mixture with nearly zero $[\alpha]_D$ and no Cotton effect, respectively. The two pure enantiomers were further separated from each other by chiral HPLC and structurally differentiated by comparing their experimental ECD spectra with the calculated ones (Figure 3).

Compounds **5/6** were determined to be monochlorinated on the basis of the ESIMS (electrospray ionization mass spectrometry) isotope ion peak at $m/z$ 397.1/399.1 ([M + Na]$^+$, ca. 3:1) and were assigned the molecular formula of $C_{19}H_{32}O_5Cl$ by (+)-HR-ESIMS analysis at *m/z* 397.1756 ([M + Na]$^+$, calcd 397.1752). The NMR data (Tables 1 and 2) for compounds **5/6** also displayed resonances for several functional groups as those in compounds **1/2**, such as two carbonyls ($\delta_C$ 174.5 and 201.0), a diene ($\delta_C$ 130.6, 141.1 and 130.3, 141.7; $\delta_H$ 6.20, 7.15 and 6.49, 6.21), and an ester methoxyl ($\delta_C$ 51.7; $\delta_H$ 3.67). Meanwhile, compounds **5/6** possessed three sp$^3$ methines ($\delta_C$ 64.6, 71.0 and 76.5; $\delta_H$ 3.97, 4.70 and 3.62) compared with only one in compounds **1/2**, which was ascribed to two hydroxyl and a chlorine substituents by analyzing the molecular composition and chemical shifts of these methines. Subsequent acquisition of 2D $^1$H-$^1$H COSY and HMBC data (Figure 2) confirmed the establishment of the planar structure of compounds **5/6** as shown, and the substitution pattern of 14-OH, 15-OH, and 16-Cl for the C-14–C-16 fragment was supported by the lower chemical shift for C-16 than those for C-14 and C-15 [16]. The relative configuration of compounds **5/6** was determined by the *J*-based configuration analysis method [17]. The magnitudes of $J_{14,15}$ (2.7 Hz) and $J_{15,16}$ (7.6 Hz) indicated a *syn*-relationship between H-14 and H-15 and an *anti*-relationship between H-15 and H-16, respectively. Alerted by the cases of compounds **1–4**, compounds **5/6** were also subjected to chiral HPLC analysis and indeed proved to be another pair of enantiomers. The absolute configurations of compounds **5/6** were further assigned by comparing their experimental ECD spectra with the computed ones (Figure 3).

Compound **9** was assigned the molecular formula of $C_{19}H_{32}O_5Cl$—same as compounds **5/6**—based on the (+)-HR-ESIMS ion peak at *m/z* 397.1757 ([M + Na]$^+$, calcd 397.1752), supportive of an isomer of the latter. Analysis of the NMR data (Tables 1 and 2) for compound **9** corroborated this conclusion, with very similar NMR data suggesting that they were diastereoisomers of the same planar structure; this was further confirmed by examination of 2D NMR correlations (Figure 2). Detailed NMR comparison between compounds **9** and **5/6** revealed nearly superimposable $^1$H and $^{13}$C NMR spectra, and the only difference was attributable to signals across CH-13 to CH-16 moiety. Obvious NMR variations were observed for resonances from H-13 to H-16, C-13 to C-14, and $J_{14,15}$ (Tables 1 and 2), which all suggested an inverted C-14 configuration in compound **9** compared with that in compounds **5/6**. The structure and relative configuration of compound **9** were thus elucidated. It was inferable that compound **9** could also be enantiomeric mixture in light of the aforementioned examples of its cometabolites. However, it was not further separated on chiral HPLC due to degradation during storage, as indicated by subsequent $^1$H NMR measurement. Moreover, the scarce amount of sample prevented us from further investigation.

In addition to the above-described molecules, compounds **7/8** were also demonstrated to be enantiomeric mixtures and separated by chiral HPLC. They had been reported in mixture as the methylation derivatives of their corresponding fatty acids [14,15], and we herein report them as enantiomerically pure isolates as new natural products. Compound **10** had been previously obtained in scalemic form [11] and (−)-form [12], respectively, from the same fungus by two Japanese research groups. In the current work, it was obtained as a nearly racemic mixture ($[\alpha]_D^{21}$ 0.3; *c* 0.10, MeOH) and was not separable on both normal-phase and reversed-phase chiral HPLC columns. The other known analogues were identified to be (9*Z*,12*Z*,14*E*)-16-oxo-octadecatrienoic acid (**11**) [18], (10*Z*,12*E*,14*Z*)-9,16-dioxo-octadecatrienoic acid (**12**) [19], linoleic acid (**13**) [20], β-(9′*Z*,12′*Z*,15′*Z*)-octadecatrienoic acid monoglyceride (**14**) [21], 1-*O*-(9*Z*,12*Z*)-octadecadienoyl glycerol (**15**) [22], α-(9′*Z*,12′*Z*,15′*Z*)-octadecatrienoic acid monoglyceride (**16**) [21], 1-*O*-(10*E*,12*E*)-9-oxo-octadecadienoyl glycerol (**17**) [23], 1-*O*-(9*Z*,11*E*)-13-oxo-octadecadienoyl glycerol (**18**) [24], and 1-*O*-(9*Z*,11*E*)-9-oxo-octadecadienoyl glycerol (**19**) [24] by spectroscopic data.

Most compounds (only those with enough amount) were screened for their antimicrobial, anti-acetylcholinesterase, and anti-inflammatory activities (Supplementary Information Tables S1 and S2); only compounds **13** and **18** displayed anti-inflammatory effect with moderate inhibition against nitric oxide (NO) production with IC$_{50}$ values of 13.08 ± 0.25 and 7.64 ± 0.21 μM, respectively.

## 3. Materials and Methods

### 3.1. General Experimental Procedures

Optical rotations were measured on a Rudolph VI polarimeter (Rudolph Research Analytical, Hackettstown, NJ, USA) with a 10 cm length cell. NMR experiments were recorded on a Bruker Avance DRX600 spectrometer (Bruker BioSpin AG, Fallanden, Switzerland) and referenced to residual solvent peaks (CD$_3$OD: $\delta_H$ 3.31, $\delta_C$ 49.00; CDCl$_3$: $\delta_H$ 7.26, $\delta_C$ 77.16). HR-ESIMS spectra were obtained on an Agilent 6545 Q-TOF mass spectrometer (Agilent Technologies Inc., Waldbronn, Germany). ESIMS analyses were carried out on an Agilent 1260-6460 Triple Quad LC-MS instrument (Agilent Technologies Inc., Waldbronn, Germany). UV spectra were obtained on a Shimadzu UV-2600 spectrophotometer (Shimadzu, Kyoto, Japan) with a 1 cm pathway cell. Normal HPLC separation was performed using an Agilent 1260 series LC instruments (Agilent Technologies Inc., Waldbronn, Germany) coupled with an Agilent SB-C$_{18}$ (9.4 × 250 mm) column (Agilent Technologies Inc., Santa Clara, CA, USA). Chiral MZ(2) RH 5u (4.6 × 250 mm) chiral column (Phenomenex, Washington, CD, USA) and CHIRALPAK AD-H (4.6 × 250 mm) chiral column (Daicel Corporation, Tokyo, Japan) were used for chiral HPLC analysis and separation. ECD spectra were acquired on a Chirascan circular dichroism spectrometer (Applied Photophysics Ltd., Surrey, UK). Column chromatography (CC) was performed on D101-macroporous absorption resin (Sinopharm Chemical Reagent Co., Ltd., Shanghai, China), MCI gel (CHP20P, Mitsubishi Chemical Corporation, Tokyo, Japan), reversed phase C18 silica gel (Merck KGaA, Darmstadt, Germany), Sephadex LH-20 (GE Healthcare Bio-Sciences AB, Uppsala, Sweden), and silica gel (300–400 mesh; Qingdao Marine Chemical Ltd., Qingdao, China). All solvents used for CC were of analytical grade (Tianjin Fuyu Fine Chemical Co., Ltd., Tianjin, China) and solvents used for HPLC were of HPLC grade (Oceanpak Alexative Chemical Ltd., Goteborg, Sweden). Pre-coated silica gel GF254 plates (Qingdao Haiyang Chemical Co., Ltd., Qingdao, China) were used for thin-layer chromatography (TLC) monitoring.

### 3.2. Plant Material

The whole plants of *P. depressa* were collected in June 2016 at Mount Kunyu, Shandong Province, and were authenticated by Prof. Jie Zhou from University of Jinan. A voucher specimen has been deposited at School of Biological Science and Technology, University of Jinan (Accession number: npmc-007).

### 3.3. Extraction and Isolation

The air-dried powder of the whole plants of *P. depressa* (15 kg) was extracted with 95% EtOH at room temperature three times to afford a crude extract (0.9 kg). The extract was then suspended in 2.0 L water and partitioned with EtOAc (2.0 L × 3). The EtOAc extract (300 g) was subjected to CC over D101-macroporous absorption resin and eluted with EtOH-H$_2$O (30%, 50%, 80% and 95%) to afford four fractions (A–D). Fraction C (80%, 87 g) was subjected to passage over an MCI gel column and eluted with MeOH-H$_2$O (50% to 100%) to give five subfractions (C1–C5). C1 was then separated by silica gel CC eluted with petroleum ether-acetone (4:1 to 1:1) to produce two eluents (C1-1 and C1-2), and C1-1 was further purified by HPLC (3.0 mL/min 80% MeOH-H$_2$O, $t_R$ = 10.0 min) to afford **7/8** (3.1 mg). Fraction C2 was then separated by silica gel CC eluted with petroleum ether-acetone (8:1 to 1:1) to produce eleven subfractions (C2-1–C2-11). C2-4 was subjected to silica gel CC eluted with CHCl$_3$-MeOH (100:1 to 10:1) to give five major eluents (C2-4-1–C2-4-5), and C2-4-4 was then purified by HPLC (3.0 mL/min 80% MeOH-H$_2$O, $t_R$ = 15.0 min) to afford **3/4** (1.0 mg). Fraction C2-6 was subjected to silica gel CC eluted with CHCl$_3$-MeOH (100:1 to 10:1) to give two major subfractions (C2-6-1 and C2-6-2), C2-6-2 was then purified by HPLC (3.0 mL/min 80% MeOH-H$_2$O, $t_R$ = 9.5 min) to afford **1/2** (3.5 mg). Fraction C2-11 was chromatographed on an RP-C18 silica gel column to give three subfractions (C2-11-1–C2-11-3), and C2-11-3 was then purified by HPLC (3.0 mL/min 80% MeOH-H$_2$O, $t_R$ = 10.0 min and 11.0 min, respectively) to afford **9** (2.5 mg) and **5/6** (2.0 mg). Fraction

C3 was then separated by silica gel CC eluted with petroleum ether-acetone (20:1 to 1:1) to produce five subfractions (C3-1–C3-5), and C3-4 was then chromatographed on an RP-C18 silica gel column to give seven eluents (C3-4-1–C3-4-7). Fraction C3-4-7 was subjected to silica gel CC eluted with CHCl$_3$-MeOH (200:1 to 10:1) to give two major fractions—C3-4-7-1 and C3-4-7-2—and C3-4-7-1 was further purified by HPLC (3.0 mL/min 90% MeOH-H$_2$O, $t_R$ = 15.5 min) to afford **12** (5.0 mg). Fraction C3-5 was chromatographed on an RP-C18 silica gel column to give seven subfractions (C3-5-1–C3-5-7). Fraction C3-5-3 was subjected to Sephadex LH-20 CC to give two subfractions (C3-5-3-1–C3-5-3-2), and C3-5-3-1 was further purified by HPLC (3.0 mL/min 90% MeOH-H$_2$O, $t_R$ = 11.0 min) to afford **10** (1.0 mg). Fraction C3-5-7 was subjected to Sephadex LH-20 CC to give one major fraction (C3-5-7-1), which was further purified by HPLC (3.0 mL/min 95% MeOH-H$_2$O, $t_R$ = 10.0 min, 11.0 min and 12.5 min, respectively) to afford **17** (1.0 mg), **18** (1.2 mg), and **19** (1.0 mg). Fraction C4 was then separated by silica gel CC eluted with petroleum ether-acetone (20:1 to 1:1) to produce six subfractions (C4-1–C4-6). Fraction C4-1 was chromatographed on an RP-C18 silica gel column to give twelve subfractions (C4-1-1–C4-1-12), and C4-1-12 was then purified by HPLC (3.0 mL/min 85% MeOH-H$_2$O, $t_R$ = 7.5 min and 10.0 min, respectively) to afford **15** (2.8 mg) and **16** (1.6 mg). Fraction C4-4 was chromatographed on an RP-C18 silica gel column to give four subfractions (C4-4-1–C4-4-4). Fraction C4-4-4 was subjected to silica gel CC eluted with CHCl$_3$-MeOH (200:1 to 10:1) to give five major eluents (C4-4-4-1–C4-4-4-5), and C4-4-4-5 was further purified by HPLC (3.0 mL/min 85% MeOH-H$_2$O, $t_R$ = 10.0 min) to afford **11** (1.8 mg). Fraction C4-6 was subjected to Sephadex LH-20 CC to give six subfractions (C4-6-1–C4-6-6). C4-6-6 was subjected to silica gel CC eluted with petroleum ether-acetone (50:1 to 1:1) to give two major eluents (C4-6-6-1–C4-6-6-2), and C4-6-6-2 was then purified by HPLC (3.0 mL/min 90% MeOH-H$_2$O, $t_R$ = 11.0 min and 17.5 min, respectively) to afford **14** (5 mg) and **13** (1 mg).

Furthermore, compounds **1–8** were separated by chiral HPLC on a Chiral MZ(2) RH column as follows: 1.0 mL/min MeCN to yield **2** (0.7 mg, $t_R$ = 1.8 min) and **1** (1.2 mg, $t_R$ = 2.4 min), 1.0 mL/min 80% MeCN-H$_2$O to afford **3** (0.2 mg, $t_R$ = 11.6 min) and **4** (0.2 mg, $t_R$ = 12.4 min), 1.0 mL/min 80% MeCN-H$_2$O to give **6** (0.3 mg, $t_R$ = 4.5 min) and **5** (0.2 mg, $t_R$ = 4.9 min), and 1.0 mL/min 80% MeCN-H$_2$O to furnish **7** (1.1 mg, $t_R$ = 6.8 min) and **8** (1.0 mg, $t_R$ = 7.3 min).

Compounds **1/2**: Colorless gum; $[\alpha]_D^{21}$ +29.1 (**1**: *c* 0.12, MeOH) and −27.1 (**2**: *c* 0.07, MeOH); UV (MeOH) $\lambda_{max}$ (log $\varepsilon$) 275 (3.2); $^1$H NMR (CD$_3$OD) see Table 1; $^{13}$C NMR (CD$_3$OD) see Table 2; (+)-ESIMS *m/z* 347.1 [M + Na]$^+$; (+)-HR-ESIMS *m/z* 325.2368 [M + H]$^+$ (calcd for C$_{19}$H$_{33}$O$_4$, 325.2373).

Compounds **3/4**: Colorless gum; $[\alpha]_D^{21}$ −10.0 (**3**: *c* 0.01, MeOH) and +11.1 (**4**: *c* 0.01, MeOH); UV (MeOH) $\lambda_{max}$ (log $\varepsilon$) 310 (3.4); $^1$H NMR (CD$_3$OD) see Table 1; $^{13}$C NMR (CD$_3$OD) see Table 2; (+)-ESIMS *m/z* 359.2 [M + Na]$^+$; (+)-HR-ESIMS *m/z* 337.2369 [M + H]$^+$ (calcd for C$_{20}$H$_{33}$O$_4$, 337.2373).

Compounds **5/6**: Colorless gum; $[\alpha]_D^{21}$ −12.6 (**5**: *c* 0.02, MeOH) and +14.1 (**6**: *c* 0.04, MeOH); UV (MeOH) $\lambda_{max}$ (log $\varepsilon$) 265 (3.5); $^1$H NMR (CDCl$_3$) see Table 1; $^{13}$C NMR (CDCl$_3$) see Table 2; (+)-ESIMS *m/z* 397.1 [M + Na]$^+$; (+)-HR-ESIMS *m/z* 397.1756 [M + Na]$^+$ (calcd for C$_{19}$H$_{32}$O$_5$Cl, 397.1752).

Compound **9**: Colorless gum; $^1$H NMR (CDCl$_3$) see Table 1; $^{13}$C NMR (CDCl$_3$) see Table 2; (+)-ESIMS *m/z* 397.1 [M + Na]$^+$; (+)-HR-ESIMS *m/z* 397.1757 [M + Na]$^+$ (calcd for C$_{19}$H$_{32}$O$_5$Cl, 397.1752).

### 3.4. Antimicrobial Assay

The antimicrobial assays were performed as we have reported earlier [25].

### 3.5. Anti-Acetylcholinesterase Assay

The anti-acetylcholinesterase assay was conducted as we have described earlier [26].

### 3.6. Anti-Inflammatory Assay

Determination of nitric oxide production. Briefly, RAW 264.7 cells were plated into 96-well plates and pretreated with a series of concentrations of compounds for 1 h before treatment with

*Molecules* **2018**, 23, 1723

1 µg/mL LPS. After 24 h incubation, detection of accumulated nitric oxide in the cell supernatants was assayed by Griess reagent kit (Beyotime Institute of Biotechnology) according to the manufacturer's instructions. Equal volumes of culture supernatant and Griess reagent were mixed, and the absorbance at 540 nm was measured using a Microplate Reader (Tecan, Switzerland).

Cell viability assay. RAW 264.7 cells were seeded into 96-well plates at $1 \times 10^4$ cells/well and allowed to attach for 24 h. The medium was replaced with 100 µL medium containing the indicated concentrations of compounds and further incubated for 24 h. 10 µL of MTT (5 mg/mL in PBS) was added into each well and the plates were incubated for 4 h at 37 °C. Supernatants were aspirated and formed formazan was dissolved in 100 µL of dimethyl sulfoxide (DMSO). The optical density (OD) was measured at an absorbance wavelength of 490 nm using a Microplate Reader (Tecan, Switzerland).

### 3.7. ECD Calculations

Conformational analysis within an energy window of 3.0 kcal/mol was performed by using the OPLS3 [27,28] molecular mechanics force field via the MacroModel [29] panel of Maestro 10.2. The conformers were then further optimized with the software package Gaussian 09 [30] at the B3LYP/6-311++G(2d,*p*) level, and the harmonic vibrational frequencies were also calculated to confirm their stability. Then, the 30 lowest electronic transitions for the obtained conformers in vacuum were calculated using TD-DFT method at the B3LYP/6-311++G(2d,*p*) level. ECD spectra of the conformers were simulated using a Gaussian function with a half-bandwidth of 0.26 eV. The overall theoretical ECD spectra were obtained according to the Boltzmann weighting of each conformer.

**Supplementary Materials:** The following materials are available online, raw spectroscopic data including chiral HPLC analyses, HR-ESIMS, and NMR ($^1$H, $^{13}$C, $^1$H-$^1$H COSY, HSQC, and HMBC) spectra for new compounds **1–6** and **9**.

**Author Contributions:** H.Z. and J.B. designed and guided the project and edited the paper. X.-Q.S. carried out the isolation, structural characterization, and the initial draft writing. K.Z. did the ECD calculations. J.-H.Y. and H.Z. participated in the spectroscopic analyses and structural characterization. Q.Z., Y.Z., F.H., Z.-Q.C., and C.-S.J. performed the biological tests. All authors read and approved the final manuscript.

**Funding:** This research was funded by Natural Science Foundation of Shandong Province [No. JQ201721], the Young Taishan Scholars Program [No. tsqn20161037] and Shandong Talents Team Cultivation Plan of University Preponderant Discipline [No. 10027].

**Acknowledgments:** We thank Jie Zhou for the identification of the plant materials.

**Conflicts of Interest:** The authors declare no conflict of interest.

## References

1.  Flora of China Editorial Committee of Chinese Academy of Sciences. *The Flora of China*; Science Press: Beijing, China, 2002; pp. 318–332.
2.  National Pharmacopoeia Committee. *Pharmacopoeia of The People's Republic of China*; Part 1; China Press of Traditional Chinese Medicine: Beijing, China, 2015; p. 69.
3.  Olennikov, D.N.; Tankhaeva, L.M.; Stolbikova, A.V.; Petrov, E.V. Phenylpropanoids and polysaccharides from *Plantago depressa* and *P. media* growing in Buryatia. *Chem. Nat. Compd.* **2011**, 47, 165–169. [CrossRef]
4.  Yu, C.-Y.; Sun, Y.-C.; Chen, G. A New Phenylethanoid glucoside from *Plantago depressa* Willd. *Nat. Prod. Res.* **2013**, 27, 609–612. [CrossRef] [PubMed]
5.  Nishibe, S.; Sasahara, M.; Jiao, Y.; Yuan, C.-L.; Tanaka, T. Phenylethanoid glycosides from *Plantago depressa*. *Phytochemistry* **1993**, 32, 975–977. [CrossRef]
6.  Yan, P.-F.; Liu, G.-Y.; Zhao, S.-M.; Song, L.-L.; Tan, L.-N.; Jin, Y.-R.; Li, X.-W. Studies on chemical constituents of *Plantago depressa* Willd. *Chin. Pharm. J.* **2009**, 44, 19–21.
7.  Zhang, Z.; Li, F.; Yuan, C.; Zheng, T.; Liu, Y. Studies on the chemical constituents of *Plantago depressa* var. *montata*. *Chin. J. Med. Chem.* **1996**, 6, 196–197.
8.  Zheng, X.-M.; Meng, F.-W.; Geng, F.; Qi, M.; Luo, C.; Yang, L.; Wang, Z.-T. Plantadeprate A, a tricyclic monoterpene zwitterionic guanidium, and related derivatives from the seeds of *Plantago depressa*. *J. Nat. Prod.* **2015**, 78, 2822–2826. [CrossRef] [PubMed]

9.   Wu, F.-H.; Liang, J.-Y.; Chen, R.; Wang, Q.-Z. Chemical constituents and hepatoprotective activity of *Plantago depressa* var. *montata* Kitag. *Chin. J. Nat. Med.* **2006**, *4*, 435–439.

10.  Smith, M.A.; Zhang, H.-X.; Purves, R.W. Identification and distribution of oxygenated fatty acids in *Plantago* seed lipids. *J. Am. Oil Chem. Soc.* **2014**, *91*, 1313–1322. [CrossRef]

11.  Amakura, Y.; Kondo, K.; Akiyama, H.; Ito, H.; Hatano, T.; Yoshida, T.; Maitani, T. Conjugated ketonic fatty acids from *Pleurocybella porrigens*. *Chem. Pharm. Bull.* **2006**, *54*, 1213–1215. [CrossRef] [PubMed]

12.  Hasegawa, T.; Ishibashi, M.; Takata, T.; Takano, F.; Ohta, T. Cytotoxic fatty acid from *Pleurocybella porrigens*. *Chem. Pharm. Bull.* **2007**, *55*, 1748–1749. [CrossRef] [PubMed]

13.  Harada, N.; Iwabuchi, J.; Yokota, Y.; Uda, H.; Nakanishi, K. A chiroptical method for determining the absolute configuration of allylic alcohols. *J. Am. Chem. Soc.* **1981**, *103*, 5590–5591. [CrossRef]

14.  Bernart, M.W.; Whatley, G.G.; Gerwick, W.H. Unprecedented oxylipins from the marine green alga *Acrosiphonia coalita*. *J. Nat. Prod.* **1993**, *56*, 245–259. [CrossRef] [PubMed]

15.  Yoshikawa, M.; Murakami, T.; Shimada, H.; Yoshizumi, S.; Saka, M.; Yamahara, J.; Matsuda, H. Medicinal foodstuffs. Xiv. on the bioactive constituents of moroheiya. (2): New fatty acids, corchorifatty acids A, B, C, D, E, and F, from the leaves of *Corchorus olitorius* L. (Tiliaceae): Structures and inhibitory effect on No production in mouse periton. *Chem. Pharm. Bull.* **1998**, *46*, 1008–1014. [CrossRef] [PubMed]

16.  Ratnayake, R.; Liu, Y.-X.; Paul, V.J.; Luesch, H. Cultivated sea lettuce is a multiorgan protector from oxidative and inflammatory stress by enhancing the endogenous antioxidant defense system. *Cancer Prev. Res.* **2013**, *6*, 989–999. [CrossRef] [PubMed]

17.  Matsumori, N.; Kaneno, D.; Murata, M.; Nakamura, H.; Tachibana, K. Stereochemical determination of acyclic structures based on carbon-proton spin-coupling constants. A method of configuration analysis for natural products. *J. Org. Chem.* **1999**, *64*, 866–876. [CrossRef] [PubMed]

18.  Wagstaff, C.; Rogers, H.J.; Thomas, B.; Feussner, I.; Griffiths, G. Characterization of a novel lipoxygenase-independent senescence mechanism in *Alstroemeria peruviana* floral tissue. *Plant Physiol.* **2002**, *130*, 273–283.

19.  Liu, H.-Y.; Liang, J.-U.; Chen, R.; Juan, Y. Study on chemical constituents of (*Hemsl.*). *Hara Herb. Chem. Ind. For. Prod.* **2008**, *28*, 8–12.

20.  Marwah, R.G.; Fatope, M.O.; Deadman, M.L.; Al-Maqbali, Y.M.; Husband, J. Musanahol: A new aureonitol-related metabolite from a *Chaetomium* sp. *Tetrahedron* **2007**, *63*, 8174–8180. [CrossRef]

21.  Ogihara, T.; Amano, N.; Mitsui, Y.; Fujino, K.; Ohta, H.; Takahashi, K.; Matsuura, H. Determination of the absolute configuration of a monoglyceride antibolting compound and isolation of related compounds from radish leaves (*Raphanus sativus*). *J. Nat. Prod.* **2017**, *80*, 872–878. [CrossRef] [PubMed]

22.  Katayama, M.; Marumo, S. (-)-*R*-Glycerol monolinolate, a minor sporogenic substance of *Sclerotinia fructicola*. *Agric. Biol. Chem.* **1978**, *42*, 1431–1433. [CrossRef]

23.  Kim, K.H.; Moon, E.; Sun, Y.K.; Kang, R.L. Antimelanogenic fatty acid derivatives from the Tuber-barks of *Colocasia antiquorum* var. *esculenta. Bull. Korean Chem. Soc.* **2010**, *31*, 2051–2053. [CrossRef]

24.  Sato, T.; Morita, I.; Shimizu, M.; Seo, S.; Watanabe, M.; Uchida, M.; Ono, M. Platelet Production Promoting Factor and Use Thereof. WO Patent WO 2008/078453 A1, 3 July 2008.

25.  Wang, P.; Yu, J.-H.; Zhu, K.-K.; Wang, Y.; Cheng, Z.-Q.; Jiang, C.-S.; Dai, J.-G.; Wu, J.; Zhang, H. Phenolic bisabolane sesquiterpenoids from a Thai mangrove endophytic fungus, *Aspergillus* sp. xy02. *Fitoterapia* **2018**, *127*, 322–327. [CrossRef] [PubMed]

26.  Li, J.-C.; Zhang, J.; Rodrigues, M.C.; Ding, D.-J.; Longo, J.P.; Azevedo, R.B.; Muehlmann, L.A.; Jiang, C.-S. Synthesis and evaluation of novel 1,2,3-triazole-based acetylcholinesterase inhibitors with neuroprotective activity. *Bioorg. Med. Chem. Lett.* **2016**, *26*, 3881–3885. [CrossRef] [PubMed]

27.  OPLS3. Schrödinger, Inc.: New York, NY, USA, 2013. Available online: https://www.schrodinger.com/opls3 (accessed on 20 June 2018).

28.  Shivakumar, D.; Harder, E.; Damm, W.; Friesner, R.A.; Sherman, W. Improving the prediction of absolute solvation free energies using the Next Generation OPLS force field. *J. Chem. Theory Comput.* **2012**, *8*, 2553–2558. [CrossRef] [PubMed]

*Molecules* **2018**, *23*, 1723

29. *MacroModel*; Schrödinger, LLC: New York, NY, USA, 2009.
30. *Gaussian 09, Revision B.01*; Gaussian, Inc.: Wallingford, CT, USA, 2010.

**Sample Availability:** Samples are available from the authors.

*molecules*

**MDPI**

*Article*

# Biosynthesis of Fluorescent β Subunits of C-Phycocyanin from *Spirulina subsalsa* in *Escherichia coli*, and Their Antioxidant Properties

Xian-Jun Wu *, Hong Yang, Yu-Ting Chen and Ping-Ping Li

College of Biology and the Environment, Nanjing Forestry University, Nanjing 210037, China; yanghong0406108@163.com (H.Y.); takeiteasy0203@163.com (Y.-T.C.); ppli@njfu.edu.cn (P.-P.L.)
* Correspondence: xjwu@njfu.edu.cn; Tel.: +86-25-8542-7210

Academic Editor: Pavel B. Drasar
Received: 29 April 2018; Accepted: 4 June 2018; Published: 6 June 2018

check for updates

**Abstract:** Phycocyanin, which covalently binds phycocyanobilin chromophores, is not only a candidate fluorescent probe for biological imaging, but also a potential antioxidative agent for healthcare. Herein, a plasmid harboring two cassettes was constructed, with *cpcB* from *Spirulina subsalsa* in one cassette and the fusion gene *cpcS::ho1::pcyA* in the other, and then expressed in *Escherichia coli*. PCB-CpcB(C-82), a fluorescent phycocyanin β subunit, was biosynthesized in *E. coli*, exhibiting an absorption maximum at 620 nm and fluorescence emission maximum at 640 nm. When *cpcS* was replaced by *cpcT*, PCB-CpcB(C-153), another fluorescent phycocyanin β subunit, was produced, exhibiting an absorption maximum at 590 nm and fluorescence emission maximum at 620 nm. These two fluorescent biliproteins showed stronger scavenging activity toward hydroxyl and DPPH free radicals than apo-CpcB. The $IC_{50}$ values for hydroxyl radical scavenging by PCB-CpcB(C-82), PCB-CpcB(C-153), and apo-CpcB were $38.72 \pm 2.48$ μg/mL, $51.06 \pm 6.74$ μg/mL, and $81.82 \pm 0.67$ μg/mL, respectively, and the values for DPPH radical scavenging were $201.00 \pm 5.86$ μg/mL, $240.34 \pm 4.03$ μg/mL, and $352.93 \pm 26.30$ μg/mL, respectively. The comparative antioxidant capacities of the proteins were PCB-CpcB(C-82) > PCB-CpcB(C-153) > apo-CpcB, due to bilin binding. The two fluorescent biliproteins exhibited a significant effect on relieving the growth of *E. coli* cells injured by $H_2O_2$. The results of this study suggest that the fluorescent phycocyanin β subunits of *S. subsalsa* were reconstructed by one expression vector in *E. coli*, and could be developed as potential antioxidants.

**Keywords:** phycocyanin; biosynthesis; antioxidant; *Spirulina*; gene expression; apo-CpcB

## 1. Introduction

Phycobiliproteins are a type of light-harvesting biliprotein involved in photosynthesis in cyanobacteria and some eukaryotic alga [1–3]. Phycobiliproteins consist of at least two dissimilar types of peptides, with linear tetrapyrrole prosthetic groups covalently attached to the apoprotein via cysteine thioether linkages [4,5]. Phycobiliproteins exhibit unique absorbance and fluorescence properties and can be classified into three types, based on their absorbance maxima: phycoerythrin (PE, $\lambda_{max}$: 540–570 nm), phycocyanin (PC, $\lambda_{max}$: 610–620 nm), and allophycocyanin (APC, $\lambda_{max}$: 650–655 nm) [6]. C-phycocyanin (CPC), one of the major phycobiliproteins, is composed of α and β subunits in the form of $(\alpha\beta)_3$ or $(\alpha\beta)_6$ aggregates. The α or β subunits of CPC can be assembled by apoproteins and chromophore groups in vivo or in vitro, with catalysis by the chromophore lyase. First, ferredoxin-dependent heme oxygenase (HO1) catalyzes the oxidative cleavage of the host heme to generate biliverdin [7], which is then reduced to phycocyanobilin (PCB) by phycocyanobilin:ferredoxin oxidoreductase (PcyA) [8]. Then, PCB is covalently attached to the apoprotein, which is catalyzed by specific lyases to generate fluorescent CPC. The lyases CpcE and CpcF are responsible for the attachment of PCB to the CPC α subunit [9]. The lyase CpcS catalyzes the attachment of PCB to the chromophore binding site Cys-β84 of CPC, while CpcT is able to bind PCB to Cys-β155 [10]. There are four genes that are considered to be essential for the formation of fluorescent CPC β subunits in *Escherichia coli*: *cpcB*, encoding apo-β-CPC (CpcB); *cpcS* or *cpcT*, encoding CpcS or CpcT, respectively; and *ho1* and *pcyA*, encoding HO1 and PcyA, respectively. Recently, a modular approach that requires only two DNA segments was reported [11]. The first segment is the *cpcB* gene from *Mastigocladus laminosus*, and the second is the fusion gene *cpcS::ho1::pcyA* or *cpcT::ho1::pcyA*, which was constructed as an end-to-end fusion of the respective genes from *Nostoc* sp. PCC 7120. The new synthetic pathway, despite requiring two plasmids, facilitates heterologous production of phycobiliproteins and the use of phycobiliproteins as fluorescent protein probes.

Spirulina is a genus of planktonic blue–green algae found in the alkaline water of volcanic lakes. Several in-vitro experiments have shown that *Spirulina* and its extracts, especially CPC, exhibit a variety of pharmacological effects, such as neuroprotective effects [12], hepatoprotective abilities [13], renoprotective effects [14], cardiovascular protective effects [15], and the elimination of cataracts [16], which is generally attributed to the antioxidant and free radical scavenging properties of CPC. Several studies have shown that the antioxidant activity is related to not only the bilin chromophore but also the apoprotein [17–19]. Some studies have been previously conducted to clone and express the relevant genes for producing α-CPC (apo-α-CPC and fluorescent α-CPC) in *E. coli* and evaluate their antioxidant activity [20,21]. Recently, recombinant apo-β-CPC has been shown to possess higher free radical scavenging activity than apo-α-CPC [22], and to be able to protect red blood cells from antioxidative damage [19]. The β-CPC subunits are considered to be highly effective antioxidants for pharmaceutical applications. However, to our knowledge, no report has been published on the antioxidant properties of recombinant fluorescent β-CPC.

In this study, the *cpcB* gene was cloned from *Spirulina subsalsa*. A modular approach with only one plasmid was employed to co-express the *cpcB* gene, as well as the fusion genes *cpcS::ho1::pcyA* or *cpcT::ho1::pcyA*, in *E. coli*, where the fluorescent biliproteins PCB-CpcB(C-82) and PCB-CpcB(C-153) were biosynthesized. Recombinant β-CPCs from *S. subsalsa* with a His tag were purified by Ni$^{2+}$-chelated affinity chromatography, and showed similar spectral characteristics to previously reported fluorescent β-CPC submits [10]. Furthermore, the antioxidant activity of PCB-CpcB(C-82) and PCB-CpcB(C-153) was evaluated and compared with that of apo-CpcB. The result suggested that PCB-CpcB(C-82) and PCB-CpcB(C-153) hold great potential to be used as therapeutic agents for healthcare.

## 2. Results

### 2.1. Cloning of the cpcB Gene and Plasmid Construction

Using designed primers, the 519-bp *cpcB* gene was amplified from *S. subsalsa* genomic DNA. A recombinant plasmid carrying the *cpcB* gene, namely, pETDuet-cpcB, was successfully constructed. Restriction enzyme digestion and DNA sequencing results showed that the *cpcB* gene was correctly inserted into the first multiple cloning sites of pETDuet, and fused in frame with the short hexahistidine tag sequence at the 5′ end. The final expression vectors, namely, pETDuet-cpcB-cpcS::*ho1::pcyA* and pETDuet-cpcB-cpcT::*ho1::pcyA*, were constructed by our methods. These vectors harbored pBR322-derived ColE1 replicons and *lacI* operator sequences for tight control of expression. Two cassettes of the final expression vectors carried *cpcB* and the genes encoding the fused enzymes under the control of separate T7 promoters (Figure 1). The *cpcB* gene followed by a His tag was in the first cassette for further purification. Upon analysis, the vectors exhibited the correct restriction enzyme patterns. Sequencing data showed the correct open reading frames in the correct orientations for *cpcB*, *cpcS*, *cpcT*, *ho1* and *pcyA* in the vector pETDuet.

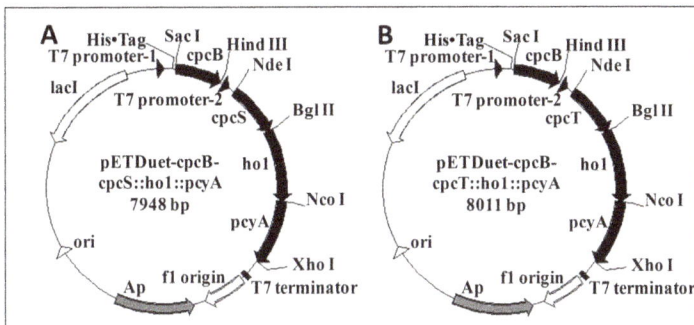

**Figure 1.** Schematic representation of the expression vectors pETDuet-cpcB-cpcS::*ho1*::*pcyA* (**A**) and pETDuet-cpcB-cpcT::*ho1*::*pcyA* (**B**).

## 2.2. Expression and Purification

After induction with IPTG, the transformants harboring pETDuet-cpcB-cpcS::*ho1*::*pcyA* and pETDuet-cpcB-cpcT::*ho1*::*pcyA* exhibited a distinct blue-green color (Figure 2a,b), which implied that both PCB-CpcB(C-82) and PCB-CpcB(C-153) might be synthesized. The recombinant *E. coli* harboring pETDuet-cpcB as a control did not exhibit any color, due to a lack of phycobilin and lyases. Visualization of the three purified proteins with a purity ratio of 0.45 for PCB-CpcB(C-82) and 0.37 for PCB-CpcB(C-153) (Table 1) on a Coomassie blue-stained SDS-PAGE gel showed only one distinct band of 20 kDa, corresponding to the calculated molecular mass of apo-CpcB plus a His tag (Figure 2c). Covalent binding of the chromophore was confirmed upon exposure to $Zn^{2+}$ and UV illumination, which resulted in a fluorescent emission (Figure 2d). The results confirmed the production of apo-CpcB and the chromophorylated derivatives of apo-CpcB in *E. coli*.

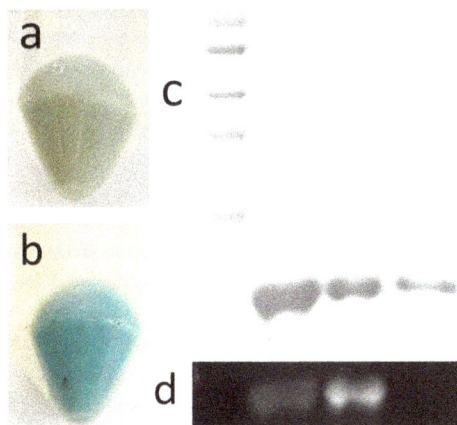

**Figure 2.** Color of the cell pellets of transformants harboring pETDuet-cpcB-cpcT::*ho1*::*pcyA* (**a**) or pETDuet-cpcB-cpcS::*ho1*::*pcyA* (**b**), and analysis of purified recombinant proteins by gradient SDS-PAGE (**c**) and chromoprotein $Zn^{2+}$ electrophoresis (**d**). (**c**) Lane 1 shows standard protein makers—the protein bands correspond to 75, 60, 45, 35, 25, 20 and 15 kDa (from top to bottom); lane 2 shows PCB-CpcB(C-153); lane 3 shows PCB-CpcB(C-82); lane 4 shows apo-CpcB. (**d**) The chromoprotein $Zn^{2+}$ electrophoresis results corresponding to lanes 1–4.

**Table 1.** IC$_{50}$ values representing the hydroxyl and DPPH free radical scavenging capacities of various proteins.[a]

| Recombinant Proteins | Absorption | Fluorescence | Chromophore-Binding Rate (A$_{max}$/A$_{280}$) | IC$_{50}$ Values (µg/mL) | |
|---|---|---|---|---|---|
| | | | | Hydroxyl Radical Scavenging Activity | DPPH Free Radical Scavenging Activity |
| apo-CpcB | 0 | 0 | 0 | 81.82 ± 0.67 | 352.93 ± 26.30 |
| PCB-CpcB(C-82) | 621 | 646 | 0.45 | 38.72 ± 2.48 | 201.00 ± 5.86 |
| PCB-CpcB(C-153) | 602 | 629 | 0.37 | 51.06 ± 6.74 | 240.34 ± 4.03 |
| Ascorbic acid | – | – | – | – | 49.91 ± 0.32 |
| Mannitol | – | – | – | 245.72 ± 9.43 | – |

[a] The values are presented as the mean ± SEM.

## 2.3. Absorbance and Fluorescence Spectrometry

Recombinant biliproteins were further analyzed for spectral characteristics of absorption and emission. The results showed that both biliproteins exhibited unique absorption and fluorescence spectra in the visible region (Figure 3). PCB-CpcB(C-82) had an absorption maximum at approximately 621 nm and an emission maximum at approximately 646 nm, whereas PCB-CpcB(C-153) exhibited an absorption maximum at approximately 602 nm, with an emission maximum at approximately 629 nm (Table 1). The absorption and emission spectra of both biliproteins were similar to those of previously reported phycocyanins [10]. The recombinant proteins also exhibited a characteristic absorption peak at approximately 280 nm in the UV region, corresponding to the total protein concentration. The ratio of the absorption maxima of biliproteins (A$_{\lambda max}$) to the absorption at 280 nm (A$_{280}$) can reflect the approximate bilin-binding rate. The results showed that the bilin-binding rate of PCB-CpcB(C-82) (A$_{621}$/A$_{280}$) was 0.45, which was slightly higher than that of PCB-CpcB(C-153) (A$_{602}$/A$_{280}$ = 0.37) (Table 1), indicating that PCB-CpcB(C-82) may have higher antioxidant activity than PCB-CpcB(C-153), due to high bilin content.

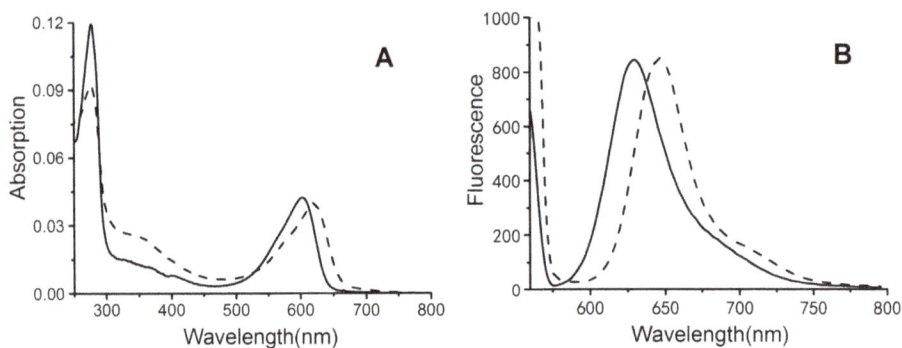

**Figure 3.** Absorption (**A**) and fluorescence (**B**) of the purified recombinant biliproteins PCB-CpcB(C-82) (dashed line) and PCB-CpcB(C-153) (solid line).

## 2.4. Antioxidant Activity of Fluorescent Phycocyanin

Different concentrations of recombinant proteins were examined for antioxidant abilities by two different antioxidant assays, including assays of hydroxyl radical and DPPH radical scavenging activities. A deoxyribose assay was conducted to study the hydroxyl radical scavenging activity of apo-CpcB, PCB-CpcB(C-82), and PCB-CpcB(C-153), plus that of mannitol as a control. The result showed that all recombinant phycocyanins exhibited distinct hydroxyl radical scavenging activity, and the scavenging rates increased with increasing protein concentration (Figure 4). The recombinant phycocyanins exhibited higher hydroxyl radical scavenging activity than mannitol (Figure 4A). Furthermore, the scavenging activity of PCB-CpcB(C-82), with an IC$_{50}$ value of 38.72 ± 2.48 µg/mL, was higher than that of PCB-CpcB(C-153), which has an IC$_{50}$ value of 51.06 ± 6.74 µg/mL (Table 1).

Apo-CpcB exhibited the lowest activity of the three recombinant proteins, with an $IC_{50}$ value of $81.82 \pm 0.67$ µg/mL, but higher activity than previously reported apo-CpcB from *Spirulina platensis* [22].

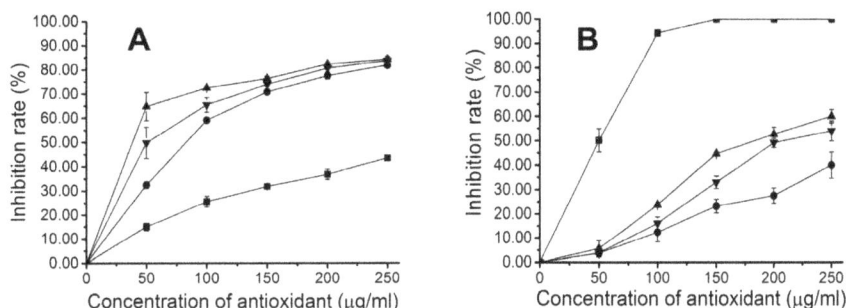

**Figure 4.** (**A**) Hydroxyl and (**B**) DPPH free radical scavenging activities of recombinant biliproteins; (▲) PCB-CpcB(C-82); (▼) PCB-CpcB(C-153); (●) apo-CpcB; (■) Mannitol (**A**) or ascorbic acid (**B**) (positive control).

The stable radical DPPH was used for determination of the antioxidant activity of apo-CpcB, PCB-CpcB(C-82), and PCB-CpcB(C-153), as well as that of ascorbic acid (Vc) as a control. The results showed that recombinant phycocyanins had DPPH free radical scavenging activity, and this activity increased with increasing protein concentration. However, the recombinant proteins exhibited significantly lower DPPH radical scavenging activity than ascorbic acid (Figure 4B). The $IC_{50}$ values of PCB-CpcB(C-82), PCB-CpcB(C-153), and apo-CpcB were $201.00 \pm 5.86$ µg/mL, $240.34 \pm 4.03$ µg/mL, and $352.93 \pm 26.30$ µg/mL, respectively, which were 4~7 times higher than the $IC_{50}$ value of ascorbic acid, which was $49.91 \pm 0.32$ µg/mL (Table 1).

The above results indicated that the three recombinant phycocyanins exhibited distinct scavenging capacities for both hydroxyl and DPPH radicals. The scavenging activity decreased in the following order: PCB-CpcB(C-82) > PCB-CpcB(C-153) > apo-CpcB. The antioxidant activity order of these phycocyanins significantly correlated with the rate of binding of bilin to the phycocyanin. Bilin and the apoprotein together contribute to the antioxidant properties of phycocyanin.

To detect the inhibition of the cellular oxidative injury by the recombinant biliproteins, the purified PCB-CpcB(C-82), PCB-CpcB(C-153), and apo-CpcB were respectively added into an *E. coli* culture system containing 3 and 6 mM $H_2O_2$ at a final concentration of 20 µg/mL. The results are shown in Figure 5. The addition of $H_2O_2$ made the growth of *E. coli* cells become slower compared to the control without $H_2O_2$, implying that *E. coli* cells were injured. The addition of the recombinant biliproteins significantly improves the survival of *E. coli* cells whether the concentration of $H_2O_2$ was 3 (Figure 5A) or 6 mM (Figure 5B). The inhibitory effect of the three recombinant phycocyanins on the oxidative injury of *E. coli* cells decreased in the following order: PCB-CpcB(C-82) > PCB-CpcB(C-153) > apo-CpcB. This is consistent with the effects of the hydroxyl and DPPH radical scavenging activities.

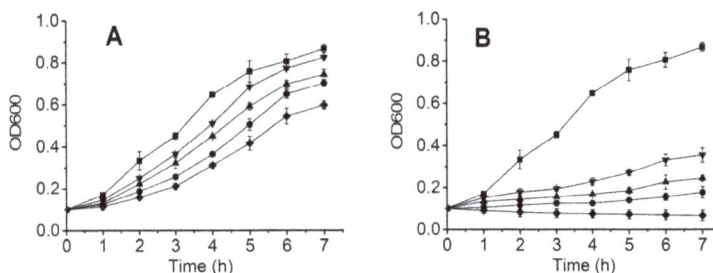

**Figure 5.** Inhibition of the cellular oxidative injury by the recombinant biliproteins (20 µg/mL). The samples were treated with 3 mM $H_2O_2$ (**A**) and 6 mM $H_2O_2$ (**B**); (▼) PCB-CpcB(C-82); (▲) PCB-CpcB(C-153); (●) apo-CpcB; (■) the control without $H_2O_2$ and the recombinant biliproteins; (♦) the control with $H_2O_2$, but without the recombinant biliproteins.

## 3. Discussion

Phycocyanins (PCs) are versatile biliproteins consisting of α and β subunits, and include fluorescent dyes [23], color additives [24], and antioxidant drugs [25]. Recombinant phycocyanins have received much attention due to their instability, the time-consuming purification process, the multimeric states of native phycocyanin, and especially because recombinant phycocyanins could potentially be used as genetically-encoded red and near-infrared fluorescent probes [26]. However, the multiple transformations involved in the process of phycocyanin biosynthesis are tedious and unpredictable. Wu et al. previously proposed a two-plasmid expression system with only two DNA segments to overcome this obstacle [11]. In this study, only one plasmid was needed to produce recombinant β subunits of C-phycocyanin, which emitted strong red fluorescence similar to that previously reported [11], further simplifying the biosynthesis of phycocyanin. Culture growth included only one resistant marker, which increased the stability of the transformants and reduced costs and antibiotic pollution in industrial applications. Guan et al. previously reported a different one-plasmid method, which employed a typical ribosomal binding site to construct an expression vector containing five genes [20]. Both one-plasmid methods could improve the rate of multigene transformation, which would be very important for the biosynthesis of phycocyanin in certain bacterial, fungal, plant, and animal cells.

Phycocyanin from *Spirulina* has attracted much attention because of the antioxidant properties and pharmacological effects of *Spirulina*. Some recombinant holo- and apophycocyanin α subunits have been reported to have higher antioxidant activities than the native phycocyanin [18,20]. The β subunits, which contain two bilin binding sites, are more complex, and might be more promising antioxidative agents than the α subunits [19]. To date, the antioxidant properties of recombinant fluorescent phycocyanin β subunits have not been reported, although those of recombinant apo-phycocyanin β subunits have been confirmed [22]. In this study, the antioxidant capacity of two recombinant fluorescent phycocyanin β subunits of *S. subsalsa* was assessed by radical scavenging assays, using hydroxyl and DPPH radicals, and by the inhibition effect of cellular oxidative injury. Our results indicated that the fluorescent phycocyanin β subunits had significantly better antioxidant potential than the apoprotein. The scavenging activity of the fluorescent phycocyanin C82-β subunit toward hydroxyl and DPPH radicals was 2.5 and 1.7 times higher than that of the apoprotein, respectively, and the scavenging activity of the fluorescent phycocyanin C153-β subunit was 1.5 and 1.4 times higher than that of the apoprotein, respectively. Meanwhile, the inhibition of the *E. coli* cells growth caused by $H_2O_2$ can be restored to some extent by the fluorescent phycocyanins β subunits, indicating that they have a good protective effect on the growth of *E. coli* cells injured by $H_2O_2$. As antioxidants, the fluorescent phycocyanin β subunits have a strong prospects for application in the food and medical industries. Moreover, Cherdkiatikul and Suwanwong reported that for both allophycocyanin and c-phycocyanin of *Spirulina*, the apo-β subunit possessed higher scavenging activity toward hydroxyl and peroxyl radicals than the apo-α subunit [22]. It is believed that the fluorescent phycocyanin β subunits have better antioxidant properties than the corresponding α subunits, but further confirmation is needed considering bilin content. In additional, more reactive oxygen or nitrogen species, such as superoxide anion, peroxynitrite, peroxyl radical, singlet oxygen, hydrogen peroxide, and hypochlorous acid [27] should be further considered for better assessment of the scavenging properties of the fluorescent phycocyanin β subunits.

The apoprotein of phycocyanin exhibited potent antioxidant activity, but bilin remained the major factor that influenced the antioxidative capability of phycocyanin. Our studies suggest that antioxidant activity increases significantly with chromophorylation of CpcB. Compared with the 153β subunit, the 82β subunit has a high chromophore binding rate ($A_{max}/A_{280}$ = 0.45 vs. 0.37), which contributes to the strong antioxidant properties of the 82β subunit ($IC_{50}$ = 38.72 vs. 51.06 μg/mL or 201 vs. 240.34 μg/mL). Moreover, the two fluorescent β subunits also showed stronger inhibition of oxidative injury to *E. coli* cells than the apo-protein. It is speculated that the differences in the bilin binding rate among recombinant phycocyanin β subunits are responsible for the differences in antioxidant activities, but the antioxidant capabilities may also be associated with exposure levels of the chromophore embedded in the binding pocket. Nevertheless, it is believed that improved chromophore binding rates could significantly enhance the antioxidant capability of phycocyanin. The rate of bilin binding to the apoprotein is associated with various factors, such as expression and culture conditions, host strains and expression vectors, and characteristics of lyases and apoproteins [28,29]. Furthermore, in addition to the differences in amino acid sequences, homologs of the α or β subunits may exhibit distinct bilin-binding abilities. Therefore, it is necessary to assess the antioxidant capabilities of homologous α or β subunits of phycocyanin from different cyanobacteria, which, in combination with genetic modification, could further improve the antioxidant activity of phycocyanin.

To increase the bilin content of the recombinant phycocyanin β subunit, we attempted to simultaneously attach phycocyanobilin to two bilin binding sites of the β subunit in *E. coli*, in the presence of the lyases CpcS and CpcT. The fluorescent 82β subunit was obtained instead of the expected holo-β subunit, which may be due to the relatively low kinetic constants of CpcT compared with those of CpcS [30]. Future studies should be directed toward synthetizing the phycocyanin holo-β subunit in vivo or in vitro, and determining the antioxidative potential of this subunit. Notably, the fluorescent phycocyanin holo-β subunit should be an excellent candidate for antioxidant therapy, due to the increased bilin-binding capacity of this subunit.

## 4. Materials and Methods

### 4.1. Cyanobacterial Cultivation and DNA Extraction

Living alga *S. subsalsa* FACHB351 was obtained from the Freshwater Algae Culture Collection at the Institute of Hydrobiology, Chinese Academy of Sciences. The alga was cultured in SP medium at 25 °C. The light intensity was 50 μmol·m$^{-2}$·s$^{-1}$, with a 12 h light/12 h dark photoperiod cycle. The filaments were collected by centrifugation (5000 rpm) at the log phase. The *Spirulina* genomic DNA was extracted using TRIzol reagent (Invitrogen, Carlsbad, CA, United States) according to the manufacturer's instructions.

### 4.2. Primer Design and PCR Amplification

The *cpcB* gene was amplified from *S. subsalsa* FACHB351 genomic DNA by polymerase chain reaction (PCR). The sequence for the primer design was obtained from the GenBank nucleotide sequence database (accession no. AY244667). The primers used to amplify the *cpcB* gene were as follows: forward primer, 5'-CAC GAG CTC TAT GTT TGA CGC ATT TAC AAG GGT TG-3'; reverse primer, 5'-TAT AAG CTT TTA GGC AAC AGC AGC AGC G-3'. These primers were synthesized by Genscript. The resulting product was digested with the restriction enzymes Sac I and Hind III, and inserted into a Sac I- and Hind III-digested pUC57 cloning vector (Thermo Scientific®, San Jose, CA, USA), to generate the plasmid pUC57-cpcB, which was then sequenced to verify the integrity of cpcB.

## 4.3. Construction of the Expression Vector and Transformation

The *cpcB* gene of Spirulina was subcloned from pUC57-*cpcB* into the first expression cassette of an expression vector, namely, pETDuet (Novagen), yielding the plasmid pETDuet-*cpcB*. The plasmids co-expressing the phycobiliprotein lyase, heme oxygenase, and ferredoxin oxidoreductase (namely, pET30a-cpcS::*ho1*::*pcyA* and pET30a-cpcT::*ho1*::*pcyA*) were constructed as described by Wu et al. [11]. The constructs *cpcS*::*ho1*::*pcyA* and *cpcT*::*ho1*::*pcyA* were digested with Nde I and Xho I and individually inserted into the second expression cassette of the plasmid pETDuet-*cpcB*, yielding two expression plasmids, namely, pETDuet-cpcB-cpcS::*ho1*::*pcyA* and pETDuet-cpcB-cpcT::*ho1*::*pcyA*. The *cpcB* gene and the fused gene were located in two different expression cassettes of pETDuet. Each cassette contained a T7 promoter followed by a ribosome-binding site. The two expression plasmids were confirmed by restriction enzyme digestion analysis and DNA sequencing, then the plasmids were individually transformed into *E. coli* BL21 (DE3) cells, according to standard procedures. The plasmid pETDuet-*cpcB*, as a control, was also transformed into *E. coli* BL21 (DE3) cells. Transformants were selected with ampicillin.

## 4.4. Protein Expression and Purification

Transformed *E. coli* containing the constructed plasmids were cultured at 37 °C in Luria-Bertani (LB) medium, supplemented with ampicillin (100 µg/mL), until the optical density at 600 nm (OD$_{600}$) was 0.5. Production of fluorescent biliprotein was induced by the addition of 1 mM isopropyl-β-D-thiogalactoside (IPTG). Cells were incubated with shaking at 150 rpm at 18 °C for 12 h in the dark. The cells were centrifuged at 9200× *g* for 5 min at 4 °C. Cell pellets were rinsed twice with distilled water and then re-suspended in ice-cold potassium phosphate buffer (KPB; 20 mM, pH 7.4) containing 0.5 M NaCl, and sonicated for 4 min at 200 W (JY92, SCIENTZ Biotechnology, Ningbo, China). The suspension was centrifuged via Ni$^{2+}$-affinity chromatography on chelating Sepharose (Amersham Biosciences, Uppsala, Sweden) equilibrated with KPB containing 0.5 M NaCl. The proteins remaining on the column were eluted with the saline KPB, which also contained imidazole (0.5 M). After collection, the protein sample was dialyzed against the saline KPB overnight at 4 °C in the dark. The protein concentration was measured using the BCA Protein Assay Reagent Kit (Thermo Scientific, Waltham, MA, USA). The purified proteins were stored at −20 °C until further use.

## 4.5. Electrophoretic Analysis

The protein samples were analyzed by polyacrylamide gel electrophoresis in the presence of SDS. The SDS-PAGE gel was composed of a 10% separation gel and 5% stacking gel. The samples were boiled with 2× SDS sample buffer containing 30 mM β-mercaptoethanol for 5 min. The purified proteins were visualized by staining with Coomassie blue, and bilins were identified by Zn$^{2+}$-induced fluorescence.

## 4.6. Spectral Analyses

Absorbance spectra were obtained on a Perkin Elmer Lambda 365 spectrophotometer. Fluorescence spectra were recorded at room temperature with a model LS 55 spectrofluorimeter (Perkin Elmer, Waltham, MA, USA). Excitation and emission slits were set at 10 nm for all measurements, with a scanning speed of 1200 nm/min.

## 4.7. Determination of Hydroxyl Radical Scavenging Activity

The hydroxyl radical scavenging ability of fluorescent phycocyanin was estimated by the deoxyribose assay. This method was performed as previously described by Bermejo et al. [31] and Cherdkiatikul and Suwanwong [22], with minor modifications. The reaction mixture contained 20 mM KPB (pH 7.4), 2.8 mM deoxyribose, 100 µM FeCl$_3$, 100 µM EDTA, 1 mM H$_2$O$_2$, 100 µM ascorbic acid, and the test sample (50, 100, 150, 200, or 250 µg/mL). Solutions of FeCl$_3$ and ascorbic acid were prepared immediately before use in de-aerated water and were mixed prior to addition into the reaction mixture. The reaction mixture (total of 500 μL) was incubated for 30 min at 37 °C. After incubation, the color was developed by adding 250 μL of thiobarbituric acid (1% $w/v$ in 50 mM KOH) and 250 μL of trichloroacetic acid (2.8% $w/v$ in deionized water), and heating the mixture in a boiling water bath for 15 min. The sample was then allowed to cool and diluted twofold with 20 mM KPB (pH 7.4). The absorbance was measured at 532 nm. KPB was the negative control. Mannitol, a classic hydroxyl radical scavenger, was used as a positive control. The hydroxyl radical scavenging activity was calculated using the following formula:

$$\text{Hydroxyl radical scavenging activity (\%)} = [(A_0 - A_1)/A_0] \times 100$$

where $A_0$ is the absorbance of the negative control, and $A_1$ is the absorbance of the sample or the positive control. The $IC_{50}$ (defined as the sample concentration at which 50% of the hydroxyl radicals were scavenged) was calculated for each sample.

*4.8. Determination of DPPH Free Radical Scavenging Activity*

The DPPH radical scavenging activity of fluorescent phycocyanin was tested according to the method of Xu et al. [32], with some modifications. The purified recombinant protein was diluted with KPB to 50, 100, 150, 200, and 250 μg/mL. Two milliliters of each dilution was added to a series of cuvettes following the addition of 500 μL of and ethanolic solution of DPPH (0.02%) and 1 mL of ethanol. The mixtures were gently mixed and incubated for 1 h at room temperature. The absorbance at 517 nm was measured after incubation. KPB was the negative control, and ascorbic acid was used as a positive control. DPPH radical scavenging activities were calculated by using the following formula:

$$\text{DPPH radical scavenging activity (\%)} = [(A_0 - A_1 + A_2)/A_0] \times 100$$

where $A_0$ is the absorbance of the negative control; $A_1$ is the absorbance of the test sample and the positive control; and $A_2$ is the absorbance of the blank, without DPPH.

*4.9. Inhibition of the Cellular Oxidative Injury by the Recombinant Biliproteins*

Inhibition of the cellular oxidative injury by the recombinant biliproteins was performed as previously described by Yu et al. [21], with some modifications. *E. coli* BL21 (DE3) were grown overnight at 37 °C in 5 mL LB media. Aliquots of *E. coli* broths were separately transferred into five tubes pre-equipped with a 5 ml LB medium until OD600 = 0.1. The recombinant biliproteins were respectively added to the above three tubes at a final concentration of 20 μg/mL. A solution of 30% $H_2O_2$ was added to the above tubes, at a final concentration of 3 or 6 mM. The *E. coli* BL21 (DE3) cells without $H_2O_2$ and with $H_2O_2$ but without the recombinant phycocyanins were used as the control group. The *E. coli* BL21 (DE3) cells were cultured at 100 rpm for 7 h at 37 °C, and then the absorbance at 600 nm was determined every other hour. The growth curves of *E. coli* BL21 (DE3) cells were plotted to determine the inhibition of E. coli cells oxidative injury.

*4.10. Statistics and Data Processing*

All experiments were repeated at least three times. Data are presented as the mean ± SEM. The difference was considered statistically significant when $p < 0.05$. Statistical analysis was performed with SPSS (SPSS 17 for Windows, SPSS Inc., Chicago, IL, USA), using the $IC_{50}$ values that were calculated.

## 5. Conclusions

We have cloned the *cpcB* gene encoding the phycocyanin β subunit from *S. subsalsa* FACHB351 and achieved the biosynthesis of two fluorescent phycocyanin β subunits, namely, PCB-CpcB(C-82) and PCB-CpcB(C-153), in *E. coli* with a one-plasmid method. The recombinant biliproteins exhibited significantly stronger scavenging activity toward hydroxyl and DPPH free radicals, and a stronger inhibitory effect on oxidative injury to *E. coli* cells than the recombinant apo-CpcB, due to covalent binding of the bilin chromophore to the apoprotein. The recombinant fluorescent phycocyanin β subunits possess high antioxidant capacities, and have potential applications in the food and pharmaceutical industries.

**Author Contributions:** X.-J.W. conceived and designed the experiments; X.-J.W., H.Y. and Y.-T.C. performed the experiments; X.-J.W. analyzed the data; P.-P.L. contributed reagents, materials, and analysis tools; X.-J.W. wrote the manuscript.

**Acknowledgments:** This work was supported by the Natural Science Foundation of Jiangsu Province, China (No. BK20150884); the China Postdoctoral Science Foundation (No. 14ZR1401500); and the Priority Academic Program Development of Jiangsu Higher Education Institutions (PAPD).

**Conflicts of Interest:** The authors declare that there is no conflict of interest.

## References

1. Glazer, A.N. Adaptive variations in phycobilisome structure. In *Advances in Molecular & Cell Biology*; Barber, J., Ed.; JAI Press: London, UK, 1994; Volume 10, pp. 119–149.
2. Gantt, B.; Grabowski, B.; Cunningham, F.X. Antenna systems of red algae: Phycobilisomes with photosystem II and chlorophyll complexes with photosystem I. In *Light-Harvesting Antennas in Photosynthesis*; Green, B., Parson, W., Eds.; Kluwer: Dordrecht, The Netherlands, 2003; pp. 307–322.
3. Grossman, A.R.; Schaefer, M.R.; Chiang, G.G.; Collier, J.L. The phycobilisome, a light-harvesting complex responsive to environmental conditions. *Microbiol. Rev.* **1993**, *57*, 725–749. [PubMed]
4. Sidler, W.A. *The Molecular Biology of Cyanobacteria*; Springer Netherlands: Dordrecht, The Netherlands, 1994; pp. 139–216.
5. Gao, X.; Wei, T.D.; Zhang, N.; Xie, B.B.; Su, H.N.; Zhang, X.Y.; Chen, X.L.; Zhou, B.C.; Wang, Z.X.; Wu, J.W.; et al. Molecular insights into the terminal energy acceptor in cyanobacterial phycobilisome. *Mol. Microbiol.* **2012**, *85*, 907–915. [CrossRef] [PubMed]
6. Glazer, A.N. Phycobiliproteins. In *Chemicals from Microalgae*; Cohen, Z., Ed.; Taylor and Francis Ltd.: Oxford, UK, 1999; pp. 262–280.
7. Beale, S.I. Biosynthesis of phycobilins. *Chem. Rev.* **1993**, *93*, 785–802. [CrossRef]
8. Frankenberg, N.; Lagarias, J.C. Biosynthesis and Biological Functions of Bilins. In *The Porphyrin Handbook*; Kadish, K.M., Smith, K.M., Guilard, R., Eds.; Academic Press: Amsterdam, The Netherlands, 2003; pp. 211–236.
9. Fairchild, C.D.; Zhao, J.; Zhou, J.; Colson, S.E.; Bryant, D.A.; Glazer, A.N. Phycocyanin alpha-subunit phycocyanobilin lyase. *Proc. Natl. Acad. Sci. USA* **1992**, *89*, 7017–7021. [CrossRef] [PubMed]
10. Zhao, K.H.; Su, P.; Tu, J.M.; Wang, X.; Liu, H.; Plöscher, M.; Eichacker, L.; Yang, B.; Zhou, M.; Scheer, H. Phycobilin: Cystein-84 biliprotein lyase, a near-universal lyase for cysteine-84-binding sites in cyanobacterial phycobiliproteins. *Proc. Natl. Acad. Sci. USA* **2007**, *104*, 14300–14305. [CrossRef] [PubMed]
11. Wu, X.J.; Chang, K.; Luo, J.; Zhou, M.; Scheer, H.; Zhao, K.H. Modular Generation of Fluorescent Phycobiliproteins. *Photochem. Photobiol. Sci.* **2013**, *12*, 1036–1040. [CrossRef] [PubMed]
12. Penton-Rol, G.; Marin-Prida, J.; Pardo-Andreu, G.; Martinez-Sanchez, G.; Acosta-Medina, E.F.; Valdivia-Acosta, A.; Lagumersindez-Denis, N.; Rodriguez-Jimenez, E.; Llopiz-Arzuaga, A.; Lopez-Saura, P.A.; et al. C-Phycocyanin is neuroprotective against global cerebral ischemia/reperfusion injury in gerbils. *Brain Res. Bull.* **2011**, *86*, 42–52. [CrossRef] [PubMed]
13. Vadiraja, B.B.; Gaikwad, N.W.; Madyastha, K.M. Hepatoprotective effect of C-phycocyanin: Protection for carbon tetrachloride and R-(+)-pulegone-mediated hepatotoxicty in rats. *Biochem. Biophys. Res. Commun.* **1998**, *249*, 428–431. [CrossRef] [PubMed]
14. Fernandez-Rojas, B.; Medina-Campos, O.N.; Hernandez-Pando, R.; Negrette-Guzman, M.; Huerta-Yepez, S.; Pedraza-Chaverri, J. C-phycocyanin prevents cisplatin-induced nephrotoxicity through inhibition of oxidative stress. *Food Funct.* **2014**, *5*, 480–490. [CrossRef] [PubMed]
15. Riss, J.; Decorde, K.; Sutra, T.; Delage, M.; Baccou, J.C.; Jouy, N.; Brune, J.P.; Oreal, H.; Cristol, J.P.; Rouanet, J.M. Phycobiliprotein C-phycocyanin from Spirulina platensis is powerfully responsible for reducing oxidative stress and NADPH oxidase expression induced by an atherogenic diet in hamsters. *J. Agric. Food Chem.* **2007**, *55*, 7962–7967. [CrossRef] [PubMed]
16. Kumari, R.P.; Sivakumar, J.; Thankappan, B.; Anbarasu, K. C-phycocyanin modulates selenite-induced cataractogenesis in rats. *Biol. Trace Elem. Res.* **2013**, *151*, 59–67. [CrossRef] [PubMed]

17. Lissi, E.; Pizarro, M.; Aspee, A.; Romay, C. Kinetics of Phycocyanine Bilin Groups Destruction by Peroxyl Radicals. *Free Radic. Biol. Med.* **2000**, *28*, 1051–1055. [CrossRef]

18. Ge, B.; Qin, S.; Han, L.; Lin, F.; Ren, Y. Antioxidant properties of recombinant allophycocyanin expressed in *Escherichia coli*. *J. Photochem. Photobiol. B* **2006**, *84*, 175–180. [CrossRef] [PubMed]

19. Pleonsil, P.; Soogarun, S.; Suwanwong, Y. Anti-oxidant activity of holo- and apo-c-phycocyanin and their protective effects on human erythrocytes. *Int. J. Biol. Macromol.* **2013**, *60*, 393–398. [CrossRef] [PubMed]

20. Guan, X.Y.; Zhang, W.J.; Zhang, X.W.; Li, Y.X.; Wang, J.F.; Lin, H.Z.; Tang, X.X.; Qin, S. A potent anti-oxidant property: Fluorescent recombinant alpha-phycocyanin of *Spirulina*. *J. Appl. Microbiol.* **2009**, *106*, 1093–1100. [CrossRef] [PubMed]

21. Yu, P.; Li, P.; Chen, X.; Chao, X. Combinatorial biosynthesis of Synechocystis PCC6803 phycocyanin holo-alpha-subunit (CpcA) in *Escherichia coli* and its activities. *Appl. Microbiol. Biotechnol.* **2016**, *100*, 5375–5388. [CrossRef] [PubMed]

22. Cherdkiatikul, T.; Suwanwong, Y. Production of the alpha and beta Subunits of *Spirulina* Allophycocyanin and C-Phycocyanin in *Escherichia coli*: A Comparative Study of Their Antioxidant Activities. *J. Biomol. Screen.* **2014**, *19*, 959–965. [CrossRef] [PubMed]

23. Glazer, A.N. Phycobiliproteins—A family of valuable, widely used fluorophores. *J. Appl. Phycol.* **1994**, *6*, 105–112. [CrossRef]

24. Rahman, D.Y.; Sarian, F.D.; van Wijk, A.; Martinez-Garcia, M.; van der Maarel, M. Thermostable phycocyanin from the red microalga *Cyanidioschyzon merolae*, a new natural blue food colorant. *J. Appl. Phycol.* **2017**, *29*, 1233–1239. [CrossRef] [PubMed]

25. Romay, C.; Gonzalez, R.; Ledon, N.; Remirez, D.; Rimbau, V. C-phycocyanin: A biliprotein with antioxidant, anti-inflammatory and neuroprotective effects. *Curr. Protein Pept. Sci.* **2003**, *4*, 207–216. [CrossRef] [PubMed]

26. Oliinyk, O.S.; Chernov, K.G.; Verkhusha, V.V. Bacterial Phytochromes, Cyanobacteriochromes and Allophycocyanins as a Source of Near-Infrared Fluorescent Probes. *Int. J. Mol. Sci.* **2017**, *18*, 1691. [CrossRef] [PubMed]

27. Trujillo, J.; Molinajijón, E.; Medinacampos, O.N.; Rodríguezmuñoz, R.; Reyes, J.L.; Loredo, M.L.; Barreraoviedo, D.; Pinzón, E.; Rodríguezrangel, D.S.; Pedrazachaverri, J. Curcumin prevents cisplatin-induced decrease in the tight and adherens junctions: Relation to oxidative stress. *Food Funct.* **2015**, *7*, 279–293. [CrossRef] [PubMed]

28. Blot, N.; Wu, X.J.; Thomas, J.C.; Zhang, J.; Garczarek, L.; Bohm, S.; Tu, J.M.; Zhou, M.; Ploscher, M.; Eichacker, L.; et al. Phycourobilin in trichromatic phycocyanin from oceanic cyanobacteria is formed post-translationally by a phycoerythrobilin lyase-Isomerase. *J. Biol. Chem.* **2009**, *284*, 9290–9298. [CrossRef] [PubMed]

29. Biswas, A.; Vasquez, Y.M.; Dragomani, T.M.; Kronfel, M.L.; Williams, S.R.; Alvey, R.M.; Bryant, D.A.; Schluchter, W.M. Biosynthesis of cyanobacterial phycobiliproteins in *Escherichia coli*: Chromophorylation efficiency and specificity of all bilin lyases from *Synechococcus* sp. strain PCC 7002. *Appl. Environ. Microbiol.* **2010**, *76*, 2729–2739. [CrossRef] [PubMed]

30. Zhao, K.H.; Zhang, J.; Tu, J.M.; Bohm, S.; Ploscher, M.; Eichacker, L.; Bubenzer, C.; Scheer, H.; Wang, X.; Zhou, M. Lyase activities of CpcS- and CpcT-like proteins from Nostoc PCC7120 and sequential reconstitution of binding sites of phycoerythrocyanin and phycocyanin beta-subunits. *J. Biol. Chem.* **2007**, *282*, 34093–34103. [CrossRef] [PubMed]

31. Bermejo, P.; Pinero, E.; Villar, A.M. Iron-chelating ability and antioxidant properties of phycocyanin isolated from a protean extract of *Spirulina platensis*. *Food Chem.* **2008**, *110*, 436–445. [CrossRef] [PubMed]

32. Xu, R.; Li, D.; Peng, J.; Fang, J.; Zhang, L.; Liu, L. Cloning, expression and antioxidant activity of a novel collagen from Pelodiscus sinensis. *World J. Microbiol. Biotechnol.* **2016**, *32*, 100. [CrossRef] [PubMed]

**Sample Availability:** Samples of the compounds are available from the authors.

**MDPI**

*Article*

# Synthesis of the Sex Pheromone of the Tea Tussock Moth Based on a Resource Chemistry Strategy

Hong-Li Zhang [1,†], Zhi-Feng Sun [2,3,†], Lu-Nan Zhou [3], Lu Liu [3], Tao Zhang [3,*] and Zhen-Ting Du [3,4,*] ®

[1] State Key Laboratory of Crop Stress Biology in Arid Areas, Northwest A&F University, Yangling 712100, China; honglizhang@126.com

[2] Shaanxi Key Laboratory for Catalysis, College of Chemical and Environment Science, Shaanxi University of Technology, Hanzhong 723001, China; sunzhifeng2018@163.com

[3] College of Chemistry and Pharmacy, Northwest A&F University, Yangling 712100, China; Zhou18792419302@163.com (L.-N.Z.); m18706733265@163.com (L.L.)

[4] Key Laboratory of Botanical Pesticide R&D in Shaanxi Province, Yangling 712100, China

* Correspondence: fuzitong@163.com (T.Z.); duzt@nwsuaf.edu.cn (Z.-T.D.); Tel.: +86-29-8709-2662 (Z.-T.D.)

† These authors contributed equally to this work.

Received: 16 May 2018; Accepted: 3 June 2018; Published: 4 June 2018

✓ check for updates

**Abstract:** Synthesis of the sex pheromone of the tea tussock moth in 33% overall yield over 10 steps was achieved. Moreover, the chiral pool concept was applied in the asymmetric synthesis. The synthesis used a chemical available on a large-scale from recycling of wastewater from the steroid industry. The carbon skeleton was constructed using the C4+C5+C8 strategy. Based on this strategy, the original chiral center was totally retained.

**Keywords:** insect sex pheromone; tea tussock moth; total synthesis; resource chemistry

---

## 1. Introduction

Today, safer agricultural produce demands that farming procedures be performed in a greener way. In this context, integrated pest management (IPM) is proposed, which is a broad-based approach that integrates practices for the economic control of pests. Among them, using low-dose pheromones to control pest has been accepted by more and more agricultural providers. In East Asia, a kind of pest called tea tussock moth (*Euproctis pseudoconspersa*) causes huge destruction in tea orchids. The bit leaves may fall off and decrease the gross product. If this kind of pest was controlled by pheromone, the tea leaves could be provided better than before. The main component of the moth's sex pheromone was first reported by Wakamura as 10,14-dimethyl-l-pentadecyl isobutyrate [1,2]. Then, a field attraction attempt was tested using crude extract from females and synthetic pheromones [3], and the result revealed that both *S* and *R* configurations have similar luring activities. As continuation of our interest in green agrochemicals [4–7], a synthesis of insect pheromone **1** based on a resource chemistry strategy was envisioned. As a matter of fact, there are several synthetic approaches to this compound in the literature. Ichikawa [8] and Zhao [2] synthesized the tussock moth pheromone (*R*)-**1** from (*S*)-citronellol or its corresponding bromide. Very recently, a synthesis of tussock pheromone based on a protective-group-free strategy [9] has been reported by our group, in which the Evans' template was adopted to control the chirality.

Some 4000 tons of sapogenin are produced every year by the Chinese steroid industry. If a new $H_2O_2$ oxidation procedure were to be applied to pregna-16,20-diol, a large quantity of (5*S*)-3-hydroxy-5-methyltetrahydro-2*H*-pyran-2-one (**2**) could become available, and 1000 tons of chiral material could be recycled. Tian et al. developed a toolkit of chiral building blocks from this resource chemical [10] and used it to achieve several syntheses of natural compounds [6,11–13]. He coined the term "resource chemical", which refers to large-scale and useful substances. Herein, we discuss the synthesis of the pheromone of the tea tussock moth based on this resource chemical.

## 2. Results and Discussions

As Scheme 1 shows, the retrosynthesis was based on two Julia coupling reactions, namely a C4+C5+C8 strategy. The left hand C4 synthon was easily obtained, and the right C8 subunit was a protected aldehyde **3**. The middle unit can be derived from a chiral methyl aldehyde, which is a large-scale resource chemical derivative.

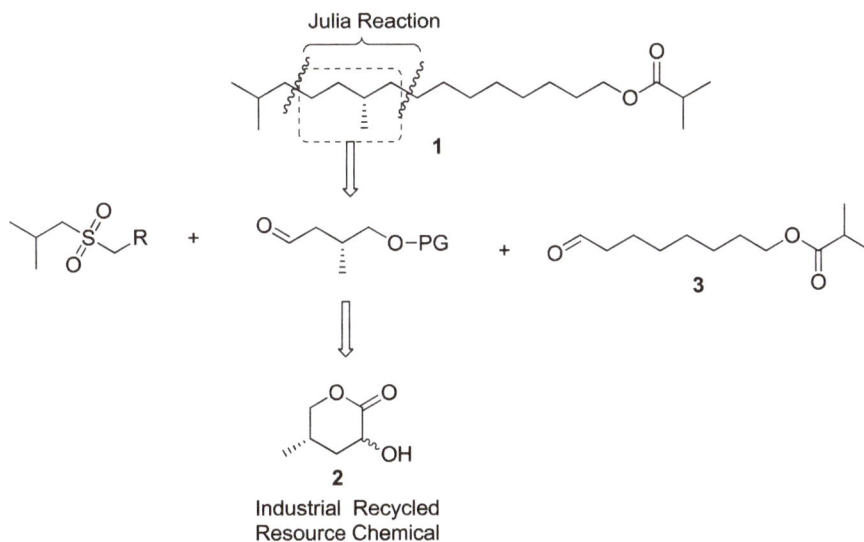

**Scheme 1.** Retrosynthesis of compound **1** based on Julia olefination and a resource-chemical.

As Scheme 2 shows, the synthesis commenced from chiral lactone **2** which was ring-opened [14,15] under basic conditions and subjected to selective benzylation to give compound **4**. In practice it proved laborious to purify the very polar hydroxyacid, so it was used directly without further purification. To reduce the carboxylic acid effectively, the ethyl ester **5** could be obtained in 75% yield for two steps under acid-catalyzed esterification conditions [11], but according to the [1]H-NMR and [13]C-NMR spectra (supplementary material), it could be observed that the hydroxyl group in the 2-position was partially epimerized. We suspect this resulted from the basic conditions used. We could observe some small peaks and high peaks with a height ratio of *c.a.* 1:3. Fortunately, the chirality of the desired methyl group didn't matter. The subsequent reduction by LiAlH$_4$ gave the vicinal diol **6** in 80% yield, which is a good substrate for an aldehyde. Therefore, the chiral aldehyde (*R*)-4-(benzyloxy)-3-methylbutanal **7** could be produced in 93% yield after a conventional oxidative diol cleavage using NaIO$_4$ in aqueous solvent. Several preparation methods of compound **7** that can be found in the literature. Aside from Wei's route [11], the rest require more chemical operations [16], an expensive chiral auxiliary such as the Evans template [17–19] or poisonous cyanide [20]. The e.e. was determined by comparison of the optical rotation value $[\alpha]_D^{25}$ + 10.6 (*c* 0.7, CHCl$_3$); lit. [17] $[\alpha]_D^{20}$ + 10.4 (*c* 0.7, CHCl$_3$). Next, at −78−−50 °C, a Julia-Kocienski reaction was performed between 5-(isobutylsulfonyl)-1-phenyl-1*H*-tetrazole (**8**), which was prepared according to the literature [21] and the chiral aldehyde **7**, affording the coupling product **9** in 87% yield (*E/Z* ratio of 1:1). The subsequent hydrogenation of the C=C double bond and removal of the benzyl protecting group was achieved in one-pot in 86% yield. (*R*)-2,6-dimethylheptan-1-ol (**10**) was converted into a Julia sulfone reagent **12** through the Mitsunobu protocol [22], followed by an oxidative reaction (83% and 99% yield, respectively). A Julia-Lythgoe reaction between aldehyde **3** and sulfone **12** gave the corresponding alkene product **13** in 79% yield with an *E*/*Z* ratio of 2:1. To prevent the isomerization and racemization of the chiral center, alkene **13** was subjected to Pt/C catalytic hydrogenation [23], giving (*S*)-**1** in 99% yield. All the analytic data for synthetic compound **1** are in accordance with that reported in the literature [8,9].

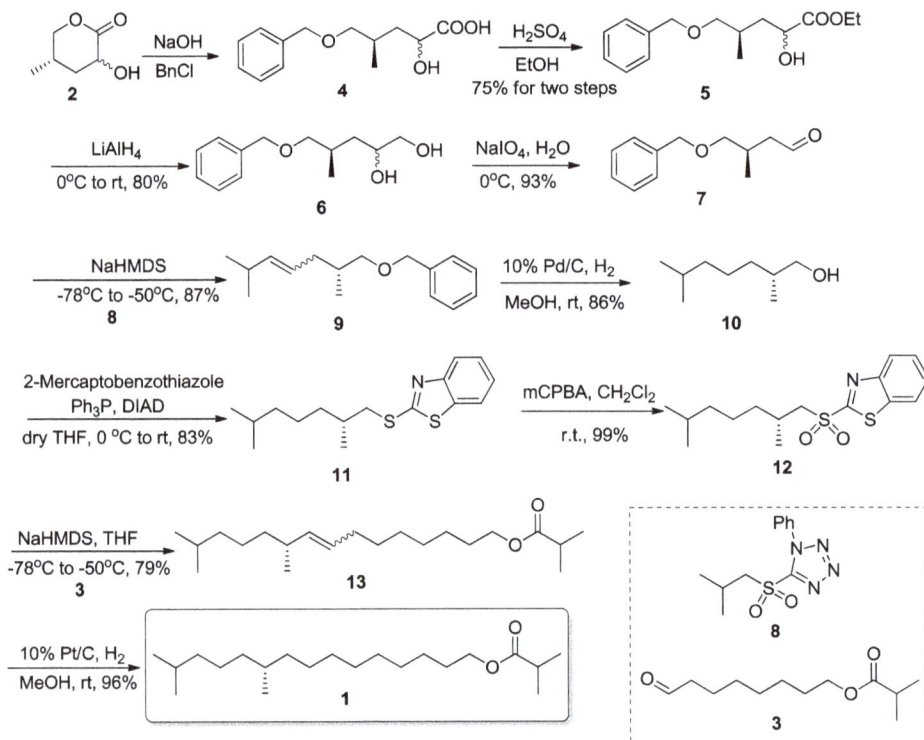

**Scheme 2.** Synthesis of (*S*)-**1** through a double Julia approach.

As Scheme 3 shows, similar to the literature [9], octane-1,8-diol was esterified selectively with 70% yield using a stoichiometric amount of isobutyric acid at presence of catalytic sulfuric acid in toluene [24]. Then, a pyridinium chlorochromate (PCC) [25] oxidation was applied, and the aldehyde **3** was produced in 90% yield (Scheme 3).

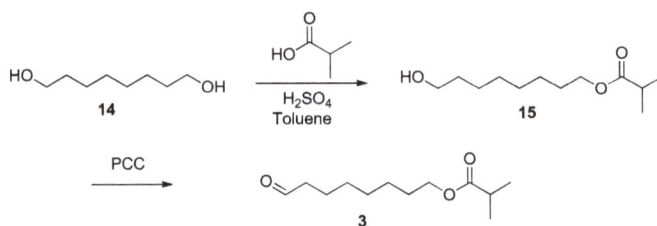

**Scheme 3.** Synthesis of compound 3.

## 3. Experimental Section

### 3.1. General Methods

THF was distilled from sodium/benzophenone and $CH_2Cl_2$ was distilled from $CaH_2$ before use. Reactions were monitored by thin-layer chromatography (TLC) on glass plates coated with silica gel with fluorescent indicator. Flash chromatography was performed on silica gel (200–300) with petroleum/EtOAc as the eluent. Optical rotations were measured on a polarimeter with a sodium lamp. HRMS spectra were measured on a LCMS-IT-TOF or LTQ-Orbitrap-XL apparatus. IR spectra were recorded using a Fourier transform infrared spectrometer. NMR spectra were recorded on a AC-500 MHz instrument (Bruker, Madison, WI, USA). Chemical shifts were reported in δ (ppm) and referenced to an internal TMS standard for $^1$H-NMR and $CDCl_3$ (77.16 ppm) for $^{13}$C- NMR (Supplementary material).

### 3.2. Ethyl (4R)-5-(Benzyloxy)-2-Hydroxy-4-Methylpentanoate (**5**)

The raw material **2** was obtained from the Pharmaceutical Factory of Shaanxi Academy of Science, Yangling, China as a strong basic, aqueous solution. If the water was removed, a large-scale raw (70–75% purity) brown solid could be obtained. In this paper, this solution was acidified to pH = 1, filtered, and the filtrate was extracted with ethyl acetate to afford crude **2**. Compound **2** was purified through column chromatography. Because the straight separation of **2** from proved laborious, the raw lactone **2** was used directly sometimes to save time.

To purified lactone **2** (6.5 g, 50 mmol), toluene (100 mL) and powdered NaOH (8 g, 200 mmol) were added. After the mixture was refluxed for 6 h, at the same time, the produced water was removed through a Dean-Stark trap, and then BnCl (15.8 g, 125 mmol) was added in 3–5 batches, and the reaction mixture was allowed to react for another 12 h. Then, water (100 mL) was added, and the aqueous phase was extracted with $Et_2O$ (70 mL × 3). The organic phases were discarded, and the aqueous phase was acidified with concentrated hydrochloric acid to pH = 1. The resulting aqueous phase was extracted with EtOAc (70 mL × 3), and the combined organic layers were dried over $MgSO_4$, filtered, and concentrated to give the crude acid without further purification. The above crude acid was dissolved in EtOH (150 mL), and then $H_2SO_4$ (5.00 mL, 98%) was added dropwise. After the mixture was refluxed for 4 h, it was monitored by TLC. Then, the reaction was dispensed into EtOAc (250 mL), and washed with water, saturated $NaHCO_3$, and brine. The solvent was evaporated, and the residue was purified through column chromatography to give the ester **5** with partly epimerized hydroxyl group 8.9 g as a yellowish oil. IR (film) $v_{max}$ 3450, 2958, 2851, 2359, 1737, 1458, 1211, 1092, 740, 701 cm$^{-1}$; $^1$H-NMR (500 MHz, $CDCl_3$) δ 7.33–7.37 (m, 5H), 4.22−4.35 (m, 3H), 3.30−3.40 (m, 1H), 3.31−3.38 (m, 2H), 2.04−2.14 (m, 1H), 1.62−1.77 (m, 2H), 1.29 (t, *J* = 7.1 Hz, 2H), 1.05 (d, *J* = 6.8 Hz, 3H); $^{13}$C-NMR (125 MHz, $CDCl_3$) δ 175.4, 138.2, 130.2, 128.4, 127.6, 75.8, 73.1, 69.3, 61.5 39.3, 30.6, 16.8, 14.2.

### 3.3. (R)-4-(Benzyloxy)-3-Methylbutanal (**7**)

To a suspension of LAH (1.11 g, 30.0 mmol) in THF (45 mL), a solution of ester **5** (5.2 g, 19.5 mmol) in THF (10 mL) was added dropwise in an ice bath. The reaction mixture was warmed to r.t after being stirred for 3 h, followed by another stirring for 10 h. The resulting mixture was carefully quenched with water (1 mL) and aqueous NaOH (5%, 2 mL), and the precipitated salt was filtered. The cake was washed with THF (30 mL), combined with the filtrate, and concentrated to give the crude diol without further purification. From the crude NMR, we could see there were two diastereomers in ca 1:3. At 0 °C, to diol **6** (1.2g 5 mmol) in water (20 mL), sulfuric acid was added to pH = 6. Then $NaIO_4$ (1.6 g, 7.5 mmol) was added portionwise and the resulting mixture was stirred for 1.5 h. After the reaction was complete, the mixture was extracted with $Et_2O$ (30 mL × 3), and the combined organic layers were washed with brine. The extract was dried, filtrated, and concentrated to give the aldehyde **7** as a colorless oil with 87% yield in two steps. $[\alpha]_D^{25}$ +10.6 (*c* 0.7, $CHCl_3$); $^1$H-NMR (500 MHz, $CDCl_3$) δ: 9.80 (t, 1H, *J* = 3.2Hz), 7.32–7.41 (m, 5H), 4.53 (s, 2H), 3.42 (dd, *J* = 9.1, 5.2 Hz, 1H), 3.26 (dd, *J* = 9.1, 7.7 Hz, 1H), 2.56 (ddd, *J* = 16.2, 6.3, 2.3 Hz, 1H), 2.30–2.44 (m, 1H), 2.28 (ddd, *J* = 16.2, 7.0, 2.1 Hz, 1H),

0.99 (d, *J* = 6.8 Hz, 3H); $^{13}$C-NMR (125 MHz, CDCl$_3$) δ: 202.4, 138.3, 128.4, 127.6, 127.6, 74.9, 73.1, 48.5, 29.2, 17.1.

### 3.4. (2R)-(((2,6-Dimethylhept-4-en-1-yl)oxy)Methyl)Benzene (9)

To a stirred solution of compound **8** (2.4 g, 8.9 mmol) in THF (30.0 mL) at –78 °C under argon, NaHMDS (2.0 M in THF, 5.60 mL, 11.3 mmol) was added dropwise. After 30 min, a solution of compound **7** (1.4 g, 7.26mmol) in THF (10.0 mL) was added dropwise. Upon stirring at −78 °C for 2 h, the reaction was warmed to −50 °C overnight. After the reaction was quenched with saturated aqueous NH$_4$Cl at −50 °C, it was extracted with ethyl acetate, washed with brine, dried over anhydrous Na$_2$SO$_4$, concentrated, and purified by flash chromatography on silica gel (hexanes: ethyl acetate = 20:1) to give *E/Z* ca. 1:1 mixture of compound **9** as a pale-yellow oil (1.9 g, 87%). [α]$_D^{25}$ + 1.32 (*c* 2.5, CHCl$_3$). $^1$H-NMR (500 MHz, CDCl$_3$) δ 7.34–7.25 (m, 5H), 5.37–5.22 (m, 2H), 4.5 (d, *J* = 2.5 Hz, 2H), 3.36–3.23 (m, 2H), 2.6–2.11 (m, 2H), 1.96–1.79 (m, 2H), 0.96–0.9 (m, 9H); $^{13}$C-NMR (125 MHz, CDCl$_3$) δ 139.4, 138.7, 128.3, 127.6, 127.6, 127.4, 125.2, 124.8, 75.5, 75.4, 73.0, 73.0, 36.6, 34.0, 33.8, 31.3, 31.1, 26.4, 23.1, 23.1, 22.7, 22.7, 17.0, 16.8.

### 3.5. (R)-2,6-dimethylheptan-1-ol (10)

To a solution of ether **9** (1.6 g, 6.89 mmol) in methanol (50 mL), Pd/C (10%, 0.3 g) was added and the atmosphere was exchanged with H$_2$. The reaction mixture was stirred for 24 h and monitored by TLC. After completion, the catalyst was filtered through a Celite pad, and the filtrate was evaporated, giving compound **10** after flash column purification as a pale-yellow oil (0.85 g, 86%). [α]$_D^{25}$ + 4.68 (*c* 2.4, CHCl$_3$). $^1$H-NMR (500 MHz, CDCl$_3$) δ 3.53-3.49 (m, 1H), 3.44-3.39 (m, 1H), 1.61 (s, 1H), 1.56–1.51 (m, 1H), 1.36–1.08 (m, 7H), 0.92 (d, *J* = 6.6 Hz, 3H), 0.87 (d, *J* = 6.4 Hz, 6H); $^{13}$C-NMR (125 MHz, CDCl$_3$) δ 68.4, 35.8, 33.4, 27.9, 24.7, 22.7, 22.6, 16.6.

### 3.6. (R)-2-((2,6-Dimethylheptyl)thio)Benzo[d]Thiazole (11)

Under argon, to a flask with (*R*)-2,6-dimethylheptan-1-ol **10** (0.7 g, 4.85mmol), THF (25 mL), benzo[*d*]thiazole-2-thiol (1.0 g, 6.0 mmol), and TPP (1.5 g, 6.0 mmol) were added. To this mixture, DIAD (1.1 mL, 6.0mmol) was added dropwise at 0 °C. After the reaction was completed, the volatile was evaporated, and the residue was purified through column chromatography to give compound **11** (1.2 g, 83% yield). [α]$_D^{25}$ –3.63 (*c* 3.1, CHCl$_3$). $^1$H-NMR (500 MHz, CDCl$_3$) δ 7.9 (d, *J* = 8.1 Hz, 1H), 7.79 (d, *J* = 8 Hz, 1H), 7.45 (t, *J* = 7.3 Hz, 1H), 7.33 (t, *J* = 7.3 Hz, 1H), 3.48–3.22 (m, 2H), 2.01–1.95 (m, 1H), 1.6–1.3 (m, 5H), 1.24–1.2 (m, 2H), 1.12 (d, *J* = 6.7Hz, 3H), 0.92 (d, *J* = 6.6 Hz, 6H); $^{13}$C-NMR (125 MHz, CDCl$_3$) δ 167.79, 153.38, 135.19, 126.0, 124.09, 121.45, 120.91, 40.78, 39.06, 36.35, 33.3, 27.95, 24.69, 22.71, 22.6, 19.4. HRMS (ESI) *m/z* calcd. for C$_{16}$H$_{24}$NO$_2$S$_2$$^+$ (M + H)$^+$: 326.1248, found 326.1245.

### 3.7. (R)-2-((2,6-Dimethylheptyl)Sulfonyl)Benzo[d]Thiazole (12)

To a solution of compound **11** (1.0 g, 3.61mmol) in CH$_2$Cl$_2$ (35 mL), *m*CPBA (4.7 g, 70%, 5.6 eq) was added, and the reaction mixture was stirred for 12 h at r.t. Then, saturated Na$_2$S$_2$O$_3$ was added to quench the reaction, and the reaction was neutralized using saturated Na$_2$CO$_3$ and extracted with CH$_2$Cl$_2$. The combined organic layers were washed with brine, and the extract was dried over MgSO$_4$, filtrated, and concentrated to give compound **12** as a colorless oil with 99% yield. [α]$_D^{25}$ − 2.99 (*c* 2.6, CHCl$_3$). $^1$H-NMR (500 MHz, CDCl$_3$) δ 8.21 (d, *J* = 8.0 Hz, 1H), 8.01 (d, *J* = 7.6 Hz, 1H), 7.66–7.58 (m, 2H), 3.56 (dd, *J* = 14.3, 4.7 Hz, 1H), 3.35 (dd, *J* = 14.3, 8.0 Hz, 1H), 2.31–2.25 (m, 1H), 1.49–1.43 (m, 2H), 1.34–1.25 (m, 3H), 1.14 (d, *J* = 6.7 Hz, 3H), 1.11–1.06 (m, 2H), 0.82(d, *J* = 6.5, 3H), 0.81(d, *J* = 6.5 Hz, 3H); $^{13}$C-NMR (125 MHz, CDCl$_3$) δ 166.83, 152.76, 136.79, 127.98, 127.64, 125.46, 122.37, 60.83, 38.7, 36.89, 28.57, 27.84, 24.08, 22.57, 22.49, 19.91. HRMS (ESI) *m/z* calcd. for C$_{16}$H$_{24}$NO$_2$S$_2$$^+$ (M + H)$^+$: 326.1248, found 326.1245.

### 3.8. 8-Hydroxyoctyl Isobutyrate **15**

To a solution of octane-1,8-diol (7.3 g, 50 mmol) and isobutyric acid (4.2 g, 48 mmol) in toluene (100 mL), four drops of concentrated sulfuric acid were added, and the reaction mixture was warmed to 75 °C for 20 h. The reaction mixture was diluted using ethyl acetate (200 mL) and washed with water, NaHCO$_3$, and brine, then dried over MgSO$_4$. The solvent was evaporated, and the residue was purified through column chromatography to give compound **15** (8.8 g, 85%) as a colorless oil. $^1$H-NMR (500 MHz, CDCl$_3$) δ 4.04 (t, *J* = 6.7 Hz, 2H), 3.62 (t, *J* = 6.6 Hz, 2H), 2.55–2.49 (m, 1H), 2.03 (s, 1H), 1.62–1.52 (m, 4H), 1.32 (s, 8H), 1.15 (d, *J* = 7.0 Hz, 6H); $^{13}$C-NMR (125 MHz, CDCl$_3$) δ 177.26, 64.31, 62.93, 34.02, 32.70, 29.24, 29.15, 28.59, 25.79, 25.62, 18.97.

### 3.9. 8-Oxo-Octyl Isobutyrate (**3**)

To a solution of 8-hydroxyoctyl isobutyrate **15** (2.8 g, 13 mmol) in CH$_2$Cl$_2$ (30 mL), silica gel (200 mesh, 2 g) and PCC (4.2 g, 19.5 mmol) were added. The above reaction mixture was stirred overnight, followed by the addition of ether (50 mL). Then, the solution of crude aldehyde was decanted and purified through flash column chromatography to give 8-oxooctyl isobutyrate **3** (2.6 g, 94%) as a colorless oil. $^1$H-NMR (500 MHz, CDCl$_3$) δ 9.76 (t, *J* = 1.8 Hz, 1H), 4.05 (t, *J* = 6.7 Hz, 2H), 2.56–2.50 (m, 1H), 2.42 (td, *J* = 1.7, 7.3 Hz, 2H), 1.64–1.61 (m, 4H), 1.34 (s, 6H), 1.15 (d, *J* = 7.0 Hz, 6H); $^{13}$C-NMR (125 MHz, CDCl$_3$) δ 202.69, 177.22, 64.22, 43.83, 34.02, 29.0, 28.95, 28.55, 25.69, 21.94, 18.99.

### 3.10. (R)-10,14-Dimethylpentadec-8-en-1-yl Isobutyrate (**13**)

Under argon, NaHMDS (1.6 mL, 2.0 M in THF, 3.2 mmol) was added dropwise to a solution of compound **12** (0.8 g, 2.5 mmol) in anhydrous THF (25 mL) at –78 °C. After half an hour, 8-oxooctyl isobutyrate **3** (0.7 g, 3.3 mmol) in THF (10 mL) was added. The reaction mixture was stirred under the same temperature for 3 h, followed by stirring for another 12 h at –50 °C. After the reaction was quenched with saturated aqueous NH$_4$Cl at –50 °C, it was extracted with ethyl acetate, washed with brine, dried over anhydrous Na$_2$SO$_4$, concentrated, and purified by flash chromatography on silica gel (hexanes: ethyl acetate = 30:1) to give compound **13** as a pale-yellow oil (0.6 g, 79%). [α]$_D^{25}$ + 3.06 (*c* 2.5, CHCl$_3$). $^1$H-NMR (500 MHz, CDCl$_3$) δ 7.33–7.25 (m, 5H), 5.36–5.08 (m, 2H), 4.49–4.44 (m, 2H), 4.06–4.03 (m, 2H), 3.49–3.39 (m, 2H), 2.65–2.5 (m, 2H), 2.07–1.93 (m, 2H), 1.7–1.43 (m, 4H), 1.3 (s, 8H), 1.16 (d, *J* = 7.0 Hz, 6H), 0.98–0.95 (m, 3H); $^{13}$C-NMR (125 MHz, CDCl$_3$) δ 177.25, 138.76, 138.73, 135.53, 129.05, 128.34, 127.66, 127.61, 127.47, 72.99, 68.82, 64.4, 37.28, 34.08, 29.85, 29.25, 29.19, 28.68, 28.61, 27.43, 25.93, 25.9, 21.51, 19.05.

### 3.11. (S)-10,14-Dimethylpentadecyl Isobutyrate (**1**)

To a solution of compound **13** (0.3 g, 9.3 mmol) in ethanol (15 mL), Pt/C (10%, 6 mg) was added and the atmosphere was exchanged with H$_2$. The reaction mixture was then stirred for 24 h and monitored by TLC. After completion, the catalyst was filtered through a Celite pad, and the filtrate was evaporated. The residue was purified by flash chromatography on silica gel (PE/EA = 30:1) to give compound **1** (0.28 g, 93%) as a colorless oil: [α]$_D^{25}$ − 0.28 (*c* 2.4, CHCl$_3$). $^1$H-NMR (500 MHz, CDCl$_3$) δ 4.05 (t, *J* = 6.8 Hz, 2H), 2.56–2.5 (m, 1H), 1.65–1.59 (m, 2H), 1.55–1.5 (m, 1H), 1.36–1.19 (m, 17H), 1.16 (d, *J* = 7.1 Hz, 6H), 1.15–1.04 (m, 4H), 0.85 (dd, *J* = 6.7, 13.2 Hz, 9H); $^{13}$C-NMR (500 MHz, CDCl$_3$) δ 177.26, 64.42, 39.4, 37.34, 37.12, 34.08, 32.79, 30.0, 29.62, 29.55, 29.27, 28.68, 28.00, 27.09, 25.93, 24.82, 22.73, 22.64, 19.73, 19.03. HRMS (ESI) *m/z* calcd. for C$_{21}$H$_{43}$O$_2^+$ (M + H)$^+$: 327.32576, found 327.32599.

## 4. Conclusions

Based on a chiral pool strategy, an efficient synthesis of (S)-10,14-dimethylpentadecyl isobutyrate, the sex pheromone of the tea tussock moth, was achieved. The longest linear synthetic step was 10 steps, and the overall yield was 33%. The key step was accomplished by a double Julia coupling. This synthesis will be helpful in research involving the differences between the biological effects of *R*-**1** and *S*-1. The scale-up and evaluation of the biological activities of the sex pheromone is currently under investigation.

**Supplementary Materials:** Supplementary Materials are available online.

**Author Contributions:** T.Z. and Z.-T.D. conceived and designed the experiments; Z.-F.S. and L.-N.Z. performed the experiments; H.-L.Z. and L.L. analyzed the data. Z.-T.D. wrote the paper.

**Acknowledgments:** Partial financial support from the National Natural Science Foundation of China (31301712, 21502151,) is greatly appreciated. Zhen-Ting Du would like to thank Opening Funds of Key Laboratory of Synthetic Chemistry of Natural Substances, Shanghai Institute of Organic Chemistry, Chinese Academy of Sciences for the partial financial support for funding this project.

**Conflicts of Interest:** The authors declare no conflict of interest.

## References

1. Wakamura, S.; Yasuda, T.; Ichikawa, A.; Fukumoto, T.; Mochizuki, F. Sex attractant pheromone of the tea tussock moth (*Euproctis pseudoconspersa*): Identification and field attraction. *Appl. Entomol. Zool.* **1994**, *29*, 403–411. [CrossRef]

2. Zhao, C.-H.; Millar, J.G.; Pan, K.-H.; Xu, C.-S. Responses of tea tussock moth, *Euproctis pseudoconspersa*, to its pheromone, (*R*)-10,14-dimethylpentadecyl isobutyrate, and to the *S*-enantiomer of its pheromone. *J. Chem. Ecol.* **1998**, *24*, 1347–1353. [CrossRef]

3. Wang, Y.; Ge, F.; Liu, X.; Feng, F.; Wang, L. Evaluation of mass-trapping for control of tea tussock moth *Euproctis pseudoconspersa* (Strand) (Lepidoptera: Lymantriidae) with synthetic sex pheromone in south China. *Int. J. Pest Manag.* **2005**, *51*, 291–298.

4. Yan, Z.; Guan, C.; Yu, Z.; Tian, W. Fluoroalkanosulfonyl fluorides-mediated cyclodehydration of beta-hydroxy sulfonamides and beta-hydroxy thioamides to the corresponding aziridines and thiazolines. *Tetrahedron Lett.* **2013**, *54*, 5788–5790. [CrossRef]

5. Zhang, T.; Feng, J.; Cai, C.; Zhang, X. Synthesis and Field Test of Three Candidates for Soybean Pod Borer's Sex Pheromone. *Nat. Prod. Commun.* **2011**, *6*, 1323–1326. [PubMed]

6. Wang, Z.; Xu, Q.; Tian, W.; Pan, X. Stereoselective synthesis of (2*S*,3*S*,7*S*)-3,7-dimethylpentadec-2-yl acetate and propionate, the sex pheromones of pine sawflies. *Tetrahedron Lett.* **2007**, *48*, 7549–7551. [CrossRef]

7. Sun, Z.-F.; Zhou, L.-N.; Zhang, T.; Du, Z.-T. Stereoselective synthesis of the Paulownia bagworm sex pheromone. *Chin. Chem. Lett.* **2017**, *28*, 558–562. [CrossRef]

8. Ichikawa, A.; Yasuda, T.; Wakamura, S. Absolute configuration of sex pheromone for tea tussock moth, Euproctis pseudoconspersa (strand) via synthesis of (*R*)- and (*S*)-10,14-dimethyl-1-pentadecyl isobutyrates. *J. Chem. Ecol.* **1995**, *21*, 627–634. [CrossRef] [PubMed]

9. Sun, Z.-F.; Zhou, L.-N.; Meng, Y.; Zhang, T.; Du, Z.-T.; Zheng, H. Concise asymmetric synthesis of the sex pheromone of the tea tussock moth. *Tetrahedron Asymmetry* **2017**, *28*, 1562–1567. [CrossRef]

10. Tian, W.S.; Wang, Z.K.; Li, B.; Xu, Q.-H. Synthesis of Thioketal and Application in Synthesis of Optical Pure Diahrotica Undecimpunctata Sex Pheromone. Patent No. CN 101050211, 10 May 2007.

11. Mao, Z.-Y.; Si, C.-M.; Liu, Y.-W.; Dong, H.-Q.; Wei, B.-G.; Lin, G.-Q. Divergent Synthesis of Revised Apratoxin E, 30-epi-Apratoxin E, and 30*S*/30*R*-Oxoapratoxin E. *J. Org. Chem.* **2017**, *82*, 10830–10845. [CrossRef] [PubMed]

12. Wang, Z.-K.; Tian, W.-S.; Pan, X.-F. Practical synthesis of the main sex pheromones of pine sawflies (2*S*,3*S*,7*S*)-3,7-dimethylpentadecan-2-ol esters by utilizing (*R*)-4-methyl-delta-valerolactone obtained from the industrial waste. *Acta Chim. Sin.* **2007**, *65*, 705–710.

13. Wang, Z.-K.; Tian, W.-S.; Pan, X.-F. A concise synthesis of the sex pheromones of Pine sawflies. *Chin. J. Org. Chem.* **2007**, *27*, 866–869.

14. Chung, J.; Kushner, A.M.; Weisman, A.C.; Guan, Z. Direct correlation of single-molecule properties with bulk mechanical performance for the biomimetic design of polymers. *Nat. Mater.* **2014**, *13*, 1055–1062. [CrossRef] [PubMed]

15. Zhang, L.; Zhu, L.; Yang, J.; Luo, J.; Hong, R. Stereoselective α-Hydroxylation of Amides Using Oppolzer's Sultam as Chiral Auxiliary. *J. Org. Chem.* **2016**, *81*, 3890–3900. [CrossRef] [PubMed]

16. Schmid, R.; Hansen, H.J. Synthesis of optically active bifunctional isoprenoid building blocks by rhodium(I)-catalyzed asymmetric allylamine to enamine isomerization. *Helv. Chim. Acta* **1990**, *73*, 1258–1275. [CrossRef]

17. Chorley, D.F.; Chen, J.L.-Y.; Furkert, D.P.; Sperry, J.; Brimble, M.A. Total synthesis of danshenspiroketallactone. *Synlett* **2012**, *23*, 128–130.

18. Hansen, D.B.; Starr, M.-L.; Tolstoy, N.; Joullie, M.M. A stereoselective synthesis of (2*S*,4*R*)-δ-hydroxyleucine methyl ester: A component of cyclomarin A. *Tetrahedron Asymmetry* **2005**, *16*, 3623–3627. [CrossRef]

19. Jones, T.K.; Reamer, R.A.; Desmond, R.; Mills, S.G. Chemistry of tricarbonyl hemiketals and application of Evans technology to the total synthesis of the immunosuppressant (−)-FK-506. *J. Am. Chem. Soc.* **1990**, *112*, 2998–3017. [CrossRef]

20. Ghosh, A.K.; Wang, Y.; Kim, J.T. Total Synthesis of Microtubule-Stabilizing Agent (−)-Laulimalide. *J. Org. Chem.* **2001**, *66*, 8973–8982. [CrossRef] [PubMed]

21. DiBlasi, C.M.; Macks, D.E.; Tan, D.S. An Acid-Stable tert-Butyldiarylsilyl (TBDAS) Linker for Solid-Phase Organic Synthesis. *Org. Lett.* **2005**, *7*, 1777–1780. [CrossRef] [PubMed]

22. Mitsunobu, O. The Use of Diethyl Azodicarboxylate and Triphenylphosphine in Synthesis and Transformation of Natural Products. *Synthesis* **1981**, *1981*, 1–28. [CrossRef]

23. Li, N.-S.; Scharf, L.; Adams, E.J.; Piccirilli, J.A. Highly Stereocontrolled Total Synthesis of β-D-Mannosyl Phosphomycoketide: A Natural Product from Mycobacterium tuberculosis. *J. Org. Chem.* **2013**, *78*, 5970–5986. [CrossRef] [PubMed]

24. Li, J.-M.; Yong, J.-P.; Huang, F.-L.; Bai, S.-Z. A facile synthesis of the sex pheromone of *Grapholitha molesta*. *Chem. Nat. Compd.* **2012**, *48*, 103–105. [CrossRef]

25. Corey, E.J.; Suggs, J.W. Pyridinium chlorochromate. An efficient reagent for oxidation of primary and secondary alcohols to carbonyl compounds. *Tetrahedron Lett.* **1975**, *16*, 2647–2650. [CrossRef]

**Sample Availability:** Samples of the compounds from **1** to **15** are available from the authors.

*molecules*

MDPI

*Review*

# Synthesis and Anticancer Activity of CDDO and CDDO-Me, Two Derivatives of Natural Triterpenoids

Rebecca Borella [1], Luca Forti [1], Lara Gibellini [2], Anna De Gaetano [1], Sara De Biasi [2], Milena Nasi [3], Andrea Cossarizza [3] and Marcello Pinti [1],*

[1]  Department of Life Sciences, University of Modena and Reggio Emilia, 41125 Modena, Italy; rebeccaborella7993@gmail.com (R.B.); luca.forti@unimore.it (L.F.); anna.degaetano@unimore.it (A.D.G.)
[2]  Department of Medical and Surgical Sciences of Children and Adults, University of Modena and Reggio Emilia, 41125 Modena, Italy; lara.gibellini@unimore.it (L.G.); debiasisara@yahoo.it (S.D.B.)
[3]  Department of Surgery, Medicine, Dentistry and Morphological Sciences, University of Modena and Reggio Emilia, 41125 Modena, Italy; milena.nasi@unimore.it (M.N.); andrea.cossarizza@unimore.it (A.C.)
*  Correspondence: marcello.pinti@unimore.it; Tel.: +39 059 205 5386; Fax: +39 059 205 5426

Received: 26 September 2019; Accepted: 10 November 2019; Published: 13 November 2019

check for updates

**Abstract:** Triterpenoids are natural compounds synthesized by plants through cyclization of squalene, known for their weak anti-inflammatory activity. 2-cyano-3,12-dioxooleana-1,9(11)-dien-28-oic acid (CDDO), and its C28 modified derivative, methyl-ester (CDDO-Me, also known as bardoxolone methyl), are two synthetic derivatives of oleanolic acid, synthesized more than 20 years ago, in an attempt to enhance the anti-inflammatory behavior of the natural compound. These molecules have been extensively investigated for their strong ability to exert antiproliferative, antiangiogenic, and antimetastatic activities, and to induce apoptosis and differentiation in cancer cells. Here, we discuss the chemical properties of natural triterpenoids, the pathways of synthesis and the biological effects of CDDO and its derivative CDDO-Me. At nanomolar doses, CDDO and CDDO-Me have been shown to protect cells and tissues from oxidative stress by increasing the transcriptional activity of the nuclear factor (erythroid-derived 2)-like 2 (Nrf2). At doses higher than 100 nM, CDDO and CDDO-Me are able to modulate the differentiation of a variety of cell types, both tumor cell lines or primary culture cell, while at micromolar doses these compounds exert an anticancer effect in multiple manners; by inducing extrinsic or intrinsic apoptotic pathways, or autophagic cell death, by inhibiting telomerase activity, by disrupting mitochondrial functions through Lon protease inhibition, and by blocking the deubiquitylating enzyme USP7. CDDO-Me demonstrated its efficacy as anticancer drugs in different mouse models, and versus several types of cancer. Several clinical trials have been started in humans for evaluating CDDO-Me efficacy as anticancer and anti-inflammatory drug; despite promising results, significant increase in heart failure events represented an obstacle for the clinical use of CDDO-Me.

**Keywords:** triterpenoids; bardoxolone methyl; anticancer drug; mitochondria

## 1. Introduction

Triterpenoids are natural compounds synthesized by plants through cyclization of squalene and represent one of the most numerous and diverse group of secondary metabolites, ubiquitously distributed in the plant kingdom [1]. So far, more than 20,000 natural triterpenoids are known [2], predominantly found in several medicinal plants, in wax-like coatings of various fruits such as apples, and in herbs including rosemary, oregano, and thyme [3–5]. The biological significance of these molecules is not completely clear: if the antibacterial and antifungal properties of some of them, such as oleananes, can explain their accumulation in fruits, most of these compounds present in plants are not toxic for herbivores or omnivores, and so have no obvious deterrent effect.

For centuries, extracts containing triterpenoids—which resemble steroids in chemical structure—have been used in Asian countries for medical purposes, as antibacterial, antifungal, antiviral, anti-inflammatory, antioxidant, antidiabetic, and hepato- and cardio-protective agents [3]. Oleanolic acid, one of these naturally-occurring triterpenoids, has a mild anti-inflammatory effect and a weak antitumorigenic activity [6–8]. However, most of the natural triterpenoids, including oleanolic acid, display their pharmacological activity at high concentration, up to 40 uM [9–11]. In an attempt to improve and enhance the biological activity of natural triterpenoids, a series of chemical modifications have been introduced to the structure of the molecules, and synthetic triterpenoids derived from oleanolic acid and ursolic acid have been obtained that exhibit optimized bioactivity such as potent anti-inflammatory and antitumorigenic activities [12]. In particular, 2-cyano-3,12-dioxooleana-1,9(11)-dien-28-oic acid (CDDO), and its C28 modified derivatives, methyl-ester (CDDO-Me, also known as bardoxolone methyl), and imidazole (CDDO-Im) have been extensively investigated for their strong antiproliferative, antiangiogenic, antimetastatic activities, and for their capability to induce apoptosis and differentiation in cancer cells. Here, we discuss the chemical properties of natural triterpenoids, the pathways of synthesis and the biological effects of CDDO and its derivative CDDO-Me.

## 2. Chemical Properties and Synthesis Pathways of Triterpenoids

Triterpenoids present a carbon skeleton containing six isoprene units, which are derived from the acyclic C30 hydrocarbon squalene, an important precursor for synthesis of all plant and animal sterols, including cholesterol and steroid hormones in humans.

In plants, terpenoids share a common biosynthetic origin: all terpenoids derive from the repetitive fusion of isoprene ($C_5H_8$) units, and the number of isoprene units determines their classification. First, an isopentenyl pyrophosphate (IPP) is generated by the mevalonate (MVA)/3-hydroxy-3-methylglutaryl-CoA reductase (HMGR) pathway or the 2-C-methyl-D-erythritol 4-phosphate (MEP)/1-deoxy-D-xylulose 5-phosphate (DOXP)/non-MVA pathway. Then, the IPP is isomerized to dimethylallyl pyrophosphate (DMAPP), and the condensation of IPP and DMAPP units, catalyzed by specific prenyltransferases, forms prenylated pyrophosphates, the precursors of different terpenoid classes. Terpenoid synthases modify these precursors to terpenoid skeletons [13], which are further modified to different terpenoids [14]. Synthesis of triterpenoids is obtained through a condensation of two IPP units with a DMAPP unit, which generates the $C_{15}$ farnesyl pyrophosphate (FPP). Two FPP are fused 'head-to-head' to generate squalene, the linear $C_{30}$ precursor of triterpenoids. Then, squalene is epoxidized to 2,3-oxidosqualene [15] and cyclized to tetra- or pentacyclic structures by specific oxidosqualene cyclases (Figure 1). Oleanolic acid, one of these pentacyclic triterpenoids, is the precursor of several synthetic or semisynthetic compounds, including CDDO, CDDO-Me, and CDDO-Im [1].

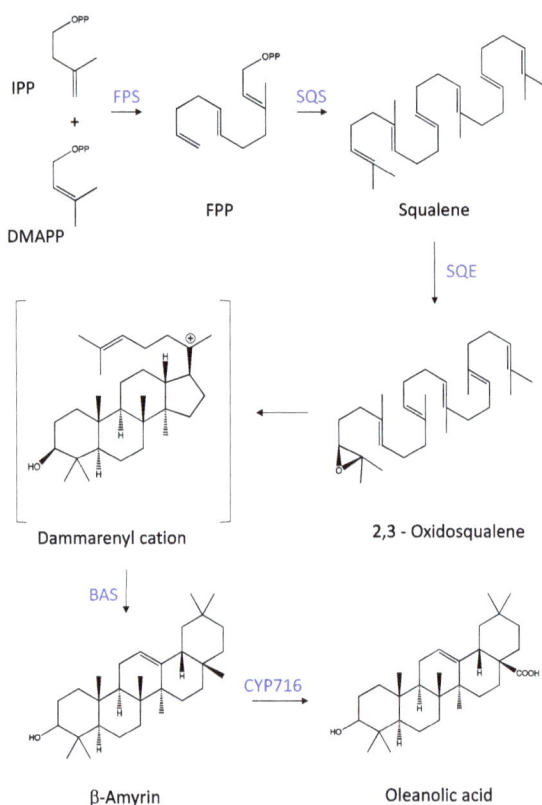

**Figure 1.** Synthesis pathway in plants and structure of oleanolic acid. See text for details. Abbreviations: DMAPP, dimethylallyl diphosphate; IPP, isopentenyl diphosphate; FPP, farnesyl pyrophosphate; FPS, farnesyl pyrophosphate synthase; SQS, squalene synthase; SQE, squalene epoxidase: BAS, β-amyrin synthase; CYP716, P450 enzymes belonging to CP71 group (CP716A12, CP716A15, CP716A17, CP716AL1). Enzyme names are in blue.

Total synthesis of naturally occurring pentacyclic triterpenes, which have at least eight chiral centers, and in particular of oleanolic acid, has been at the center of intense research activity for decades. The first enantioselective total synthesis of oleanolic acid was reported in 1993 [16]. Oleanolic acid derivatives were firstly synthesized in 1998, in an attempt to identify new inhibitors of nitric oxide (NO) production in macrophages [17]. Honda et al. randomly modified the structure of oleanolic acid and obtained about 60 compounds, which were tested in vitro as inhibitors of the production of NO induced by IFN-γ. Several compounds showed significant inhibitory activity and one of them (3,12-dioxoolean-1,9-dien-28-oic acid) displayed the highest activity (IC$_{50}$ = 7 µM) [17]. In a following study, the same research group reported the synthesis of a much more potent compound, named CDDO, whose inhibitory activity was comparable to that of dexamethasone [18]. A further efficient and multi-gram level synthesis procedure of CDDO-Me was described in 2013 [19]. The synthesis pathway of CDDO, as described in [18] is illustrated in the Figure 2A; Figure 2B shows the structure of CDDO and of its methyl derivative, CDDO-Me.

2-cyano-3,12-dioxoolean-1,9-dien-28-oic acid (CDDO)

CDDO-Me

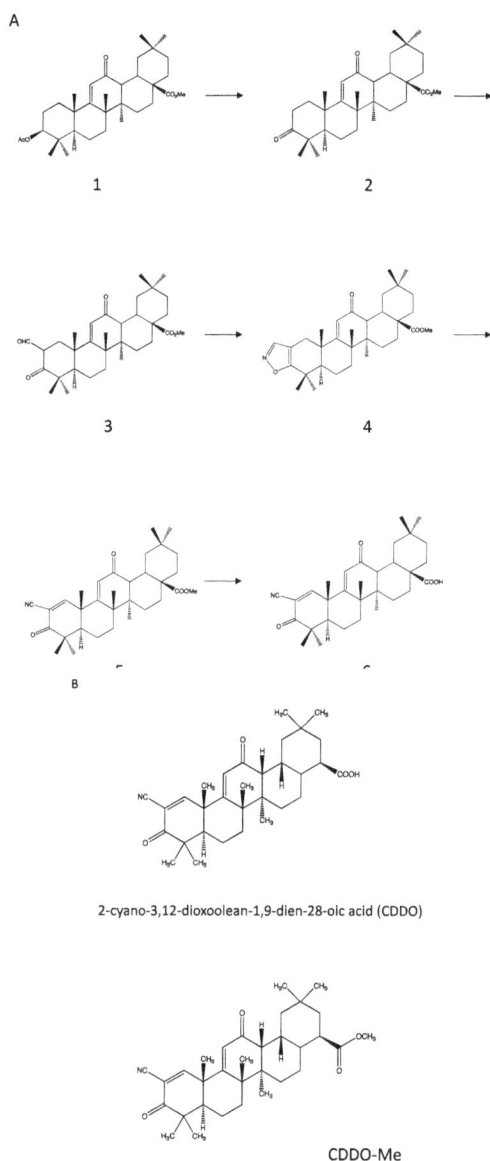

**Figure 2.** Structures and synthesis of CDDO and CDDO-Me, two synthetic derivatives of oleanolic acid. (**A**) Synthesis pathway of CDDO, as described by Honda et al. [18]. Compound **2** was obtained from the already known compound **1** by alkali hydrolysis and Jones oxidation. Compound **3** was obtained by formylation of **2** with ethyl formate; compound **4** was obtained from **3** by addition of hydroxylamine, and compound **5** by cleavage of isoxazole 16 with sodium methoxide and subsequent double bond introduction at C1 with PhSeCl–H$_2$O$_2$. CDDO (compound **6**) was obtained from **5** by halogenolysis of **5** with lithium iodide in dimethylformamide. (**B**) Structure of CDDO and CDDO-Me. CDDO-Me is the C28 methyl ester of CDDO.

## 3. Anticancer Effects of Triterpenoids

Triterpenoids have pleiotropic effects. At low doses they display anti-inflammatory, antioxidative stress effects, while at intermediate doses they are able to induce cell differentiation and at high doses they exert cytotoxic, antiproliferative, and proapoptotic effects. Thus, they can theoretically have an anticancer activity at multiple levels: the low doses of these molecules can prevent the process of carcinogenesis and mitigate the damage of procarcinogens, while at intermediate–high levels they can slow down the proliferation of cancer cells, and/or cause cell death by apoptosis.

The antiproliferative activity of CDDO-Me is generally observed at concentrations ranging from 0.1 to 1.0 µM, and the proapoptotic activity can be seen with concentrations above 0.5 µM. CDDO is slightly less effective than CDDO-Me and is active at concentrations 5–10 times higher.

As far as antiproliferative effect is concerned, the most striking features of these molecules are their capability to inhibit cell growth independently from p53 status, and the fact that inhibition is often observed in neoplastic cells, but not in their normal counterpart. The latter point has particularly drawn the attention of researchers for using CDDO and its derivatives in cancer treatment.

It has been proposed that the effects of CDDO and CDDO-Me are largely mediated by the presence of two electrophilic Michael acceptor sites in the A and C rings, which allow the formation of adducts with proteins containing redox-sensitive Cys residues; this effect has been directly demonstrated in the case of relevant targets such as the IκB kinase (IKK), the ubiquitin-specific-processing protease 7 (USP7) or the erythroblastic oncogene B2 (ErbB2) [20–23]. All these proteins contain specific Cys crucial for their functions, and CDDO and CDDO-Me interact only with some of them, in a specific manner. This mechanism of action has been directly proved in the case of IKK: the Cys179 is specifically targeted in this protein, and when this residue is mutated, the effect of CDDO-Me on NF-κB pathway is abrogated.

Thus, different concentrations of CDDO and CDDO-Me can have different—even opposite—effects on target cells because of the diverse binding affinity for target proteins when forming Michael adducts. At low concentrations, these synthetic triterpenoids interact with kelch-like ECH-associated protein 1 (Keap1) and activate a cytoprotective pathway, while at higher concentrations other proteins (such as PPAR-γ, USP7, IKK, Lonp1) are targeted to inhibit proliferation or induce apoptosis. It must be noted that the formation of Michael adducts by CDDO and CDDO-Me is reversible. Thus, it is possible that CDDO and CDDO-Me induce a biological response in a cell by binding a specific target, but they may not remain bound to this target, making the effect transient.

### 3.1. Effects In Vitro at Low Doses

At nanomolar concentrations, CDDO and CDDO-Me have been shown to protect cells and tissues from oxidative stress by increasing the transcriptional activity of the nuclear factor (erythroid-derived 2)-like 2 (Nrf2). Nrf2 is the principal regulator of the phase II cellular antioxidant response and represents an endogenous defense mechanism against toxic cell stress [24]. Under physiological conditions, Nrf2 is sequestered in the cytosol by its repressor protein, Keap1, and is subsequently ubiquitinated and degraded [25,26]. During cell stress, ubiquitination of Nrf2 by Keap1 is disrupted, allowing Nrf2 to translocate to the nucleus, and up-regulate genes containing an antioxidant response element (ARE) in their promoter regions [27]. Nrf2- regulated genes facilitate a variety of functions, including antioxidative activity, detoxification and transport of xenobiotics, proteasome activity, and well as glutathione homeostasis [28]. Electrophilic compounds have been shown to induce the activation of the Nrf2 pathway and many of them (including CDDO), have demonstrated cytoprotective responses against oxidative and inflammatory stress in vitro [29–31]. Along with the activation of the antioxidative response, nanomolar doses of CDDO and its derivatives have also an antioxidant effect. The mechanisms at the basis of this effect is not completely understood, but is at least in part due to the capability of these molecules to suppress iNOS in innate immune cells [18,32], and to reduce the expression of proinflammatory cytokines, including (but not limited to) tumor necrosis factor (TNF)-α, interleukin (IL)-1β, IL-6, and interferon (IFN)-γ in a variety of cell types [33–40]. It is interesting to note that CDDO-Me has an opposite effect in M2 macrophages, as it reduces anti-inflammatory cytokines like IL-10 and increases the production of TNF-α and IL-6 [40].

While the activation of Keap1/Nrf2/ARE pathway by CDDO and its derivatives has been shown to be beneficial in several experimental models of human diseases, such as neurodegenerative diseases [30,31,41–44], eye diseases [45,46], and lung pathologies [47,48], the lines of evidence of a possible effect on cancer development are less convincing. Indeed, several studies have provided evidence that CDDO, CDDO-Me can act as a chemopreventer in vivo (see Section 5) but few of them have shown a possible role of the Nrf2 pathway in this process. This is probably because in some tumor models, activation of the Nrf2 pathway does not prevent cancer development but can be detrimental by aiding carcinogenesis or giving rise to resistance to chemotherapeutic drugs.

Synthetic triterpenoids can also have an antiproliferative and proapoptotic effect at these low doses. CDDO-me inhibits the activation of the PI3K/Akt/mTOR pathway, which is often dysregulated in cancer [49,50]. In human prostate cancer cells, CDDO-Me inhibits the activity of *p*-Akt and mTOR and of their downstream targets [51,52] and overexpression of Akt leads to resistance to CDDO-Me [52]. The inhibition of this pathway is likely ROS-dependent, as the inhibition of ROS generation by antioxidants like *N*-acetylcysteine (NAC) prevents the inhibition of constitutively active Akt, nuclear factor κB (NF-κB), and mTOR by CDDO-Me. A similar mechanism of action has been demonstrated in several other cancer cell lines, such as pancreatic cancer cells [53,54], colorectal cancer cells [55,56], ovarian cancer cells [57,58], human glioblastoma and neuroblastoma cell lines [55], suggesting that this is a general phenomenon.

## 3.2. Effects In Vitro at Intermediate Doses

At doses higher than those needed for activating Nrf2 pathways (>100 nM), CDDO and CDDO-Me are able to modulate the differentiation of a variety of cell types, both tumor cell lines and primary culture cells. This effect has been observed in human leukemia cell lines, monocytic cell lines, osteosarcoma and adipocytes, while an effect on stem cells or progenitor cells has been observed with CDDO-Im, but not with CDDO or CDDO-Me [59–64].

The mechanisms that underpin the induction of cell differentiation by the CDDO and CDDO-Me are far from being clear, and several factors involved in cell differentiation have been identified as targets of synthetic triterpenoid activity.

The most convincing data have been obtained in preadipocytes, where CDDO has been shown to induce adipocyte differentiation by modulating PPAR-γ activity. In these cells, CDDO shows a biphasic activity. When used at doses ranging from 10 to 100 nM, it induces differentiation of 3T3-L1 preadipocytes. At the dose of 1 uM, CDDO fails to induce differentiation, and inhibits that caused by all other known differentiating molecules, such as rosiglitazone. While the mechanism of inhibition is unknown, the differentiation effect is due to the binding of CDDO to PPAR-γ, which is an agonist. Unlike the cases of IKKβ, JAK, or STAT3, CDDO binds to PPAR-γ in a non-covalent and reversible manner, as no direct adduct formation has been observed. It is interesting to note that, even if CDDO-Me can bind to PPAR-γ with similar affinity, CDDO-Ma acts as an antagonist of the transactivator; this striking difference is due to the fact that CDDO releases the nuclear receptor corepressor (NCoR) from PPAR-γ, while CDDO-Me does not [65]. The involvement of PPAR-γ in the differentiation activity of synthetic triterpenoids has been independently confirmed in other cell models, such as the acute promyelocytic leukemia (AML) HL-60, NB4, and MR2 cell lines, and in patient-derived primary AML blasts. In HL-60 cells, CDDO induces PPAR-γ activation, and enhances binding of the vitamin D-interacting protein (DRIP205) coactivator to PPAR-γ. Accordingly, the PPAR-γ antagonist T007 blocks differentiation in HL-60 cells treated with CDDO, and the differentiation induced by CDDO is enhanced in the same cells overexpressing DRIP205 [66].

CDDO is able to enhance the differentiating effect of all-trans-retinoic acid (ATRA) in two models of acute promyelocytic leukemia (APL): in the ATRA-sensitive cell line NB4, CDDO enhances the differentiating effect of ATRA, while in the ATRA-resistant MR2 cell line it partially reverses ATRA resistance [63]. These effects are mediated by CDDO-induction of PPAR-γ, and by enhancing the ability of ATRA to induce retinoic acid receptor (RAR) β2 gene expression in APL cells. Independently from the effect on PPAR-γ , the combination of ATRA and CDDO partially increases histone acetylation in the RARβ2 promoter, which in turn allows the recruitment of RARE to the RARβ2 promoter [63].

CDDO is also able to induce differentiation in a PPAR-γ independent manner. In osteosarcoma cells, differentiation induced by CDDO is abrogated by the overexpression of the extrinsic caspase-8 inhibitor cytokine response modifier A (CrmA), suggesting that CDDO-induced differentiation is mediated by caspase-8 activity [67].

### 3.3. Effects In Vitro at High Doses

At high doses (about 1–10 μM, depending on the molecule and cell model) CCDO and its derivatives exert an anticancer activity through several mechanisms, often selective for proliferating cancer cells but not for non-transformed cells. Although most of the mechanisms described below act directly on cancer cells, CDDO and its derivatives can also redirect and reactivate the immune response versus cancer cells [40].

### 3.3.1. Induction of Apoptosis

Apoptosis can be triggered via either the caspase-mediated extrinsic or intrinsic pathways [68]. In the extrinsic or death receptor mediated pathway, binding of death ligands (e.g., FasL, TRAIL) with their death receptors activates initiator caspase-8, which cleaves and activates effector caspases 3, 6, and 7, which ultimately lead to apoptosis [68]. The intrinsic apoptotic pathway is triggered by stimuli that trigger permeabilization of mitochondria and release of cytochrome-c (cyt-c) into the cytoplasm [69]. Cyt-c binds to Apaf-1 and forms the apoptosome, which in turn recruits and activates caspase-9, the first caspase of the intrinsic pathway effector caspases 3, 6, and 7 [68].

The literature discussing how CDDO and CDDO-Me trigger apoptosis is extensive, with notable discrepancies in the different cell models. As a general rule, CDDO has been shown to activate extrinsic apoptosis, while CDDO-Me activates the intrinsic pathway [67,70–73]. CDDO exerts its activity on extrinsic pathway at multiple levels. First, it activates caspase-8, which in turns leads to the activation of caspase-3 and, via cleavage of Bid, to the release of cytochrome c from mitochondria. Second, treatment with CDDO upregulates death receptors (DR) 4 and 5 and downregulates the antiapoptotic protein cellular FLICE-like inhibitory protein (c-FLIP) [74,75]. Conversely, convincing lines of evidence have been reported that CDDO-Me preferentially activates the intrinsic pathway, by upregulating the proapoptotic protein Bax, or by permeabilizing the inner mitochondrial membrane, so favoring cyt-c release [72,76]. Nevertheless, some models in which CDDO-Me activates the extrinsic pathway have been reported: by depleting GSH and causing reticulum stress, CDDO-Me activates c-Jun $NH_2$-terminal kinase (JNK), which activates CCAAT/enhancer-binding protein homologous protein (CHOP). CHOP in turn upregulates decoy receptor (DR) 5, so favoring apoptosis [77,78].

The mechanisms by which CDDO and CDDO-Me initiate the apoptotic cascade are far from being clear. One proposed mechanism is that these molecules disrupt the oxidative balance of cells, but it is difficult to establish if increase in cellular ROS is an initiating event, or if it is rather a consequence of mitochondrial perturbation [79]. CDDO and CDDO-Me decrease intracellular GSH levels [79], one of the most important scavengers of reactive oxygen species (ROS). The maintenance of correct GSH levels is important not only for scavenging ROS but also for preventing apoptosis triggered by mitochondria, which do not synthetize GSH but import it from the cytosol [80]. The cytotoxic effects of CDDOs are mediated by a rapid and selective decrease of mitochondrial GSH (mGSH) which leads to caspase-independent apoptosis [76]. In this model, CDDO-Im induced depletion of mGSH occurs prior to the onset of apoptosis and results in the generation of ROS, mitochondrial dysfunction, and intracellular glutathione pool oxidation. Treatment with the triterpenoid induced rapid alterations in the cytoplasmic morphology that were insensitive to the pharmacological inhibition of caspases. Similarly, the loss of mitochondrial membrane potential was also not prevented by the inhibition of caspases. Notably, cotreatment with sulfhydryl antioxidants prevents the depletion of mGSH, the loss of mitochondrial membrane potential, and the shrinking of cytoplasm, suggesting that redox stress can mediate the activation of caspase-independent apoptosis [76].

### 3.3.2. Induction of Autophagy

Less attention has been paid to the effects of these compounds on autophagy. CDDO-Me is able to induce autophagy in chronic myeloid leukemia cells, in which the toxic effect on mitochondria is rapidly followed by engulfment in autophagosomes of damaged organelles or by mitochondrial-induced apoptosis [81]. In K562 cells, CDDO-Me induces autophagy by suppressing the PI3K/Akt/mTOR signaling pathway [82]. The same mechanism has been independently observed in esophageal squamous cancer cell lines Ec109 and KYSE70, suggesting that this effect of CDDO-Me could be a general phenomenon [83].

### 3.3.3. Inhibition of Janus-Activated Kinases (JAKs)

Signal transducer and activator of transcription (STAT) proteins are involved in the regulation of cell proliferation and differentiation, and of apoptosis [84]. Among them, STAT3 is often constitutively active in cancer cells, and can contribute to neoplastic transformation, invasion, and metastasis. After binding of a ligand to its cognate growth factor receptor, such as the IL-6 receptor, a JAK kinase phosphorylates the receptor, allowing recruitment and phosphorylation of a STAT. STATs then dimerize, translocate to the nucleus, and induce the transcription of several STAT targets, including cyclin D1, myc, and survivin proteins [85,86]. CDDO-Me has been shown to inhibit JAK/STAT3 pathways in different cell models, such as in ovarian cancer breast cancer and osteosarcoma cancer cells [37,87,88]. Studies on the mechanisms of action demonstrated that CDDO-Me inhibits this pathway at multiple levels and in a specific way. At micromolar concentrations, CDDO-Me suppresses JAK1 phosphorylation by direct binding to Cys1077. Thus, Jak1 is unable to phosphorylate STAT3, which is crucial for its dimerization and activation [87]. Independently from the effect on Jak1, CDDO-Me can also form adducts with STAT3 that are dependent on Cys259, as mutations of this aminoacidic residue abrogates the inhibitory effect [87].

### 3.3.4. Inhibition of NF-κB Activity

Synthetic triterpenoids, and in particular CDDO-Me, can exert their activity also by inhibiting the NF-κB pathway. NF-κB is a transcription factor crucial for activating a variety of genes involved in inflammation, proliferation, and survival. Since the proinflammatory microenvironment is a feature of most cancers, its inhibition can play a role in inhibiting tumorigenesis and cancer proliferation. This capability has been shown in a variety of cells, including PC-3 and C4-2 cells, prostate cancer cells, and colorectal cancer cells, where CDDO-Me inhibits the growth and causes cells death at intermediate–high concentrations (0.625–2.5 μM) through the inhibition of NF-κB, p-Akt, and mTOR [52].

### 3.3.5. Inhibition of Telomerase Activity

CDDO-Me can inhibit cancer proliferation and provoke apoptosis by inhibiting telomerase activity, which is associated with promotion of tumorigenesis [53]. When telomeres, which shorten at any cell division, becomes too short, they trigger senescence or apoptosis. The enzyme telomerase can counteract this process by elongating telomeres and cancer cells often upregulate telomerase to circumvent cell senescence due to telomere shortening. In a model of human pancreatic cancer cells, CDDO-Me is able to inhibit telomerase gene and protein expression, as well as its enzymatic activity [54], through a ROS-dependent mechanism [89]. The action of CDDO-Me on telomerase activity is also indirect, as it inhibits c-Myc, Sp1, NF-κB, p-STAT3, and p-Akt, a group of factors regulating telomerase, and causes decreased histone deacetylation and histone demethylation at the promoter of the human telomerase gene [90]. This inhibitory effect on telomerase is likely a general phenomenon, as it has been demonstrated in other cancer cell lines [58,91].

### 3.3.6. Inhibition of Mitochondrial Protease Lonp1

More recently, we and others have shown that CCDO and CDDO-Me can exert their antiproliferative, proapoptotic activity by inhibiting the mitochondrial protease Lonp1. This is a ubiquitous serine protease present in the mitochondrial matrix, but encoded by the nucleus, which has three main functions: (i) it degrades oxidized or damaged proteins, with an ATP-dependent mechanism; (ii) it acts as a chaperone for the folding of imported proteins into the mitochondrial matrix; (iii) it binds mtDNA and contributes to the maintenance of the normal levels of this molecule in the mitochondria [92–94]. Lonp1 plays a critical role in antioxidant stress response acting as a chaperone for the degradation of mutant and abnormal proteins, such as toxic aggregates of oxidized mitochondrial proteins, and regulates the maintenance of mitochondria DNA, morphology, and dynamics [93]. CDDO and its derivatives mediate apoptosis in lymphoma cells through a mitochondria-mediated mechanism, by which CDDO leads to mitochondrial protein thiol modification and the generation of mitochondrial protein aggregates, which in turn contribute to the increase in the permeability of mitochondrial membranes and lead to the initiation of apoptosis [95]. The CDDO-induced mitochondrial protein aggregation can be the consequence of the inhibition of the mitochondrial ATP-dependent Lon protease. Then, CDDO and its derivatives have been demonstrated to directly and selectively inhibit Lonp1: CDDO blocks Lonp1-mediated proteolysis in biochemical and cellular assays but does not inhibit the 20S proteasome. Furthermore, a biotinylated-CDDO conjugate modifies mitochondrial Lonp1. As Lonp1 protein levels are increased in malignant lymphoma cells if compared with B cells, and considering that Lonp1 knockdown causes lymphoma cell death, the pharmacological inhibition of Lonp1 by CDDO could represent a promising therapeutic approach for B-cell lymphoma [95]. Our group expanded and deepened such idea. Using colon cancer cellular models, we observed that CDDO and CDDO-Me decrease proliferation and induce apoptosis in a dose-dependent manner. Furthermore, they are able to determine an increase in mitochondrial hydrogen peroxide and mitochondrial superoxide anion, which in turn induces the increase of carbonylation of mitochondrial proteins, causing mitochondrial depolarization, reduction of mitochondrial mass, and alteration of the organelle morphology [96]. In particular, both molecules determine an evident fragmentation of the mitochondria and the loss of the normal morphology of the matrix and of the cristae. This effect is due, at least in part, to the inhibition of Lonp1 functions, since the expression of this enzyme is not significantly altered by CDDO and CDDO-Me, but the levels of Lonp1 enzymatic activity targets, such as aconitase or TFAM, significantly increase. In line with these observations, Lonp1 overexpression abrogates the effects of CDDO and CDDO-Me, protecting cells from apoptosis [97,98].

### 3.3.7. Inhibition of Ubiquitin-Specific-Processing Protease 7 (USP7)

The last mechanism demonstrated for the anticancer activity of CDDO-Me is the inhibition of USP7, a deubiquitylating enzyme that cleaves ubiquitin from its substrates [22]. USP7 is the antagonist of MDM2, the regulator of p53 levels, and is involved in the pathogenesis and progression of several types of cancers. CDDO-Me directly binds to USP7 in cells, likely in its ubiquitin carboxyl terminus-binding pocket, and inhibits its activity, leading to the decrease of its substrates MDM2, MDMX, and UHRF1. This effect has been demonstrated in an in vitro model of ovarian cancer, and further confirmed by suppression of tumor growth in a xenograft model.

The anticancer effects described in the following paragraphs are summarized in Table 1.

**Table 1.** In vitro evidence of CDDO and CDDO-Me anticancer activity.

| Compound | Cell Line(s) | Effect(s) | Reference(s) |
|---|---|---|---|
| CDDO-Me | LNCaP, DU145, and PC3 prostate cancer cell lines | Inhibition of proliferation and induction of apoptosis; Inhibition of Akt, mTOR, NF-κB, and NF-κB-regulated antiapoptotic and proangiogenic proteins | [51] |
| CDDO-Me | PC-3 (AR(−)) and C4-2 (AR(+)) prostate cancer cells | Growth inhibition and induction of apoptosis; Inhibition of p-AKT and mTOR | [52] |
| CDDO-Me | MiaPaCa-2 and Panc-1 pancreatic cancer cells | Downregulation of p-Akt, p-mTOR and NF-kappaB; Generation of hydrogen peroxide and superoxide anion | [54] |
| CDDO-Me | U87MG, U251MG glioblastoma, and SK-N-MC neuroblastoma cell lines | Inhibition of antiapoptotic and prosurvival p-Akt, NF-kappaB (p65), and Notch1 molecules; Induction of apoptosis | [55] |
| CDDO | U87MG, U251MG glioblastoma and SK-N-MC neuroblastoma cell lines | Induction of apoptosis | [55] |
| CDDO-Me | HCT 8, HCT-15, HT-29, and Colo 205 colorectal cancer cells | Growth inhibition and induction of apoptosis; Generation of reactive oxygen species; Inhibition of Akt, mTOR, and NF-κB | [56] |
| CDDO-Me | OVCAR-3, OVCAR-5, and SK-OV3 ovarian cancer cell lines | Growth inhibition and induction of apoptosis; Inhibition of p-AKT, NF-κB, and p-mTOR | [57] |
| CDDO-Me | OVCAR-5 and MDAH 2774 ovarian cancer cells | Growth inhibition and induction of apoptosis; Inhibition of p-AKT, NF-κB, and p-mTOR; Inhibition of BCL-2, BCL-xL, c-IAP1 | [58] |
| CDDO-Me | Saos-2 osteosarcoma cells | Osteoblastic differentiation; Induction of apoptosis by caspase-8-dependent mechanisms | [67] |
| CDDO-Me | H460, A549, and H1944, H522, H157, and H1792 non-small-cell lung carcinoma cell lines | Induction of apoptosis via DR5 expression and caspase-8 activation | [69] |
| CDDO-Me | H460 and H1792 non-small-cell lung carcinoma cell lines | Trigger of ER stress; JNK-dependent, CHOP-mediated DR5 upregulation | [70] |
| CDDO-Me | H157 and A549 non-small-cell lung carcinoma cell lines | Induction of ubiquitin/proteasome-dependent c-FLIP degradation | [71] |
| CDDO | U-937 leukemia cells. | Induction of apoptosis via intrinsic pathway; Higher levels of ROS and lower levels of intracellular glutathione (GSH). | [73] |
| CDDO-Me | U-937 leukemia cells. | Induction of apoptosis via intrinsic pathway; Higher levels of ROS and lower levels of intracellular glutathione (GSH) | [73] |
| CDDO-Me | KBM5 chronic myeloid leukemia cells. | Induction of apoptosis and autophagic cell death | [76] |
| CDDO-Me | K562 chronic myeloid leukemia | Cell cycle arrest, apoptosis, and autophagy via PI3K/Akt/mTOR and p38 MAPK/Erk1/2 | [77] |

**Table 1.** *Cont.*

| Compound | Cell Line(s) | Effect(s) | Reference(s) |
|---|---|---|---|
| **CDDO-Me** | Ec109 and KYSE70 esophageal squamous cancer cells | Cell cycle arrest in G2/M phase; induction of apoptosis; Induction of autophagy by suppressing PI3K/Akt/mTOR pathway | [78] |
| **CDDO-Me** | MDA-MB-468 breast cancer cells | Inhibition of JAK1/STAT3 pathway | [82] |
| **CDDO-Me** | HeLa cervical cancer cells | Inhibition of JAK1/STAT3 pathway | [82] |
| **CDDO-Me** | KHOS, U-2OS, SaOS osteosarcoma cells | Induction of apoptosis via inhibition of STAT3 nuclear translocation and Bcl-X$_L$, survivin, and MCL-1 downregulation | [83] |
| **CDDO-Me** | MiaPaCa-2 and Panc-1 pancreatic cancer cell lines | Inhibition of telomerase activity though a ROS-dependent mechanism; Decrease of histone deacetylation and histone demethylation at hTERT promoter | [84,85] |
| **CDDO-Me** | LNCaP and PC-3 prostate cancer cell lines | Inhibition of hTERT gene expression and of hTERT telomerase activity | [86] |
| **CDDO, CDDO-Me** | OCI-Ly7, OCI-Ly19, OCI-Ly3, and OCI-Ly1 diffuse large B-cell lymphoma cell lines | Inhibition of Lonp1 protease activity | [90] |
| **CDDO, CDDO-Me** | RKO colorectal cancer cells | Impairment of mitochondrial proteome and block of mitochondrial respiration via Lonp1 inhibition | [91–93] |
| **CDDO, CDDO-Me** | SKOV3, OVCAR3, A2780, A2780/CP70, and HeyC2 ovarian cancer cell lines | Inhibition of deubiquitinating enzyme USP7 | [22] |

## 4. Anticancer Effects In Vivo

Several studies have been performed to test if CDDO and CDDO-Me anticancer effects observed in vitro could be mirrored by a similar effect in animal models. Most of the studies in animal models have been focused on CDDO-Me, which was considered the most potent molecule, and the most promising as a candidate for tests in humans. CDDO-Me demonstrated its efficacy as anticancer drugs in different mouse models, and versus several types of cancer. Doses of CDDO-Me tested were between 7.5–60 mg/kg/day (see Table 2).

CDDO-Me has been shown to inhibit lung carcinogenesis in vivo. Treatment with vinyl carbamate, a potent mutagenic agent, induces lung adenocarcinoma in female A/J mice in 16 weeks, but treatment with CDDO-Me together with vinyl carbamate markedly reduced number, size, and severity of tumors [99]. The same group observed that in another model of carcinoma, i.e., the mouse mammary tumor MMTV-neu transgenic model, CDDO-Me plus the rexinoid LG100268 significantly delayed the onset of estrogen receptor (ER)-negative mammary tumors if compared to controls [100]. The effects of the two drugs were synergic, as mice treated with both compounds showed a much higher reduction of tumor development than mice treated with individual drugs.

**Table 2.** Anticancer effects of CDDO and CDDO-Me in vivo described in the text.

| Compound | Animal Model | Treatment | Effect(s) | Reference(s) |
|---|---|---|---|---|
| **CDDO-Me** | Female A/Jm mice | Oral assumption; 40 mg/kg from the 8th week of age | CDDO-Me reduced number size and severity of lung carcinomas induced by vinyl carbamate; acts synergistically with the rexinoid LG100268 | [94] |
| **CDDO -Me** | FVB/N-Tg(MMTVneu)202Mul/J female mice | Oral assumption; 60 mg/kg from the 10th week of age for up to 45 weeks | CDDO-Me delays development of ER-negative tumors of 14 weeks; acts synergistically with the rexinoid LG100268 | [95] |
| **CDDO-Me** | FVB/N-Tg(MMTV-PyVT)634Mul/J mice | Oral assumption; 50 mg/kg | CDDO-Me delays mammary carcinogenesis in PyMT breast ER-negative cancer by 4.3 weeks | [97] |
| **CDDO-Me** | $Brca1^{Co/Co}$; $MMTV$-$Cre;p53^{+/-}$ mice | Oral assumption; 50 mg/kg | CDDO-Me delays breast cancer development by an average of 5.2 weeks | [96] |
| **CDDO-Me** | C57BL/6-Tg(TRAMP)8247Ng/J mice | Oral assumption; 7.5 mg/kg from the 5th week of age; treatment for 7 or 20 weeks. | CDDO-Me inhibits the progression of the preneoplastic lesions to prostate adenocarcinoma; inhibits metastasis | [98,99] |
| **CDDO-Me** | $LSL$-$Kras^{G12D/+}$; $LSL$-$Trp53^{R127H/+}$; $Pdx$-$1$-$Cre$ (KPC) mice | Oral assumption; 60 mg/kg from the 4th week of age | CDDO-Me increases mice survival by 3–4 weeks; acts synergistically with rexinoid LG268 | [100] |
| **CDDO-Me** | Female C57BL/6 mice | Intravenous injections of CDDO-Me nanoparticles; intraperitoneal injections of CDDO-Me every other day (5 mg/kg) | CDDO-Me enhances efficacy of vaccine therapy for melanoma | [101] |

CDDO-Me also delays mammary carcinogenesis in the aggressive PyMT model of estrogen receptor-negative breast cancer. In this model, the PyMT gene is under the control of the MMTV promoter, and the mice developed a tumor that recapitulates the key features of the human disease [101]. CDDO-Me, at the dose of 50 mg/kg/day, significantly delays tumor onset. This increase in survival is mediated by different mechanisms: inhibition of EGFR and STAT3 pathways, reduction in the infiltration of tumor-associated macrophages in the tumor microenvironment, reduction of levels of chemokines able to attract and activate lymphocytes and monocytes, such as CXCL12 and CCL2, and decreased secretion of matrix metalloproteinases, crucial for invasion and metastasis [102].

CDDO-Me delays tumor development in a mouse model with ablation of breast cancer-associated gene (BRCA1) and single allele mutation of p53 (Brca1Co/Co; MMTV-Cre; p53$^{+/-}$ mice). In this model, supplementation of CDDO-Me in the diet from 12 weeks of age delayed breast cancer development by an average of 5.2 weeks [23].

CDDO-Me inhibits the progression of preneoplastic lesions to adenocarcinoma in a transgenic mouse model of prostate adenocarcinoma [103]. The delayed progression has been observed in more than 70% of the mice and importantly, no evident toxicity of the drug was observed [103]. Not surprisingly, studies on primary cell culture from the same model showed that the anticancer effect was due to antiproliferative, proapoptotic effect of CDDO-Me, mediated by the downregulation of Akt, mTOR, NF-κB, and of the NF-κB-regulated antiapoptotic and proangiogenic proteins [104], as well as to the reduction of telomerase reverse transcriptase activity [91]. A similar effect has been observed in a transgenic model of pancreatic cancer that recapitulates the genetic mutations, clinical symptoms, and histopathology of the human disease [105]. In this model, CDDO-Me, alone or in combination with the drug LG268, increases survival of mice by 3–4 weeks.

Finally, CDDO-Me enhances the efficacy of vaccine therapy for melanoma [106]. In an experimental model of melanoma, the efficacy of Trp2 vaccination in female C57BL/6 mice inoculated with inoculated with B16F10 melanoma cells was significantly increased by CDDO-Me. The enhanced efficacy of the vaccine is due to the remodeling of the tumor microenvironment, in which CDDO-Me, delivered by nanoparticle to the tumor mass, remodels the tumor-associated fibroblasts, collagen and vessel and enhances the Fas signaling pathway, which in turn sensitizes cancer cells for killing by cytotoxic T lymphocytes.

## 5. Anticancer Effects in Humans: Clinical Trials

Since pre-clinical studies showed beneficial activity of CDDO-Me in animal models as an antitumor compound, several clinical trials have been conducted in humans to test its efficacy to evaluate its activity against solid tumors and lymphoid malignancies. In these trials, CDDO-Me is usually referred to as bardoxolone methyl or RTA-4012. So far, 33 clinical trials have been registered in clinicaltrials.gov.

The first phase I clinical trial of CDDO-Me was conducted in patients with advanced solid tumor and lymphoma to identify the determine the dose-limiting toxicity (DLT) and the maximum tolerated dose (MTD). CDDO-Me was administered orally once a day for 21 days. The MTD was established as 900 mg/day and was associated with the antitumor activity, with complete response in a patient with mantle cell lymphoma, and partial response in a patient with anaplastic thyroid carcinoma [107]. In this first trial, an increase in estimated glomerular filtration rate (eGFR) was also noted. This observation led to the proposal to use CDDO-Me for treatment of patients with chronic kidney disease (CKD) and prompted a phase II trial in patients with moderate to severe CKD and type 2 diabetes. In this trial, patients received placebo or oral CDDO-Me at a dose of 25, 75, or 150 mg once daily for 52 weeks. Kidney function improvements were observed, and only mild to moderate adverse effects occurred, with muscle spasms, hypomagnesemia, mild elevations in alanine aminotransferase levels, and gastrointestinal effects being the most common [108].

Then, a phase III trial, named BEACON (NCT01351675) was designed to test the efficacy of CDDO-Me on patients with stage 4 CKD and type 2 diabetes [109,110]. BEACON was a randomized, double-blind, parallel-group, international, multicenter trial of once-daily administration of 20 mg of CDDO-Me, compared with placebo. Patients enrolled in BEACON were adults with T2DM and stage 4 CKD. Patients received background conventional therapy (inhibitors of the renin-angiotensin-aldosterone system, insulin or hypoglycemic agents) and were randomized 1:1 for administration of CDDO-Me or placebo. BEACON was stopped early because of a significant increase in heart failure events within the first 4 weeks of treatment [109]. These events were caused by fluid retention and occurred in patients with prior history of heart failure and elevated baseline B-type natriuretic peptide, while no evidence of direct cardiotoxicity was observed [111,112]. Thus, trials are ongoing, focused on the use of CDDO-Me for treating CKD or pulmonary hypertension, rather than for cancer treatment.

## 6. Conclusions and Future Perspectives

CDDO and CDDO-Me represent interesting examples of molecules derived from natural compounds that potentiate the effects of the natural counterpart. Nevertheless, the difficulty to identify all the targets and mechanisms of action of these compounds, as well as the toxic effects observed in clinical trials limit their potential as candidates for cancer treatment in humans. As the potential of natural terpenoids remains largely unexplored, it is likely that other triterpenoids and derivatives hold potential as future therapeutics.

**Author Contributions:** R.B., L.F., A.D.G. and L.G. performed bibliography research, prepared figures and tables, L.F., R.B., S.D.B., M.N., A.C. and M.P. wrote the manuscript, M.P. conceived and supervised the work.

**Funding:** This study has been funded by Associazione Italiana per la Ricerca sul Cancro (AIRC), IG 19876 to MP.

**Conflicts of Interest:** The authors declare no conflicts of interest.

## References

1. Phillips, D.R.; Rasbery, J.M.; Bartel, B.; Matsuda, S.P. Biosynthetic diversity in plant triterpene cyclization. *Curr. Opin. Plant Biol.* **2006**, *9*, 305–314. [CrossRef]
2. Ovesna, Z.; Vachalkova, A.; Horvathova, K.; Tothova, D. Pentacyclic triterpenoic acids: New chemoprotective compounds. Minireview. *Neoplasma* **2004**, *51*, 327–333.
3. Tang, W.; Eisenbrand, G. *Chinese Drugs of Plant Origin: Chemistry, Pharmacology, and Use in Traditional and Modern Medicine*; Springer: Berlin/Heidelberg, Germany; New York, NY, USA, 1992; p. ix. 1056 p.
4. Huang, M.T.; Ho, C.T.; Wang, Z.Y.; Ferraro, T.; Lou, Y.R.; Stauber, K.; Ma, W.; Georgiadis, C.; Laskin, J.D.; Conney, A.H. Inhibition of skin tumorigenesis by rosemary and its constituents carnosol and ursolic acid. *Cancer Res.* **1994**, *54*, 701–708.
5. Jager, S.; Trojan, H.; Kopp, T.; Laszczyk, M.N.; Scheffler, A. Pentacyclic Triterpene Distribution in Various Plants - Rich Sources for a New Group of Multi-Potent Plant Extracts. *Molecules* **2009**, *14*, 2016–2031. [CrossRef]
6. Zhu, Y.Y.; Huang, H.Y.; Wu, Y.L. Anticancer and apoptotic activities of oleanolic acid are mediated through cell cycle arrest and disruption of mitochondrial membrane potential in HepG2 human hepatocellular carcinoma cells. *Mol. Med. Rep.* **2015**, *12*, 5012–5018. [CrossRef]
7. Jesus, J.A.; Lago, J.H.; Laurenti, M.D.; Yamamoto, E.S.; Passero, L.F. Antimicrobial activity of oleanolic and ursolic acids: An update. *Evid. Based Complement. Alterna. Med. eCAM* **2015**, *2015*, 620472. [CrossRef]
8. Kim, S.; Lee, H.; Lee, S.; Yoon, Y.; Choi, K.H. Antimicrobial action of oleanolic acid on Listeria monocytogenes, Enterococcus faecium, and Enterococcus faecalis. *PLoS ONE* **2015**, *10*, e0118800. [CrossRef]
9. Liby, K.T.; Sporn, M.B. Synthetic oleanane triterpenoids: Multifunctional drugs with a broad range of applications for prevention and treatment of chronic disease. *Pharmacol. Rev.* **2012**, *64*, 972–1003. [CrossRef]
10. Wang, X.; Bai, H.; Zhang, X.; Liu, J.; Cao, P.; Liao, N.; Zhang, W.; Wang, Z.; Hai, C. Inhibitory effect of oleanolic acid on hepatocellular carcinoma via ERK-p53-mediated cell cycle arrest and mitochondrial-dependent apoptosis. *Carcinogenesis* **2013**, *34*, 1323–1330. [CrossRef]
11. Zhao, X.; Liu, M.; Li, D. Oleanolic acid suppresses the proliferation of lung carcinoma cells by miR-122/Cyclin G1/MEF2D axis. *Mol. Cell. Biochem.* **2015**, *400*, 1–7. [CrossRef]
12. Honda, T.; Rounds, B.V.; Bore, L.; Finlay, H.J.; Favaloro, F.G., Jr.; Suh, N.; Wang, Y.; Sporn, M.B.; Gribble, G.W. Synthetic oleanane and ursane triterpenoids with modified rings A and C: A series of highly active inhibitors of nitric oxide production in mouse macrophages. *J. Med. Chem.* **2000**, *43*, 4233–4246. [CrossRef]
13. Chen, F.; Tholl, D.; Bohlmann, J.; Pichersky, E. The family of terpene synthases in plants: A mid-size family of genes for specialized metabolism that is highly diversified throughout the kingdom. *Plant J.* **2011**, *66*, 212–229. [CrossRef]
14. Vranova, E.; Coman, D.; Gruissem, W. Structure and dynamics of the isoprenoid pathway network. *Mol. Plant* **2012**, *5*, 318–333. [CrossRef]
15. Augustin, J.M.; Kuzina, V.; Andersen, S.B.; Bak, S. Molecular activities, biosynthesis and evolution of triterpenoid saponins. *Phytochemistry* **2011**, *72*, 435–457. [CrossRef]

16. Corey, E.J.; Lee, J. Enantioselective Total Synthesis of Oleanolic Acid, Erythrodiol, Beta-Amyrin, and Other Pentacyclic Triterpenes from a Common Intermediate. *J. Am. Chem. Soc.* **1993**, *115*, 8873–8874. [CrossRef]

17. Honda, T.; Finlay, H.J.; Gribble, G.W.; Suh, N.; Sporn, M.B. New enone derivatives of oleanolic acid and ursolic acid as inhibitors of nitric oxide production in mouse macrophages. *Bioorg. Med. Chem. Lett.* **1997**, *7*, 1623–1628. [CrossRef]

18. Honda, T.; Rounds, B.V.; Gribble, G.W.; Suh, N.; Wang, Y.; Sporn, M.B. Design and synthesis of 2-cyano-3,12-dioxoolean-1,9-dien-28-oic acid, a novel and highly active inhibitor of nitric oxide production in mouse macrophages. *Bioorg. Med. Chem. Lett.* **1998**, *8*, 2711–2714. [CrossRef]

19. Fu, L.; Gribble, G.W. Efficient and scalable synthesis of bardoxolone methyl (cddo-methyl ester). *Org. Lett.* **2013**, *15*, 1622–1625. [CrossRef]

20. Couch, R.D.; Browning, R.G.; Honda, T.; Gribble, G.W.; Wright, D.L.; Sporn, M.B.; Anderson, A.C. Studies on the reactivity of CDDO, a promising new chemopreventive and chemotherapeutic agent: Implications for a molecular mechanism of action. *Bioorg. Med. Chem. Lett.* **2005**, *15*, 2215–2219. [CrossRef]

21. Ahmad, R.; Raina, D.; Meyer, C.; Kharbanda, S.; Kufe, D. Triterpenoid CDDO-Me blocks the NF-kappaB pathway by direct inhibition of IKKbeta on Cys-179. *J. Biol. Chem.* **2006**, *281*, 35764–35769. [CrossRef]

22. Qin, D.; Wang, W.W.; Lei, H.; Luo, H.; Cai, H.Y.; Tang, C.X.; Wu, Y.Z.; Wang, Y.Y.; Jin, J.; Xiao, W.L.; et al. CDDO-Me reveals USP7 as a novel target in ovarian cancer cells. *Oncotarget* **2016**, *7*, 77096–77109. [CrossRef]

23. Kim, E.H.; Deng, C.; Sporn, M.B.; Royce, D.B.; Risingsong, R.; Williams, C.R.; Liby, K.T. CDDO-methyl ester delays breast cancer development in BRCA1-mutated mice. *Cancer Prev. Res. (Phila)* **2012**, *5*, 89–97. [CrossRef]

24. Nguyen, T.; Sherratt, P.J.; Huang, H.C.; Yang, C.S.; Pickett, C.B. Increased protein stability as a mechanism that enhances Nrf2-mediated transcriptional activation of the antioxidant response element. Degradation of Nrf2 by the 26 S proteasome. *J. Biol. Chem.* **2003**, *278*, 4536–4541. [CrossRef]

25. Itoh, K.; Wakabayashi, N.; Katoh, Y.; Ishii, T.; O'Connor, T.; Yamamoto, M. Keap1 regulates both cytoplasmic-nuclear shuttling and degradation of Nrf2 in response to electrophiles. *Genes Cells* **2003**, *8*, 379–391. [CrossRef]

26. Kobayashi, M.; Yamamoto, M. Nrf2-Keap1 regulation of cellular defense mechanisms against electrophiles and reactive oxygen species. *Adv. Enzyme Regul.* **2006**, *46*, 113–140. [CrossRef]

27. Dinkova-Kostova, A.T.; Holtzclaw, W.D.; Cole, R.N.; Itoh, K.; Wakabayashi, N.; Katoh, Y.; Yamamoto, M.; Talalay, P. Direct evidence that sulfhydryl groups of Keap1 are the sensors regulating induction of phase 2 enzymes that protect against carcinogens and oxidants. *Proc. Natl. Acad. Sci. USA* **2002**, *99*, 11908–11913. [CrossRef]

28. Nguyen, T.; Sherratt, P.J.; Pickett, C.B. Regulatory mechanisms controlling gene expression mediated by the antioxidant response element. *Annu. Rev. Pharmacol. Toxicol.* **2003**, *43*, 233–260. [CrossRef]

29. Li, J.; Johnson, D.; Calkins, M.; Wright, L.; Svendsen, C.; Johnson, J. Stabilization of Nrf2 by tBHQ confers protection against oxidative stress-induced cell death in human neural stem cells. *Toxicol. Sci.* **2005**, *83*, 313–328. [CrossRef]

30. Tran, T.A.; McCoy, M.K.; Sporn, M.B.; Tansey, M.G. The synthetic triterpenoid CDDO-methyl ester modulates microglial activities, inhibits TNF production, and provides dopaminergic neuroprotection. *J. Neuroinflammation* **2008**, *5*, 14. [CrossRef]

31. Yang, L.; Calingasan, N.Y.; Thomas, B.; Chaturvedi, R.K.; Kiaei, M.; Wille, E.J.; Liby, K.T.; Williams, C.; Royce, D.; Risingsong, R.; et al. Neuroprotective effects of the triterpenoid, CDDO methyl amide, a potent inducer of Nrf2-mediated transcription. *PLoS ONE* **2009**, *4*, e5757. [CrossRef]

32. Suh, N.; Honda, T.; Finlay, H.J.; Barchowsky, A.; Williams, C.; Benoit, N.E.; Xie, Q.W.; Nathan, C.; Gribble, G.W.; Sporn, M.B. Novel triterpenoids suppress inducible nitric oxide synthase (iNOS) and inducible cyclooxygenase (COX-2) in mouse macrophages. *Cancer Res.* **1998**, *58*, 717–723.

33. Thimmulappa, R.K.; Scollick, C.; Traore, K.; Yates, M.; Trush, M.A.; Liby, K.T.; Sporn, M.B.; Yamamoto, M.; Kensler, T.W.; Biswal, S. Nrf2-dependent protection from LPS induced inflammatory response and mortality by CDDO-Imidazolide. *Biochem. Biophys Res. Commun.* **2006**, *351*, 883–889. [CrossRef]

34. Segal, B.H.; Han, W.; Bushey, J.J.; Joo, M.; Bhatti, Z.; Feminella, J.; Dennis, C.G.; Vethanayagam, R.R.; Yull, F.E.; Capitano, M.; et al. NADPH oxidase limits innate immune responses in the lungs in mice. *PLoS ONE* **2010**, *5*, e9631. [CrossRef]

35. Choi, S.H.; Kim, B.G.; Robinson, J.; Fink, S.; Yan, M.; Sporn, M.B.; Markowitz, S.D.; Letterio, J.J. Synthetic triterpenoid induces 15-PGDH expression and suppresses inflammation-driven colon carcinogenesis. *J. Clin. Investig.* **2014**, *124*, 2472–2482. [CrossRef]

36. Fitzpatrick, L.R.; Stonesifer, E.; Small, J.S.; Liby, K.T. The synthetic triterpenoid (CDDO-Im) inhibits STAT3, as well as IL-17, and improves DSS-induced colitis in mice. *Inflammopharmacology* **2014**, *22*, 341–349. [CrossRef]

37. Duan, Z.; Ames, R.Y.; Ryan, M.; Hornicek, F.J.; Mankin, H.; Seiden, M.V. CDDO-Me, a synthetic triterpenoid, inhibits expression of IL-6 and Stat3 phosphorylation in multi-drug resistant ovarian cancer cells. *Cancer Chemother. Pharmacol.* **2009**, *63*, 681–689. [CrossRef]

38. Kulkarni, A.A.; Thatcher, T.H.; Hsiao, H.M.; Olsen, K.C.; Kottmann, R.M.; Morrissette, J.; Wright, T.W.; Phipps, R.P.; Sime, P.J. The triterpenoid CDDO-Me inhibits bleomycin-induced lung inflammation and fibrosis. *PLoS ONE* **2013**, *8*, e63798. [CrossRef]

39. Wang, Y.Y.; Zhang, C.Y.; Ma, Y.Q.; He, Z.X.; Zhe, H.; Zhou, S.F. Therapeutic effects of C-28 methyl ester of 2-cyano-3,12-dioxoolean-1,9-dien-28-oic acid (CDDO-Me; bardoxolone methyl) on radiation-induced lung inflammation and fibrosis in mice. *Drug Des. Devel. Ther.* **2015**, *9*, 3163–3178.

40. Ball, M.S.; Shipman, E.P.; Kim, H.; Liby, K.T.; Pioli, P.A. CDDO-Me Redirects Activation of Breast Tumor Associated Macrophages. *PLoS ONE* **2016**, *11*, e0149600. [CrossRef]

41. Dumont, M.; Wille, E.; Calingasan, N.Y.; Tampellini, D.; Williams, C.; Gouras, G.K.; Liby, K.; Sporn, M.; Nathan, C.; Flint Beal, M.; et al. Triterpenoid CDDO-methylamide improves memory and decreases amyloid plaques in a transgenic mouse model of Alzheimer's disease. *J. Neurochem.* **2009**, *109*, 502–512. [CrossRef]

42. Neymotin, A.; Calingasan, N.Y.; Wille, E.; Naseri, N.; Petri, S.; Damiano, M.; Liby, K.T.; Risingsong, R.; Sporn, M.; Beal, M.F.; et al. Neuroprotective effect of Nrf2/ARE activators, CDDO ethylamide and CDDO trifluoroethylamide, in a mouse model of amyotrophic lateral sclerosis. *Free Radic. Biol Med.* **2011**, *51*, 88–96. [CrossRef] [PubMed]

43. Pareek, T.K.; Belkadi, A.; Kesavapany, S.; Zaremba, A.; Loh, S.L.; Bai, L.; Cohen, M.L.; Meyer, C.; Liby, K.T.; Miller, R.H.; et al. Triterpenoid modulation of IL-17 and Nrf-2 expression ameliorates neuroinflammation and promotes remyelination in autoimmune encephalomyelitis. *Sci. Rep.* **2011**, *1*, 201. [CrossRef] [PubMed]

44. Wei, H.J.; Pareek, T.K.; Liu, Q.; Letterio, J.J. A unique tolerizing dendritic cell phenotype induced by the synthetic triterpenoid CDDO-DFPA (RTA-408) is protective against EAE. *Sci. Rep.* **2017**, *7*, 9886. [CrossRef] [PubMed]

45. Cano, M.; Thimmalappula, R.; Fujihara, M.; Nagai, N.; Sporn, M.; Wang, A.L.; Neufeld, A.H.; Biswal, S.; Handa, J.T. Cigarette smoking, oxidative stress, the anti-oxidant response through Nrf2 signaling, and Age-related Macular Degeneration. *Vision Res.* **2010**, *50*, 652–664. [CrossRef]

46. Wei, Y.; Gong, J.; Yoshida, T.; Eberhart, C.G.; Xu, Z.; Kombairaju, P.; Sporn, M.B.; Handa, J.T.; Duh, E.J. Nrf2 has a protective role against neuronal and capillary degeneration in retinal ischemia-reperfusion injury. *Free Radic. Biol. Med.* **2011**, *51*, 216–224. [CrossRef]

47. Sussan, T.E.; Rangasamy, T.; Blake, D.J.; Malhotra, D.; El-Haddad, H.; Bedja, D.; Yates, M.S.; Kombairaju, P.; Yamamoto, M.; Liby, K.T.; et al. Targeting Nrf2 with the triterpenoid CDDO-imidazolide attenuates cigarette smoke-induced emphysema and cardiac dysfunction in mice. *Proc. Natl. Acad. Sci. USA* **2009**, *106*, 250–255. [CrossRef]

48. Reddy, N.M.; Suryanaraya, V.; Yates, M.S.; Kleeberger, S.R.; Hassoun, P.M.; Yamamoto, M.; Liby, K.T.; Sporn, M.B.; Kensler, T.W.; Reddy, S.P. The triterpenoid CDDO-imidazolide confers potent protection against hyperoxic acute lung injury in mice. *Am. J. Respir. Crit. Care Med.* **2009**, *180*, 867–874. [CrossRef]

49. Fruman, D.A.; Rommel, C. PI3K and cancer: Lessons, challenges and opportunities. *Nat. Rev. Drug Discov.* **2014**, *13*, 140–156. [CrossRef]

50. Janku, F.; Yap, T.A.; Meric-Bernstam, F. Targeting the PI3K pathway in cancer: Are we making headway? *Nat. Rev. Clin. Oncol.* **2018**, *15*, 273–291. [CrossRef]

51. Deeb, D.; Gao, X.; Dulchavsky, S.A.; Gautam, S.C. CDDO-me induces apoptosis and inhibits Akt, mTOR and NF-kappaB signaling proteins in prostate cancer cells. *Anticancer Res.* **2007**, *27*, 3035–3044.

52. Deeb, D.; Gao, X.; Jiang, H.; Dulchavsky, S.A.; Gautam, S.C. Oleanane triterpenoid CDDO-Me inhibits growth and induces apoptosis in prostate cancer cells by independently targeting pro-survival Akt and mTOR. *Prostate* **2009**, *69*, 851–860. [CrossRef] [PubMed]

53. Deeb, D.; Gao, X.; Arbab, A.S.; Barton, K.; Dulchavsky, S.A.; Gautam, S.C. CDDO-Me: A Novel Synthetic Triterpenoid for the Treatment of Pancreatic Cancer. *Cancers* **2010**, *2*, 1779–1793. [CrossRef] [PubMed]

54. Deeb, D.; Gao, X.; Liu, Y.B.; Gautam, S.C. Inhibition of cell proliferation and induction of apoptosis by CDDO-Me in pancreatic cancer cells is ROS-dependent. *J. Exp. Ther. Oncol.* **2012**, *10*, 51–64. [PubMed]

55. Gao, X.; Deeb, D.; Jiang, H.; Liu, Y.; Dulchavsky, S.A.; Gautam, S.C. Synthetic triterpenoids inhibit growth and induce apoptosis in human glioblastoma and neuroblastoma cells through inhibition of prosurvival Akt, NF-kappaB and Notch1 signaling. *J. Neurooncol.* **2007**, *84*, 147–157. [CrossRef] [PubMed]

56. Gao, X.; Deeb, D.; Liu, P.; Liu, Y.; Arbab-Ali, S.; Dulchavsky, S.A.; Gautam, S.C. Role of reactive oxygen species (ROS) in CDDO-Me-mediated growth inhibition and apoptosis in colorectal cancer cells. *J. Exp. Ther. Oncol.* **2011**, *9*, 119–127. [PubMed]

57. Gao, X.; Liu, Y.; Deeb, D.; Arbab, A.S.; Guo, A.M.; Dulchavsky, S.A.; Gautam, S.C. Synthetic oleanane triterpenoid, CDDO-Me, induces apoptosis in ovarian cancer cells by inhibiting prosurvival AKT/NF-kappaB/mTOR signaling. *Anticancer Res.* **2011**, *31*, 3673–3681. [PubMed]

58. Gao, X.; Liu, Y.; Deeb, D.; Liu, P.; Liu, A.; Arbab, A.S.; Gautam, S.C. ROS mediate proapoptotic and antisurvival activity of oleanane triterpenoid CDDO-Me in ovarian cancer cells. *Anticancer Res.* **2013**, *33*, 215–221.

59. Suh, N.; Wang, Y.; Honda, T.; Gribble, G.W.; Dmitrovsky, E.; Hickey, W.F.; Maue, R.A.; Place, A.E.; Porter, D.M.; Spinella, M.J.; et al. A novel synthetic oleanane triterpenoid, 2-cyano-3,12-dioxoolean-1,9-dien-28-oic acid, with potent differentiating, antiproliferative, and anti-inflammatory activity. *Cancer Res.* **1999**, *59*, 336–341.

60. Ikeda, T.; Kimura, F.; Nakata, Y.; Sato, K.; Ogura, K.; Motoyoshi, K.; Sporn, M.; Kufe, D. Triterpenoid CDDO-Im downregulates PML/RARalpha expression in acute promyelocytic leukemia cells. *Cell Death Differ.* **2005**, *12*, 523–531. [CrossRef]

61. Ji, Y.; Lee, H.J.; Goodman, C.; Uskokovic, M.; Liby, K.; Sporn, M.; Suh, N. The synthetic triterpenoid CDDO-imidazolide induces monocytic differentiation by activating the Smad and ERK signaling pathways in HL60 leukemia cells. *Mol. Cancer Ther.* **2006**, *5*, 1452–1458. [CrossRef]

62. Koschmieder, S.; D'Alo, F.; Radomska, H.; Schoneich, C.; Chang, J.S.; Konopleva, M.; Kobayashi, S.; Levantini, E.; Suh, N.; Di Ruscio, A.; et al. CDDO induces granulocytic differentiation of myeloid leukemic blasts through translational up-regulation of p42 CCAAT enhancer binding protein alpha. *Blood* **2007**, *110*, 3695–3705. [CrossRef] [PubMed]

63. Tabe, Y.; Konopleva, M.; Kondo, Y.; Contractor, R.; Tsao, T.; Konoplev, S.; Shi, Y.; Ling, X.; Watt, J.C.; Tsutsumi-Ishii, Y.; et al. PPARgamma-active triterpenoid CDDO enhances ATRA-induced differentiation in APL. *Cancer Biol. Ther.* **2007**, *6*, 1967–1977. [CrossRef] [PubMed]

64. Suh, N.; Paul, S.; Lee, H.J.; Yoon, T.; Shah, N.; Son, A.I.; Reddi, A.H.; Medici, D.; Sporn, M.B. Synthetic triterpenoids, CDDO-Imidazolide and CDDO-Ethyl amide, induce chondrogenesis. *Osteoarthr. Cartil.* **2012**, *20*, 446–450. [CrossRef] [PubMed]

65. Wang, Y.; Porter, W.W.; Suh, N.; Honda, T.; Gribble, G.W.; Leesnitzer, L.M.; Plunket, K.D.; Mangelsdorf, D.J.; Blanchard, S.G.; Willson, T.M.; et al. A synthetic triterpenoid, 2-cyano-3,12-dioxooleana-1,9-dien-28-oic acid (CDDO), is a ligand for the peroxisome proliferator-activated receptor gamma. *Mol. Endocrinol.* **2000**, *14*, 1550–1556. [PubMed]

66. Tsao, T.; Kornblau, S.; Safe, S.; Watt, J.C.; Ruvolo, V.; Chen, W.; Qiu, Y.; Coombes, K.R.; Ju, Z.; Abdelrahim, M.; et al. Role of peroxisome proliferator-activated receptor-gamma and its coactivator DRIP205 in cellular responses to CDDO (RTA-401) in acute myelogenous leukemia. *Cancer Res.* **2010**, *70*, 4949–4960. [CrossRef]

67. Ito, Y.; Pandey, P.; Sporn, M.B.; Datta, R.; Kharbanda, S.; Kufe, D. The novel triterpenoid CDDO induces apoptosis and differentiation of human osteosarcoma cells by a caspase-8 dependent mechanism. *Mol. Pharmacol.* **2001**, *59*, 1094–1099. [CrossRef]

68. Hengartner, M.O. The biochemistry of apoptosis. *Nature* **2000**, *407*, 770–776. [CrossRef]

69. Troiano, L.; Ferraresi, R.; Lugli, E.; Nemes, E.; Roat, E.; Nasi, M.; Pinti, M.; Cossarizza, A. Multiparametric analysis of cells with different mitochondrial membrane potential during apoptosis by polychromatic flow cytometry. *Nature protocols* **2007**, *2*, 2719–2727. [CrossRef]

70. Ito, Y.; Pandey, P.; Place, A.; Sporn, M.B.; Gribble, G.W.; Honda, T.; Kharbanda, S.; Kufe, D. The novel triterpenoid 2-cyano-3,12-dioxoolean-1,9-dien-28-oic acid induces apoptosis of human myeloid leukemia cells by a caspase-8-dependent mechanism. *Cell Growth Differ.* **2000**, *11*, 261–267.

71. Stadheim, T.A.; Suh, N.; Ganju, N.; Sporn, M.B.; Eastman, A. The novel triterpenoid 2-cyano-3,12-dioxooleana-1,9-dien-28-oic acid (CDDO) potently enhances apoptosis induced by tumor necrosis factor in human leukemia cells. *J. Biol. Chem.* **2002**, *277*, 16448–16455. [CrossRef]

72. Konopleva, M.; Tsao, T.; Ruvolo, P.; Stiouf, I.; Estrov, Z.; Leysath, C.E.; Zhao, S.; Harris, D.; Chang, S.; Jackson, C.E.; et al. Novel triterpenoid CDDO-Me is a potent inducer of apoptosis and differentiation in acute myelogenous leukemia. *Blood* **2002**, *99*, 326–335. [CrossRef] [PubMed]

73. Samudio, I.; Konopleva, M.; Pelicano, H.; Huang, P.; Frolova, O.; Bornmann, W.; Ying, Y.; Evans, R.; Contractor, R.; Andreeff, M. A novel mechanism of action of methyl-2-cyano-3,12 dioxoolean-1,9 diene-28-oate: Direct permeabilization of the inner mitochondrial membrane to inhibit electron transport and induce apoptosis. *Mol. Pharmacol.* **2006**, *69*, 1182–1193. [CrossRef] [PubMed]

74. Hyer, M.L.; Croxton, R.; Krajewska, M.; Krajewski, S.; Kress, C.L.; Lu, M.; Suh, N.; Sporn, M.B.; Cryns, V.L.; Zapata, J.M.; et al. Synthetic triterpenoids cooperate with tumor necrosis factor-related apoptosis-inducing ligand to induce apoptosis of breast cancer cells. *Cancer Res.* **2005**, *65*, 4799–4808. [CrossRef] [PubMed]

75. Hyer, M.L.; Shi, R.; Krajewska, M.; Meyer, C.; Lebedeva, I.V.; Fisher, P.B.; Reed, J.C. Apoptotic activity and mechanism of 2-cyano-3,12-dioxoolean-1,9-dien-28-oic-acid and related synthetic triterpenoids in prostate cancer. *Cancer Res.* **2008**, *68*, 2927–2933. [CrossRef] [PubMed]

76. Samudio, I.; Konopleva, M.; Hail, N., Jr.; Shi, Y.X.; McQueen, T.; Hsu, T.; Evans, R.; Honda, T.; Gribble, G.W.; Sporn, M.; et al. 2-Cyano-3,12-dioxoolena-1,9-dien-28-imidazolide (CDDO-Im) directly targets mitochondrial glutathione to induce apoptosis in pancreatic cancer. *J. Biol.Chem.* **2005**, *280*, 36273–36282. [CrossRef] [PubMed]

77. Zou, W.; Liu, X.; Yue, P.; Zhou, Z.; Sporn, M.B.; Lotan, R.; Khuri, F.R.; Sun, S.Y. c-Jun NH2-terminal kinase-mediated up-regulation of death receptor 5 contributes to induction of apoptosis by the novel synthetic triterpenoid methyl-2-cyano-3,12-dioxoolena-1, 9-dien-28-oate in human lung cancer cells. *Cancer Res.* **2004**, *64*, 7570–7578. [CrossRef] [PubMed]

78. Zou, W.; Yue, P.; Khuri, F.R.; Sun, S.Y. Coupling of endoplasmic reticulum stress to CDDO-Me-induced up-regulation of death receptor 5 via a CHOP-dependent mechanism involving JNK activation. *Cancer Res.* **2008**, *68*, 7484–7492. [CrossRef]

79. Ikeda, T.; Sporn, M.; Honda, T.; Gribble, G.W.; Kufe, D. The novel triterpenoid CDDO and its derivatives induce apoptosis by disruption of intracellular redox balance. *Cancer Res.* **2003**, *63*, 5551–5558.

80. Fernandez-Checa, J.C.; Kaplowitz, N.; Garcia-Ruiz, C.; Colell, A. Mitochondrial glutathione: Importance and transport. *Semin. Liver Dis.* **1998**, *18*, 389–401. [CrossRef]

81. Samudio, I.; Kurinna, S.; Ruvolo, P.; Korchin, B.; Kantarjian, H.; Beran, M.; Dunner, K., Jr.; Kondo, S.; Andreeff, M.; Konopleva, M. Inhibition of mitochondrial metabolism by methyl-2-cyano-3,12-dioxoolena-1,9-diene-28-oate induces apoptotic or autophagic cell death in chronic myeloid leukemia cells. *Mol. Cancer Ther.* **2008**, *7*, 1130–1139. [CrossRef]

82. Wang, X.Y.; Zhang, X.H.; Peng, L.; Liu, Z.; Yang, Y.X.; He, Z.X.; Dang, H.W.; Zhou, S.F. Bardoxolone methyl (CDDO-Me or RTA402) induces cell cycle arrest, apoptosis and autophagy via PI3K/Akt/mTOR and p38 MAPK/Erk1/2 signaling pathways in K562 cells. *Am. J. Transl. Res.* **2017**, *9*, 4652–4672. [PubMed]

83. Wang, Y.Y.; Yang, Y.X.; Zhao, R.; Pan, S.T.; Zhe, H.; He, Z.X.; Duan, W.; Zhang, X.; Yang, T.; Qiu, J.X.; et al. Bardoxolone methyl induces apoptosis and autophagy and inhibits epithelial-to-mesenchymal transition and stemness in esophageal squamous cancer cells. *Drug Des. Devel. Ther.* **2015**, *9*, 993–1026. [PubMed]

84. Johnson, D.E.; O'Keefe, R.A.; Grandis, J.R. Targeting the IL-6/JAK/STAT3 signalling axis in cancer. *Nat. Rev. Clin. Oncol.* **2018**, *15*, 234–248. [CrossRef] [PubMed]

85. Yu, H.; Pardoll, D.; Jove, R. STATs in cancer inflammation and immunity: A leading role for STAT3. *Nat. Rev. Cancer* **2009**, *9*, 798–809. [CrossRef]

86. Grivennikov, S.I.; Karin, M. Dangerous liaisons: STAT3 and NF-kappaB collaboration and crosstalk in cancer. *Cytokine Growth Factor Rev.* **2010**, *21*, 11–19. [CrossRef]

87. Ahmad, R.; Raina, D.; Meyer, C.; Kufe, D. Triterpenoid CDDO-methyl ester inhibits the Janus-activated kinase-1 (JAK1)–>signal transducer and activator of transcription-3 (STAT3) pathway by direct inhibition of JAK1 and STAT3. *Cancer Res.* **2008**, *68*, 2920–2926. [CrossRef]

88. Ryu, K.; Susa, M.; Choy, E.; Yang, C.; Hornicek, F.J.; Mankin, H.J.; Duan, Z. Oleanane triterpenoid CDDO-Me induces apoptosis in multidrug resistant osteosarcoma cells through inhibition of Stat3 pathway. *BMC Cancer* **2010**, *10*, 187. [CrossRef]

89. Deeb, D.; Gao, X.; Liu, Y.; Varma, N.R.; Arbab, A.S.; Gautam, S.C. Inhibition of telomerase activity by oleanane triterpenoid CDDO-Me in pancreatic cancer cells is ROS-dependent. *Molecules* **2013**, *18*, 3250–3265. [CrossRef]

90. Deeb, D.; Brigolin, C.; Gao, X.; Liu, Y.; Pindolia, K.R.; Gautam, S.C. Induction of Apoptosis in Pancreatic Cancer Cells by CDDO-Me Involves Repression of Telomerase through Epigenetic Pathways. *J. Carcinog. Mutagen.* **2014**, *5*, 177. [CrossRef]

91. Liu, Y.; Gao, X.; Deeb, D.; Arbab, A.S.; Gautam, S.C. Telomerase reverse transcriptase (TERT) is a therapeutic target of oleanane triterpenoid CDDO-Me in prostate cancer. *Molecules* **2012**, *17*, 14795–14809. [CrossRef]

92. Pinti, M.; Gibellini, L.; Liu, Y.; Xu, S.; Lu, B.; Cossarizza, A. Mitochondrial Lon protease at the crossroads of oxidative stress, ageing and cancer. *Cell. Mol. Life Sci. CMLS* **2015**, *72*, 4807–4824. [CrossRef]

93. Pinti, M.; Gibellini, L.; Nasi, M.; De Biasi, S.; Bortolotti, C.A.; Iannone, A.; Cossarizza, A. Emerging role of Lon protease as a master regulator of mitochondrial functions. *Biochim. Biophys. Acta* **2016**, *1857*, 1300–1306. [CrossRef]

94. Gibellini, L.; Bianchini, E.; De Biasi, S.; Nasi, M.; Cossarizza, A.; Pinti, M. Natural Compounds Modulating Mitochondrial Functions. *Evid. Based Complement. Alterna. Med. eCAM* **2015**, *2015*, 527209. [CrossRef] [PubMed]

95. Bernstein, S.H.; Venkatesh, S.; Li, M.; Lee, J.; Lu, B.; Hilchey, S.P.; Morse, K.M.; Metcalfe, H.M.; Skalska, J.; Andreeff, M.; et al. The mitochondrial ATP-dependent Lon protease: A novel target in lymphoma death mediated by the synthetic triterpenoid CDDO and its derivatives. *Blood* **2012**, *119*, 3321–3329. [CrossRef] [PubMed]

96. Gibellini, L.; Pinti, M.; Boraldi, F.; Giorgio, V.; Bernardi, P.; Bartolomeo, R.; Nasi, M.; De Biasi, S.; Missiroli, S.; Carnevale, G.; et al. Silencing of mitochondrial Lon protease deeply impairs mitochondrial proteome and function in colon cancer cells. *FASEB J. Off. Publ. Fed. Am. Soc. Exp. Biol.* **2014**, *28*, 5122–5135. [CrossRef] [PubMed]

97. Gibellini, L.; Pinti, M.; Bartolomeo, R.; De Biasi, S.; Cormio, A.; Musicco, C.; Carnevale, G.; Pecorini, S.; Nasi, M.; De Pol, A.; et al. Inhibition of Lon protease by triterpenoids alters mitochondria and is associated to cell death in human cancer cells. *Oncotarget* **2015**, *6*, 25466–25483. [CrossRef]

98. Gibellini, L.; Losi, L.; De Biasi, S.; Nasi, M.; Lo Tartaro, D.; Pecorini, S.; Patergnani, S.; Pinton, P.; De Gaetano, A.; Carnevale, G.; et al. LonP1 Differently Modulates Mitochondrial Function and Bioenergetics of Primary Versus Metastatic Colon Cancer Cells. *Front. Oncol.* **2018**, *8*, 254. [CrossRef]

99. Liby, K.; Risingsong, R.; Royce, D.B.; Williams, C.R.; Ma, T.; Yore, M.M.; Sporn, M.B. Triterpenoids CDDO-methyl ester or CDDO-ethyl amide and rexinoids LG100268 or NRX194204 for prevention and treatment of lung cancer in mice. *Cancer Prev. Res. (Phila)* **2009**, *2*, 1050–1058. [CrossRef]

100. Liby, K.; Risingsong, R.; Royce, D.B.; Williams, C.R.; Yore, M.M.; Honda, T.; Gribble, G.W.; Lamph, W.W.; Vannini, N.; Sogno, I.; et al. Prevention and treatment of experimental estrogen receptor-negative mammary carcinogenesis by the synthetic triterpenoid CDDO-methyl Ester and the rexinoid LG100268. *Clin Cancer Res.* **2008**, *14*, 4556–4563. [CrossRef]

101. Lin, E.Y.; Jones, J.G.; Li, P.; Zhu, L.; Whitney, K.D.; Muller, W.J.; Pollard, J.W. Progression to malignancy in the polyoma middle T oncoprotein mouse breast cancer model provides a reliable model for human diseases. *Am. J. Pathol.* **2003**, *163*, 2113–2126. [CrossRef]

102. Tran, K.; Risingsong, R.; Royce, D.; Williams, C.R.; Sporn, M.B.; Liby, K. The synthetic triterpenoid CDDO-methyl ester delays estrogen receptor-negative mammary carcinogenesis in polyoma middle T mice. *Cancer Prev. Res. (Phila)* **2012**, *5*, 726–734. [CrossRef] [PubMed]

103. Gao, X.; Deeb, D.; Liu, Y.; Arbab, A.S.; Divine, G.W.; Dulchavsky, S.A.; Gautam, S.C. Prevention of Prostate Cancer with Oleanane Synthetic Triterpenoid CDDO-Me in the TRAMP Mouse Model of Prostate Cancer. *Cancers* **2011**, *3*, 3353–3369. [CrossRef] [PubMed]

104. Deeb, D.; Gao, X.; Dulchavsky, S.A.; Gautam, S.C. CDDO-Me inhibits proliferation, induces apoptosis, down-regulates Akt, mTOR, NF-kappaB and NF-kappaB-regulated antiapoptotic and proangiogenic proteins in TRAMP prostate cancer cells. *J. Exp. Ther. Oncol.* **2008**, *7*, 31–39. [PubMed]

105. Liby, K.T.; Royce, D.B.; Risingsong, R.; Williams, C.R.; Maitra, A.; Hruban, R.H.; Sporn, M.B. Synthetic triterpenoids prolong survival in a transgenic mouse model of pancreatic cancer. *Cancer Prev Res (Phila)* **2010**, *3*, 1427–1434. [CrossRef]

106. Zhao, Y.; Huo, M.; Xu, Z.; Wang, Y.; Huang, L. Nanoparticle delivery of CDDO-Me remodels the tumor microenvironment and enhances vaccine therapy for melanoma. *Biomaterials* **2015**, *68*, 54–66. [CrossRef]

107. Hong, D.S.; Kurzrock, R.; Supko, J.G.; He, X.; Naing, A.; Wheler, J.; Lawrence, D.; Eder, J.P.; Meyer, C.J.; Ferguson, D.A.; et al. A phase I first-in-human trial of bardoxolone methyl in patients with advanced solid tumors and lymphomas. *Clin. Cancer Res.* **2012**, *18*, 3396–3406. [CrossRef]

108. Pergola, P.E.; Raskin, P.; Toto, R.D.; Meyer, C.J.; Huff, J.W.; Grossman, E.B.; Krauth, M.; Ruiz, S.; Audhya, P.; Christ-Schmidt, H.; et al. Bardoxolone methyl and kidney function in CKD with type 2 diabetes. *N. Engl. J. Med.* **2011**, *365*, 327–336. [CrossRef]

109. de Zeeuw, D.; Akizawa, T.; Audhya, P.; Bakris, G.L.; Chin, M.; Christ-Schmidt, H.; Goldsberry, A.; Houser, M.; Krauth, M.; Lambers Heerspink, H.J.; et al. Bardoxolone methyl in type 2 diabetes and stage 4 chronic kidney disease. *N. Engl. J. Med.* **2013**, *369*, 2492–2503. [CrossRef]

110. De Zeeuw, D.; Akizawa, T.; Agarwal, R.; Audhya, P.; Bakris, G.L.; Chin, M.; Krauth, M.; Lambers Heerspink, H.J.; Meyer, C.J.; McMurray, J.J.; et al. Rationale and trial design of Bardoxolone Methyl Evaluation in Patients with Chronic Kidney Disease and Type 2 Diabetes: The Occurrence of Renal Events (BEACON). *Am. J. Nephrol.* **2013**, *37*, 212–222. [CrossRef]

111. Chin, M.P.; Reisman, S.A.; Bakris, G.L.; O'Grady, M.; Linde, P.G.; McCullough, P.A.; Packham, D.; Vaziri, N.D.; Ward, K.W.; Warnock, D.G.; et al. Mechanisms contributing to adverse cardiovascular events in patients with type 2 diabetes mellitus and stage 4 chronic kidney disease treated with bardoxolone methyl. *Am. J. Nephrol.* **2014**, *39*, 499–508. [CrossRef]

112. Chin, M.P.; Wrolstad, D.; Bakris, G.L.; Chertow, G.M.; de Zeeuw, D.; Goldsberry, A.; Linde, P.G.; McCullough, P.A.; McMurray, J.J.; Wittes, J.; et al. Risk factors for heart failure in patients with type 2 diabetes mellitus and stage 4 chronic kidney disease treated with bardoxolone methyl. *J. Card. Fail.* **2014**, *20*, 953–958. [CrossRef] [PubMed]

*molecules*

MDPI

*Review*

# Advances in Biosynthesis, Pharmacology, and Pharmacokinetics of Pinocembrin, a Promising Natural Small-Molecule Drug

Xiaoling Shen, Yeju Liu, Xiaoya Luo and Zhihong Yang *

Institute of Medicinal Plant Development, Chinese Academy of Medical Sciences and Peking Union Medical College, Beijing 100193, China; 15189806892@163.com (X.S.); lyj18811358958@163.com (Y.L.); 18804897936@163.com (X.L.)
* Correspondence: zhyang@implad.ac.cn; Tel.: +86-10-57833219

Academic Editors: Pavel B. Drasar and Vladimir A. Khripach
Received: 31 May 2019; Accepted: 23 June 2019; Published: 24 June 2019

check for updates

**Abstract:** Pinocembrin is one of the most abundant flavonoids in propolis, and it may also be widely found in a variety of plants. In addition to natural extraction, pinocembrin can be obtained by biosynthesis. Biosynthesis efficiency can be improved by a metabolic engineering strategy and a two-phase pH fermentation strategy. Pinocembrin poses an interest for its remarkable pharmacological activities, such as neuroprotection, anti-oxidation, and anti-inflammation. Studies have shown that pinocembrin works excellently in treating ischemic stroke. Pinocembrin can reduce nerve damage in the ischemic area and reduce mitochondrial dysfunction and the degree of oxidative stress. Given its significant efficacy in cerebral ischemia, pinocembrin has been approved by China Food and Drug Administration (CFDA) as a new treatment drug for ischemic stroke and is currently in progress in phase II clinical trials. Research has shown that pinocembrin can be absorbed rapidly in the body and easily cross the blood–brain barrier. In addition, the absorption/elimination process of pinocembrin occurs rapidly and shows no serious accumulation in the body. Pinocembrin has also been found to play a role in Parkinson's disease, Alzheimer's disease, and specific solid tumors, but its mechanisms of action require in-depth studies. In this review, we summarized the latest 10 years of studies on the biosynthesis, pharmacological activities, and pharmacokinetics of pinocembrin, focusing on its effects on certain diseases, aiming to explore its targets, explaining possible mechanisms of action, and finding potential therapeutic applications.

**Keywords:** pinocembrin; microbial biosynthesis; pharmacological activities; pharmacokinetic features; research progress

## 1. Introduction

Pinocembrin is a pharmacologically active flavonoid that is mainly found in propolis, with a content reaching up to 606–701 mg/g in balsam which extracted from propolis with 70% ethanol [1]. Besides, it can be isolated from a variety of medicinal plants, such as *Peperomia* and *Piper* genera and *Asteraceae* families [2–4]. In addition to extraction from natural products, pinocembrin can be synthesized by biological and chemical methods. Biosynthesis plays an important role in the synthesis of pinocembrin owing to its high yield and low production cost. In terms of pharmacological effects, pinocembrin can exhibit anti-inflammatory [5], anti-oxidant [6], antibacterial [7] and neuroprotective activities [8]. Research has shown that pinocembrin exhibits a positive effect on the treatment of ischemic stroke. Pinocembrin can reduce the area of cerebral infarction in rats with cerebral ischemia and reduce the degree of cerebral edema and apoptosis of nerve cells. In addition, pinocembrin can protect the integrity of the blood–brain barrier (BBB), thereby reducing mortality and improving neurobehavioral scores in rats [9]. Due to its significant pharmacological activity, pinocembrin has been approved by China Food and Drug Administration (CFDA) as a new drug for treatment of ischemic stroke, now in Phase II clinical trials. Recent studies have also revealed that pinocembrin possesses an anti-tumor effect, such as against melanoma [10]. It also shows anti-fibrosis [11] effect. An excellent drug not only possesses good pharmacological activities but also good pharmacokinetic (PK) parameters. Pinocembrin features a small molecular weight (Figure 1) and good liposolubility, allowing it to easily pass through the BBB. The transport mode of pinocembrin may primarily transpire by passive transport. This property suggests that this candidate drug can be used for treatment of brain diseases [12]. Both clinical and preclinical experiments have shown that pinocembrin can be absorbed quickly and distributed widely without notable accumulation of residues, indicating that this compound possesses good PK profiles. Thus, pinocembrin is a potential natural small-molecule drug with good prospects for further development.

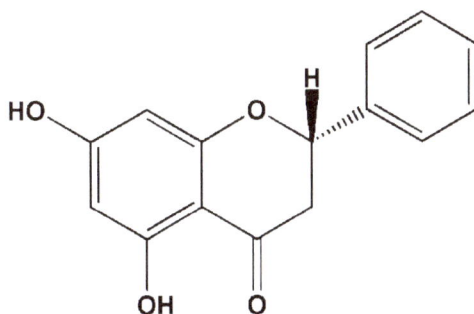

**Figure 1.** Chemical structure of (2S)-pinocembrin. Overall, this article reviews and expounds in detail the progress on pinocembrin biosynthesis, its pharmacological actions and partial mechanisms, and certain drug metabolism characteristics in vitro and in vivo, which will provide advantageous information for the comprehensive study of the pharmacokinetic (PK) features and pharmacological mechanisms of pinocembrin.

## 2. Microbial Biosynthesis

### 2.1. Synthesis of Pinocembrin from Glucose

Pinocembrin can be extracted from natural products but at a high production cost and insufficient yield. Microbial biosynthesis features the advantages of low cost and large output, which compensate for the lack of natural sources of pinocembrin. *Escherichia coli* is widely used in producing pinocembrin. In recent years, researchers have been working on how to produce pinocembrin efficiently. Biosynthesis of pinocembrin often require supplementation of expensive phenylpropanoic precursors (Figure 2), presenting a major problem in the past studies. To solve this issue, genetic engineering is used to construct engineering bacteria in order to synthesize pinocembrin from glucose. In order to produce the flavonoid precursor (2S)-pinocembrin directly from glucose, 3-deoxy-D-arabinoheptulosonate-7-phosphate synthase, chorismate mutase/prephenate dehydratase, phenylalanine ammonia lyase (PAL), 4-coumarate:CoA ligase (4CL), chalcone synthase (CHS), chalcone isomerase (CHI), malonate synthetase, and malonate carrier protein have been assembled in four vectors. Synthesizing pinocembrin from glucose can be realized with the adjustment of other corresponding conditions [13,14].

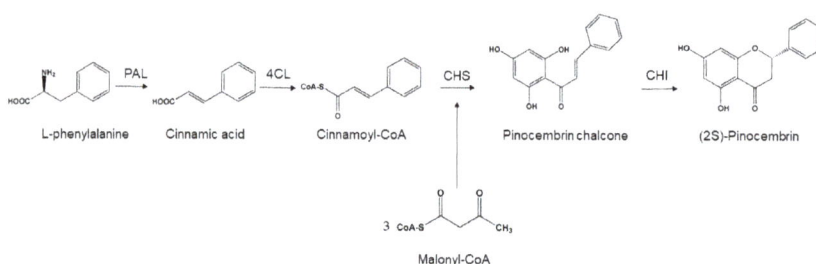

**Figure 2.** Classic biosynthetic pathway of pinocembrin [15]. L-Phenylalanine as a precursor compound produces cinnamic acid under the action of phenylalanine ammonia lyase, which generates cinnamoyl-CoA under the action of CoA ligase, and adds malonyl-CoA to the reaction system to form pinocembrin chalcone under the action of chalcone synthase. Finally, pinocembrin is generated under the action of chalcone isomerase.

## 2.2. Production Optimization Measures of Pinocembrin

Several key factors may affect the production efficiency of pinocembrin. First, *E. coli* metabolites affect the synthesis of pinocembrin. The production efficiency of pinocembrin is limited by the content of malonyl-CoA. However, in *E. coli*, both the biosynthesis of pinocembrin and self-fatty acid biosynthesis of the bacterium consume malonyl-CoA. Thus, limiting *E. coli* self-fatty acid synthesis can improve the synthetic efficiency of pinocembrin. Research has shown that overexpressing enzymes β-ketoacyl-acyl carrier protein synthase III (FabH), FabF, or both enzymes in *E. coli* BL21 (DE3) decreased fatty acid synthesis and increased cellular malonyl-CoA levels, thereby up-regulating the production of pinocembrin [16]. In addition, the accumulation of cinnamic acid adversely affects the production of pinocembrin. Screening gene sources and optimizing gene expression are employed to regulate the synthetic pathway of cinnamic acid. Then, site-directed mutagenesis of chalcone synthase and cofactor engineering are used to optimize the downstream pathway of cinnamic acid consumption. These strategies reduce the accumulation of cinnamic acid and increase the yield of pinocembrin [17].

Adenosine triphosphate (ATP) is the source of energy for various activities in living organisms. Therefore, ATP concentration in *E. coli* must be controlled. To screen several ATP-related candidate genes, a clustered regularly interspaced short palindromic repeats (CRISPR) interference system has been established. MetK and proB have been found to show potential in improving ATP level and increasing the production of pinocembrin [18]. Different culture pH values during microbial fermentation also affect microbial fermentation. Studies have shown that in the biosynthesis of pinocembrin, high pH values favor upstream pathway catalysis, whereas low pH values favor downstream pathway catalysis. Thus, a two-stage pH control strategy has been proposed [19].

In one-step (2*S*)-pinocembrin production, expensive malonyl-CoA precursor malonate is needed, and morpholinopropane sulfonate is required to provide buffering capacity. To solve this problem, a CRISPR interference system has been established to effectively guide carbon flux to malonyl coenzyme A. In addition, by adjusting the pH value of the fermentation system, the yield of pinocembrin can be significantly improved [19]. In summary, the increase in pinocembrin production can be achieved by a stepwise metabolic engineering strategy in combination with malonyl-CoA engineering. In addition to the above strategies, a two-stage pH fermentation strategy and optimized strain culture should be combined to increase the production of pinocembrin. The synthesis of (2*S*)-pinocembrin can be achieved by assembling the Oc4CL1, OcCHS2, and MsCHI genes obtained in alfalfa into *E. coli* by a gene manufacturer [20].

## 3. Pharmacological Effects of Pinocembrin

Pinocembrin is mainly used for ischemic stroke treatment. However, recent studies have indicated that pinocembrin may exert therapeutic effects on Parkinson's disease (PD) and Alzheimer's disease (AD). Pinocembrin also exhibits anti-pulmonary fibrosis and vasodilating activities. Pinocembrin undergoes multiple mechanisms to perform its pharmacological effects. Furthermore, pinocembrin may alleviate BBB disruption and neurological injury via reducing the levels of reactive oxygen species (ROS) and inflammatory factors. In addition, pinocembrin may preserve mitochondrial integrity by activating the extracellular signal-regulated kinase/nuclear factor erythroid 2-related factor 2 (Erk1/2-Nrf2) pathway [21]. Pinocembrin also attenuates apoptosis by affecting the p53 pathway, thereby influencing the Bax-Bcl-2 ratio and the release of cytochrome C [22]. Antioxidant and anti-inflammatory activities serve as the basis for various pharmacological effects of pinocembrin (Figure 3). An in-depth understanding of its pharmacological activities and mechanisms of actions can aid in discovering new targets and potential therapeutic applications of pinocembrin.

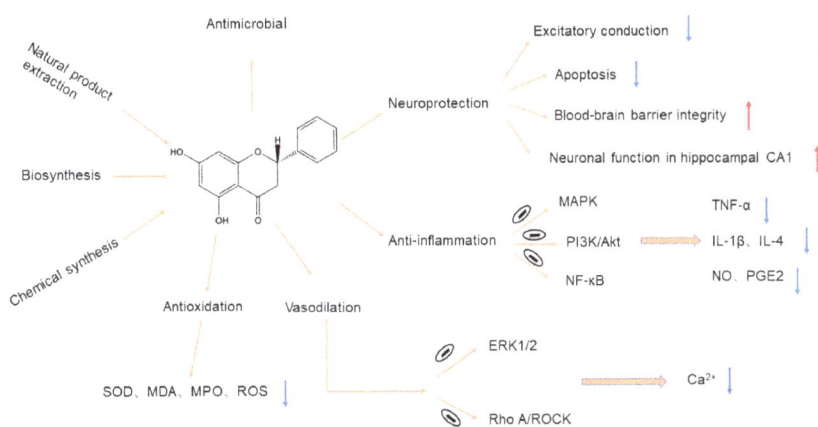

**Figure 3.** Pharmacological action and possible mechanisms of pinocembrin. Pinocembrin features a variety of pharmacological activities. This compound can inhibit the expression of pro-inflammatory factors by inhibiting mitogen-activated protein kinase (MAPK), phosphoinositide 3-kinase (PI3K)/AKT, and nuclear factor kappa B (NF-κB) signaling pathways, thereby exerting anti-inflammatory effects. Its vasodilation effect is achieved by inhibiting the ERK1/2 and Rho-associated protein kinase (ROCK) signaling pathways and then downregulating calcium ion concentration. The neuroprotective effects of pinocembrin mainly include the reduction in nerve excitability and neuronal apoptosis and enhances activity of cells in hippocampal CA1 region. Pinocembrin can down-regulate the contents of superoxide dismutase (SOD), malondialdehyde (MDA), myeloperoxidase (MPO), and ROS to achieve antioxidant effects.

### 3.1. Neuroprotective Activity

#### 3.1.1. Neuroprotective Effect in Cerebral Ischemia

Research shows that pinocembrin possesses the potential to become a drug for the treatment of ischemic stroke. Against the background trend of an increasingly aging population worldwide, the risk of cerebral ischemia is increasing, with 70% of survivors presenting physical disabilities [23]. Thus, developing drugs to treat cerebral ischemia is crucial.

In vitro, pinocembrin can inhibit the reactivity of SN/L7 to 5-HT by reducing the excitatory conduction of synapses in co-cultured Aplysia SN/L7 neurons reversibly. This phenomenon is related to glutamate receptors in the postsynaptic membrane [24]. These events indicate that pinocembrin exerts certain effects on the nervous system. In an oxygen-glucose deprivation/reoxygenation model, pinocembrin can increase neuronal survival rates, decrease the amount of lactate dehydrogenase, and alleviate neurite length and apoptosis during reoxygenation [8]. These studies have initially shown that pinocembrin can alleviate nerve damage caused by cerebral ischemia in vitro. To verify whether pinocembrin also features neuroprotective effects in vivo, a rat model of focal cerebral ischemia has been applied. The result has shown that pinocembrin reduced the leakage of Evans Blue and sodium fluorescein, manifesting a protective action on BBB integrity [25]. Continued research has revealed that pinocembrin can improve the morphology of brain cortex, striatum, and hippocampal neurons in rats with acute focal cerebral ischemia/reperfusion. After ischemia reperfusion in rats, pinocembrin significantly suppressed the levels of neuronal specific enolase and S-100β protein in blood. In addition, the morphology of neurons in the hippocampal CA1 area improved, and the survival rate of neurons increased. All this evidence indicate that pinocembrin exhibits neuroprotective effects that prevent brain ischemia/reperfusion acute injury [5,26]. Moreover, the treatment time window of pinocembrin is wider than that of tissue plasminogen activator (t-PA). Pretreatment with pinocembrin shortly before t-PA infusion can significantly protect BBB function and improve neurological function after long-term ischemia, thereby improving the therapeutic effects of t-PA [27]. These data have shown that pinocembrin presents desirable neuroprotective effects and may be beneficial for the treatment of stroke in combination with t-PA.

What are the mechanisms of pinocembrin for neuroprotection? Studies have shown that the neuroprotective effects of pinocembrin include mitochondrial protection, inhibition of autophagy, anti-oxidation, anti-apoptosis, and other pharmacological effects. Pinocembrin reduces production of ROS, nitric oxide (NO), neuronal NO synthase (nNOS) and induces NO synthase (iNOS). Pinocembrin also increases the content of glutathione, thereby exerting an antioxidant effect and achieving neuroprotection [24]. The anti-inflammatory effect of pinocembrin also plays an important role in neuroprotection. Pinocembrin suppresses expression of inflammatory markers, such as tumor necrosis factor-alpha (TNF-α), interleukin-1 (IL)-beta, intercellular adhesion molecule-1, vascular cell adhesion molecule-1, iNOS, and aquaporin-4. In addition, pinocembrin inhibits the activation of microglials and astrocytes and downregulates the expression of matrix metalloproteinases (MMPs) in ischemic brain area. These events result in the protective action of pinocembrin on neurovascular units [28,29]. Anti-apoptosis and anti-autophagy of pinocembrin directly affect the total number of nerve cells. Treatment with pinocembrin can increase the viability of cells and attenuate apoptosis in a dose-dependent manner. Part of the mechanism of pinocembrin is to inhibit the release of p53, thereby affecting the Bax-Bcl-2 ratio and released amount of cytochrome C [20]. Pinocembrin decreases the expression of autophagy protein LC3 II and Beclin 1 and increases the level of p62, which are key proteins of autophagy in the hippocampus CA1 area. These results suggest that pinocembrin may achieve neuroprotection by inhibiting autophagy activity [30].

Pinocembrin can provide mitochondrial protection. Cerebral ischemia often causes mitochondrial damage in the brain. Thus, restoring mitochondrial function is important to fight this disease. $Ca^{2+}$ overload consistently results in mitochondrial structure and function impairment. Pinocembrin can reduce the content of $Ca^{2+}$ in mitochondria, thereby alleviating mitochondrial membrane swelling and reducing Mn-SOD activity. In addition, pinocembrin can reduce ATP synthesis and energy metabolism disorders caused by $Ca^{2+}$ overload. Pinocembrin can also reduce ROS production in mitochondria and reduce electronic leakage of the NADH respiratory chain, improve the efficiency of oxidative phosphorylation, and promote mitochondrial respiratory function and synthesis of mitochondrial ATP [31–33]. The role of pinocembrin in mitochondrial protection depends in part on the activation of the Erk1/2-Nrf2 axis [20].

3.1.2. Neuroprotective Effect of Pinocembrin in Alzheimer's Disease and Parkinson's Disease

AD is a central nervous system degenerative disease that seriously threatens the health of the elderly. Modern pharmacological studies have shown that common pathological features of AD include β-amyloid (Aβ) plaques, excessive phosphorylation of Tau protein to form neurofibrillary tangles, reduction of acetylcholine transmitters and inflammatory responses, and neuronal loss [34,35].

Pinocembrin has been found to affect cognitive function and protect nerve cells against Aβ-induced toxicity. Administration of pinocembrin in Aβ25–35-induced mice can preserve ultrastructural nerve fibers in the mouse brain and reduce neurodegeneration in the cerebral cortex, thereby improving cognitive function. The function of the nervous system is also affected by the interaction between Aβ and receptors of advanced glycation end products (RAGE). Pinocembrin significantly inhibits the upregulation of RAGE transcription and protein translation in vivo and in vitro and the inflammatory response following Aβ-RAGE interaction. These effects are achieved by inhibition of p38 MAPK–MAPK-activated protein kinase-2–heat shock protein 27 and stress-activated protein kinase/c-Jun N-terminal kinase–c-Jun pathway and NF-κB signaling pathway. Pinocembrin also improves the cholinergic system by conserving the ERK–cAMP-response element-binding protein–brain-derived neurotrophic factor pathway [36–38]. Treatment with pinocembrin reduced the damage induced by fibrillar Aβ (1–40) in human brain microvascular endothelial cells (hBMECs), and this finding may be related to inhibition of inflammation by pinocembrin. The related mechanisms may include the inhibited activation of MAPK, down-regulated the level of IκB kinase, decreased degradation of IκBα, and blocked nuclear translocation of NF-κB p65 by pinocembrin, thereby reducing the release of pro-inflammatory factors. Further studies have shown that pinocembrin can protect SH-SY5Y cells from Aβ (25–35)-induced neurotoxicity by activating the Nrf2/heme oxygenase-1 (HO-1) pathway [39]. The protective effect on mitochondria is also one of the anti-AD mechanisms of pinocembrin.

PD is characterized by the loss of dopaminergic neurons in the substantia nigra. Oxidative stress and mitochondrial dysfunction are possibly involved in the etiology of PD [40]. In a methyl-4-phenyl pyridinium (MPP$^+$) -induced PD model, pinocembrin significantly reduced the loss of cell viability, production of intracellular ROS, apoptosis rate of cells, and activation of caspase-3 in SH-SY5Y cells. The related mechanisms may include the activation of ERK1/2 signaling pathways, enhancement of HO-1 expression, and suppression of MPP$^+$-induced oxidative damage by pinocembrin. In addition, pinocembrin can alleviate MPP$^+$-induced mitochondrial dysfunction by reducing mitochondrial membrane potential, down-regulating Bcl-2/Bax ratio, and inhibiting the release of cytochrome C [41,42]. In 6-hydroxydopamine (6-OHDA)-induced PD model, pinocembrin activates the expression of HO-1 and γ-glutamylcysteine synthetase via the Nrf2/antioxidant response element pathway in SH-SY5Y cells. As a result, the loss of cell viability and apoptosis rate induced by 6-OHDA in SH-SY5Y cells decreased, thereby reducing nerve damage [43].

### 3.2. Anti-Inflammation Activity

Pinocembrin significantly reduces the degree of cerebral edema and serum TNF-α and IL-1β levels in rats with focal cerebral ischemia reperfusion. The anti-inflammation effect may partly account for the mechanisms in ischemic stroke treatment [5]. In diabetic mice, pinocembrin protects neurons from inflammatory damage, thereby reducing their cognitive deficits [44]. Moreover, pinocembrin inhibits the inflammation of allergic airways induced by ovalbumin (OVA) in mice and significantly reduces the content of Th2 cytokines, IL-4, IL-5, and IL-13 in broncho-alveolar lavage fluid and OVA-specific antibody IgE in serum. A possible mechanism involves the inhibition of IκBα and NF-κB p65 phosphorylation. These findings indicate that pinocembrin features the potential to become a natural antiallergic drug [45]. The degradation of extracellular matrix induced by MMPs is an important cause of cartilage destruction. Pinocembrin inhibits the expression of MMP-1, MMP-3, and MMP-13 at both mRNA and protein levels in human chondrocytes [46]. In lipopolysaccharide-induced inflammation, pinocembrin inhibits the production of TNF-α, IL-1β, NO, and PGE2 by suppressing PI3K/Akt/NF-κB signaling pathway [47,48]. The mechanism by which pinocembrin inhibits inflammation includes inhibition of the MAPK and NF-κB signaling pathway. Through the above routes, pinocembrin can reduce the release of pro-inflammatory cytokines [37].

*3.3. Antioxidation Activity*

The anti-oxidation effect of pinocembrin is the basis for treatment of numerous diseases using this compound. The antioxidation activity of pinocembrin includes neuroprotection and mitochondrial protection against cerebral ischemia, AD, PD, and other diseases. In the global cerebral ischemia model, pinocembrin can reduce brain tissue damage by reducing the compensatory activity of SOD and reducing MDA levels and MPO activity in a dose-dependent manner [49]. In oxidative stress injury evoked by $CCl_4$, pinocembrin restores liver transaminases and total cholesterol to normal levels through the inhibition of reduced glutathione depletion and lipid peroxidation and elevation of superoxide dismutase (SOD) [6]. Pinocembrin can also protect human aortic endothelial cells from ox-low-density lipoprotein (LDL)-induced injury. The mechanisms may relate to ROS reduction induced by ox-LDL [50]. The nephrotoxicity induced by gentamicin can be alleviated by pinocembrin due in part to its antioxidant effect [51]. These findings have indicated that pinocembrin reduces the degree of atherosclerosis and is a promising antioxidant.

*3.4. Antimicrobial Activity*

Pinocembrin possesses antibacterial, antifungal, and antiparasitic effects. A study has shown that pinocembrin features anti-*Staphylococcus aureus* action both in vitro and in vivo [52]. Pinocembrin presents significant inhibition of zoospore mobility and mildew development and thus could be used as a natural antifungal product [53]. Pinocembrin significantly suppressed parasitemia in *Plasmodium berghei*-infected mice [54]. More importantly, pinocembrin plays a highly reversible role as antimicrobial. Substituent groups will affect the pharmacological activities of pinocembrin. Introducing oleyl or linoleoyl in the seventh carbon, the derivatives showed high inhibitory effects on bacterial proliferation, with minimum inhibitory concentration values of 32 µg/mL against *Staphylococcus aureus* [55].

*3.5. Vasodilation Activity*

Pinocembrin has been observed to inhibit angiotensin II (Ang II)-induced vasoconstriction in aortic rings of rats. In the docking model, pinocembrin binds effectively to the active site of Angiotensin II receptor type 1(AT1R), thereby inhibiting Ang II induced $Ca^{2+}$ release and $Ca^{2+}$ influx. These inhibitory effects may be related to the reduction of Ang II-induced increase in $Ca^{2+}$ and ERK1/2 activation via blocking of AT1R [56]. Pinocembrin can induce endothelium-independent relaxation in rat aortic rings, and the mechanism is at least partly due to the blockade of the Rho A/ROCK pathway [57,58].

*3.6. Hepatoprotection Activity*

In treatment of liver fibrosis, inactivating hepatic stellate cells (HSCs) has been an effective therapeutic strategy. It has been discovered that pinocembrin inhibits the expressions of fibrotic markers in LX-2 cells and rat HSCs (HSC-T6). Pinocembrin can reduce ROS accumulation by elevating the expression and activity of silent mating type information regulation 2 homolog 3 (SIRT3) and then activating SOD2. In addition, pinocembrin inhibits the PI3K/Akt signaling pathway, resulting in decreased production of transforming growth factor-beta and inhibition of transcriptional factors Sma- and Mad-related protein (Smad) nuclear translocation. Moreover, pinocembrin activates glycogen synthase kinase 3β by acting on SIRT3, thereby enhancing the degradation of Smad protein [10]. Pinocembrin-7-*O*-[3″-*O*-galloyl-4″,6″-hexahydroxydiphenoyl]-β-glucose is the derivative of pinocembrin, and it significantly contributes to the hepatoprotective effects of the latter [59]. This finding indicates that pinocembrin can be used as a drug candidate for the treatment of liver diseases.

*3.7. Others*

Recent studies have also reported that pinocembrin features an anti-tumor effect. Pinocembrin up-regulates the levels of caspase 3 and LC3-II in MDA-MB-231 cells, induces apoptosis and autophagy, and exhibits potential anticancer effect [60]. In melanoma, pinocembrin can induce endoplasmic reticulum (ER) stress-mediated apoptosis and suppress autophagy, showing its potentiality for melanoma treatment. Pinocembrin induces ER stress via the inositol-requiring endonuclease 1 $\alpha$/X-box binding protein 1 pathway and then triggers caspase-12/-4 mediated apoptosis by suppressing autophagy through the activation of PI3K/Akt/mTOR pathway [9]. Pinocembrin also reduced ventricular arrhythmias in I/R rats by enhancing $Na^+$-$K^+$ ATPase and $Ca^{2+}$-$Mg^{2+}$ ATPase activity and up-regulating Cx43 and Kir2.1 protein expression [61].

Furthermore, studies have shown that pinocembrin can prevent kidney damage caused by diabetes, but when the kidney is damaged, pinocembrin will aggravate the organ's condition [62]. This finding indicates that pinocembrin can be used as a preventive agent for kidney damage before injury. Pinocembrin may enhance the activities of hexokinase and pyruvate kinase via the Akt/mTOR signaling pathway, thereby improving insulin resistance [63]. In daily life, given its antibacterial and antioxidant activities, pinocembrin can be used as mouth cleaning agent and sunscreen [7,64].

## 4. Pharmacokinetic Profiles of Pinocembrin

### 4.1. Transport Features Across Blood–Brain Barrier (BBB) In Vitro

Cultured rat BMECs have been used as an in vitro BBB model. The findings have shown the uptake of pinocembrin in a time- and concentration-dependent manner. Passive transport may be the main process for pinocembrin to pass through the BBB, whereas P-gp is likely to cause little effect on the transport process of pinocembrin. Furthermore, pinocembrin may show no effect on the functional activity and protein expression of the P-gp transporter at the BBB [11].

### 4.2. Pharmacokinetic (PK) Profiles In Vivo

#### 4.2.1. PK Profiles in Rats

In rats, pinocembrin exhibits a large volume of distribution ($V_d$) and a short half-life ($T_{1/2}$) and is easily metabolized in the body (Table 1). After intravenous injection of pinocembrin in rats, the cumulative excretion scores of drug prototypes excreted from urine and excrement in 72 h totaled 6.99% ± 5.97% and 59.17% ± 22.13%, respectively. In rats, the drug is metabolized rapidly, and 40.6% ± 23.52% of the metabolites are excreted by urine [65]. In racemic delivery (±) of pinocembrin to rats (20 mg/kg, intravenously (iv.)), the concentration–time profile of pinocembrin followed a biexponential pattern, which indicates differences in the in vivo process of different configurations of pinocembrin. S-Pinocembrin and R-pinocembrin could be detected in serum. Similar values of $V_d$ are observed between enantiomers, and both enantiomers exhibit a serum half-life ($T_{1/2}$) of about 15 min in rats. After oral administration, pinocembrin is rapidly glucuronidated. The peak concentration of S-pinocembrin glucuronide and R-pinocembrin glucuronide measure 140 and 160 µg/mL, respectively. The main metabolic mode of pinocembrin is phase II metabolism. The large $V_d$ coupled with the short serum $T_{1/2}$ suggests the extensive distribution of pinocembrin into the tissues [66–68].

#### 4.2.2. PK Profiles in Humans

Pinocembrin is well absorbed and widely distributed in the human body. In a single-dose study, five dose groups have been established, and the mean peak plasma pinocembrin concentration has been obtained at the end of 30 min infusion. $T_{1/2}$ is similar in the five dose groups and ranges from 40 min to 55 min. At 4 h after administration, the cumulative excretion rate of pinocembrin in the urine reaches a plateau, and the level of urine and fecal excretion of pinocembrin is extremely low, with each dose group yielding similar values. The data show that pinocembrin is easily metabolized in vivo. The compound is mainly metabolized into two metabolites in the human body, sulfonic products and glucuronide products, among which 5-hydroxy-flavanone-7-O-sulfonate and 5-hydroxy-flavanone-7-O-β-D glucuronic acid can be synthesized by artificial synthesis. The $V_d$ of pinocembrin approximates 136.6 ± 52.8 L, and the clearance (CL) rate equals 2.0 ± 0.31 L/min, indicating that pinocembrin is well absorbed and widely distributed in human body. The PK features of pinocembrin under multiple dose are similar to those observed in single-dose studies, showing no evidence of accumulation. Pinocembrin is well tolerated when administered iv. to healthy adults [69–72]. Although the absorption/elimination process of pinocembrin occurs rapidly, and no serious accumulation exists in the body, the related drug interaction still deserves attention. In a study, 6β-hydroxylation of testosterone was used as a labeling reaction for CYP3A4 activity. The resulting product was determined by high-performance liquid chromatography in conjunction with diode array detector. Metabolism, time dependence, and direct inhibition were tested to determine if inhibition of CYP3A4 activity is reversible or irreversible. The result showed that pinocembrin irreversibly inhibited the metabolic activity of the CYP3A4 enzyme, decreasing the enzyme activity by 50% [73]. In addition, pinocembrin has been shown to inhibit hOATP2B1 and hOATP1A2, with $IC_{50}$ of 37.3 ± 1.3 and 2.0 ± 1.7 µM, respectively, affecting the intake of statins [74]. In addition, racemic pinocembrin reveals the inhibitory activity of CYP2D6 at low concentrations. At 0.01 and 0.1 µM, the inhibition of CYP2D6 approximates 50% compared with the positive control [67]. Given the inhibitory effect of pinocembrin on drug-metabolizing enzymes, drug interactions should be considered when using the compound.

**Table 1.** Summarizes the PK parameters of pinocembrin.

| Subject | Mode of Administration | Dose (mg/kg) | AUC (h*µg/mL) | $V_d$ (L/kg) | $CL_{total}$ (L/h/kg) | $T_{1/2}$ Serum (h) | References |
|---|---|---|---|---|---|---|---|
| SD rats | iv. | 10 | S-1.821 ± 0.211; R-1.876 ± 0.427 | S-1.758 ± 1.313; R-1.793 ± 0.805 | S-5.527 ± 0.641; R-5.535 ± 1.217 | S-0.212 ± 0.140; R-0.223 ± 0.083 | [67] |
| SD rats | iv. | 10 | S-1.83 ± 0.092; R-1.876 ± 0.312 | S-1.46 ± 0.591; R-1.80 ± 0.271 | S-5.44 ± 0.287; R-5.83 ± 0.865 | S-0.262 ± 0.071; R-0.263 ± 0.027 | [66] |
| SD rats | po. | 100 | S-570 ± 21.7; R-531 ± 82.1 | S-3.80 ± 1.34; R-5.14 ± 1.81 | S-2.82 ± 0.084; R-2.83 ± 0.844 | S-20.3 ± 8.41; R-27.1 ± 18.8 | [66] |
| SD rats | iv. | 10 | 0.686.1 ± 0.0651 | 48.7 ± 19.6 | 15.5 ± 1.4 | 2.14 ± 0.68 | [68] |
| SD rats | po. | 50 | 0.518 ± 0.170 | 478 ± 213 | 110 ± 31.4 | 3.11 ± 1.21 | [68] |
| Human | iv. | 20 (mg) | 10.3381 ± 1.5394 (min µg/mL) | 136.6 ± 52.8 (L) | 2.0 ± 0.3 (L/min) | 0.79 ± 0.23 | [71] |

AUC: area under the curve; CL: clearance; iv.: intravenous; $T_{1/2}$: half-life; $V_d$: volume of distribution.

## 5. Conclusions and Prospects

Although pinocembrin is widely found in honey and various plants, the yield of natural extraction remains insufficient. Biosynthesis fills the gap in this area. In this review, we summarized the progress in biosynthesis of pinocembrin, relying on genetic engineering technology to construct engineered bacteria to achieve the synthesis of pinocembrin from glucose. The production of pinocembrin can be considerably improved by regulating the metabolism of engineered bacteria and regulating pH and energy supply of the fermentation system.

Pinocembrin exerts certain effects on ischemic stroke, PD, AD, solid tumors, and some other diseases. In central nervous system diseases, pinocembrin can reduce the release of inflammatory factors by inhibiting multiple signaling pathways, such as MAPK and PI3K/AKT. Pinocembrin can also reduce the release of NO, ROS, nNOS, and iNOS, thus playing an antioxidant role. The activation of ERK1/2-Nrf2 axis is one mechanism of mitochondrial protection by pinocembrin. In addition, this compound can increase the number of nerve cells by inhibiting their autophagy. Given its wide range of pharmacological activities, pinocembrin can be linked to disease production mechanisms, such as network pharmacology and molecular target docking, to explore its application in some other diseases.

Pinocembrin may easily cross the BBB due to its low molecular weight and good liposolubility. Several studies have shown that pinocembrin features an anti-tumor effect, undergoing passive transport when passing the BBB; this property can be used for the treatment of drug-resistant brain tumors. However, the anti-tumor mechanism of pinocembrin remains unclear, thus requiring further research. Numerous diseases are often associated with inflammation and oxidative damage during production and development. Pinocembrin performs significant anti-oxidant and anti-inflammatory activities, indicating that it possesses the potential to treat a variety of diseases and thus needs further research. In addition, pinocembrin exhibits an inhibitory effect on various drug-metabolizing enzymes and transporters. Thus, drug interactions should be considered when using this compound to ensure drug safety. Pinocembrin is in a phase II clinical trial and requires more in-depth studies.

**Author Contributions:** Writing—original draft preparation, X.S.; Z.Y.; writing—review and editing, X.S.; Y.L.; X.L.; Z.Y.; funding acquisition—Z.Y.

**Funding:** This work was supported by grants from Beijing Natural Science Foundation (No. 7173267), National Natural Science Foundation of China (Nos. 81473579, 81273654 and 81102879), and National Science and National Science and Technology Major Projects for "Major New Drugs Innovation and Development" (No. 2013ZX09103002-022).

**Conflicts of Interest:** The authors declare no conflict of interest.

## References

1. Escriche, I.; Juan-Borrás, M. Standardizing the analysis of phenolic profile in propolis. *Food Res. Int.* **2018**, *106*, 834–841. [CrossRef] [PubMed]
2. López, A.; Ming, D.S.; Towers, G.H. Antifungal activity of benzoic acid derivatives from Piper lanceaefolium. *J. Nat. Prod.* **2002**, *65*, 62–64. [CrossRef] [PubMed]
3. Danelutte, A.P.; Lago, J.H.; Young, M.C.; Kato, M.J. Antifungal flavanones and prenylated hydroquinones from Piper crassinervium Kunth. *Phytochemistry* **2003**, *64*, 555–559. [CrossRef]
4. Feng, R.; Guo, Z.K.; Yan, C.M.; Li, E.G.; Tan, R.X.; Ge, H.M. Anti-inflammatory flavonoids from Cryptocarya chingii. *Phytochemistry* **2012**, *76*, 98–105. [CrossRef] [PubMed]
5. Wu, C.X.; Du, G.H. Effects of pinocembrin on inflammation in rats with focal cerebral ischemia reperfusion injury. *J. Shandong Med. College* **2015**, *37*, 247–250.
6. Said, M.M.; Azab, S.S.; Saeed, N.M.; El-Demerdash, E. Antifibrotic Mechanism of Pinocembrin: Impact on Oxidative Stress, Inflammation and TGF-β/Smad Inhibition in Rats. *Ann. Hepatol.* **2018**, *17*, 307–317. [CrossRef]
7. Celerino de Moraes Porto, I.C.; Chaves Cardoso de Almeida, D.; Vasconcelos Calheiros de Oliveira Costa, G.; Sampaio Donato, T.S.; Moreira Nunes, L.; Gomes do Nascimento, T.; Dos Santos Oliveira, J.M.; Batista da Silva, C.; Barbosa Dos Santos, N.; de Alencar E Silva Leite, M.L.; et al. Mechanical and aesthetics compatibility of Brazilian red propolis micellar nanocomposite as a cavity cleaning agent. *BMC Complement. Altern. Med.* **2018**, *18*, 219. [CrossRef] [PubMed]
8. Liu, R.; Gao, M.; Yang, Z.H.; Du, G.H. Pinocembrin protects rat brain against oxidation and apoptosis induced by ischemia-reperfusion both in vivo and in vitro. *Brain Res.* **2008**, *1216*, 104–115. [CrossRef]
9. Meng, F.R.; Liu, R.; Gao, M.; Wang, Y.H.; Yu, X.T.; Xuan, Z.H.; Sun, J.L.; Yang, F.; Wu, C.F.; Du, G.H. Pinocembrin attenuates blood-brain barrier injury induced by global cerebral ischemia-reperfusion in rats. *Brain Res.* **2011**, *1391*, 93–101. [CrossRef]
10. Zheng, Y.; Wang, K.; Wu, Y.; Chen, Y.; Chen, X.; Hu, C.W.; Hu, F. Pinocembrin induces ER stress mediated apoptosis and suppresses autophagy in melanoma cells. *Cancer Lett.* **2018**, *431*, 31–42. [CrossRef]
11. Zhou, F.Y.; Wang, A.Q.; Li, D.; Wang, Y.T.; Lin, L.G. Pinocembrin from Penthorumchinense Pursh suppresses hepatic stellate cells activation through a unified SIRT3-TGF-β-Smad signaling pathway. *Toxicol. Appl. Pharmacol.* **2018**, *341*, 38–50. [CrossRef] [PubMed]
12. Yang, Z.H.; Sun, X.; Qi, Y.; Mei, C.; Sun, X.B.; Du, G.H. Uptake characteristics of pinocembrin and its effect on p-glycoprotein at the blood-brain barrier in in vitro cell experiments. *J. Asian Nat. Prod. Res.* **2012**, *14*, 14–21. [CrossRef] [PubMed]
13. Wu, J.J.; Du, G.C.; Zhou, J.W.; Chen, J. Metabolic engineering of Escherichia coli for (2S)-pinocembrin production from glucose by a modular metabolic strategy. *Metab. Eng.* **2013**, *16*, 48–55. [CrossRef] [PubMed]
14. Kim, B.G.; Lee, H.; Ahn, J.H. Biosynthesis of pinocembrin from glucose using engineered escherichia coli. *J. Microbiol. Biotechnol.* **2014**, *24*, 1536–1541. [CrossRef] [PubMed]
15. Guo, L.; Kong, J.Q. Progress in synthetic biology of pinocembrin. *Chin. J. Biotechnol.* **2015**, *31*, 451–460.
16. Cao, W.J.; Ma, W.C.; Zhang, B.; Wang, X.; Chen, K.Q.; Li, Y.; Ouyang, P.K. Improved pinocembrin production in Escherichia coli by engineering fatty acid synthesis. *J. Ind. Microbiol. Biotechnol.* **2016**, *43*, 557–566. [CrossRef] [PubMed]

17. Cao, W.J.; Ma, W.C.; Wang, X.; Zhang, B.W.; Cao, X.; Chen, K.Q.; Li, Y.; Ouyang, P.Y. Enhanced pinocembrin production in Escherichia coli by regulating cinnamic acid metabolism. *Sci. Rep.* **2016**, *6*, 32640. [CrossRef] [PubMed]

18. Tao, S.; Qian, Y.; Wang, X.; Cao, W.J.; Ma, W.C.; Chen, K.Q.; Ouyang, P.K. Regulation of ATP levels in Escherichia coli using CRISPR interference for enhanced pinocembrin production. *Microb. Cell Fact.* **2018**, *17*, 147–159. [CrossRef]

19. Wu, J.J.; Zhang, X.; Zhou, J.W.; Dong, M.S. Efficient biosynthesis of (2S)-pinocembrin from d-glucose by integrating engineering central metabolic pathways with a pH-shift control strategy. *Bioresour. Technol.* **2016**, *218*, 999–1007. [CrossRef]

20. Guo, L.; Chen, X.; Li, L.N.; Tang, W.; Pan, Y.T.; Kong, J.Q. Transcriptome-enabled discovery and functional characterization of enzymes related to (2S)-pinocembrin biosynthesis from Ornithogalum caudatum and their application for metabolic engineering. *Microb. Cell Fact.* **2016**, *15*, 27–45. [CrossRef]

21. De Oliveira, M.R.; Peres, A.; Gama, C.S.; Bosco, S.M.D. Pinocembrin Provides Mitochondrial Protection by the Activation of the Erk1/2-Nrf2 Signaling Pathway in SH-SY5Y Neuroblastoma Cells Exposed to Paraquat. *Mol. Neurobiol.* **2017**, *54*, 6018–6031. [CrossRef] [PubMed]

22. Gao, M.; Zhang, W.C.; Liu, Q.S.; Hu, J.J.; Liu, G.T.; Du, G.H. Pinocembrin prevents glutamate-induced apoptosis in SH-SY5Y neuronal cells via decrease of bax/bcl-2 ratio. *Eur. J. Pharmacol.* **2008**, *91*, 73–79. [CrossRef]

23. Feigin, V.L.; Mensah, G.A.; Norrving, B.; Murray, C.J.; Roth, G.A.; GBD 2013 Stroke Panel Experts Group. Atlas of the Global Burden of Stroke (1990-2013): The GBD 2013 Study. *Neuroepidemiology* **2015**, *45*, 230–236. [CrossRef]

24. Ying, J.; Jiang, Y.D.; Chen, Y.; Samuel, S.; Du, G.H. Electrophysiological effects of pinocembrin on Aplysia SN/L7 co-cultures. *Chin. Pharm. Bull.* **2011**, *27*, 755–759.

25. Gao, M.; Liu, R.; Zhu, S.Y.; Du, G.H. Acute neurovascular unit protective action of pinocembrin against permanent cerebral ischemia in rats. *J. Asian Nat. Prod. Res.* **2008**, *10*, 551–558. [CrossRef] [PubMed]

26. Wu, C.X.; Du, G.H. Pinocembrin prevented brain acute injury induced by focal cerebral ischemia-reperfusion. *Chin. J. Clin. Pharmacol. Ther.* **2015**, *20*, 1208–1211, 1220.

27. Ma, Y.Z.; Li, L.; Kong, L.L.; Zhu, Z.M.; Zhang, W.; Song, J.K.; Chang, J.L.; Du, G.H. Pinocembrin Protects Blood-Brain Barrier Function and Expands the Therapeutic Time Window for Tissue-Type Plasminogen Activator Treatment in a Rat Thromboembolic Stroke Model. *Biomed. Res. Int.* **2018**, *2018*, 1–14. [CrossRef]

28. Gao, M.; Zhu, S.Y.; Tan, C.B.; Xu, B.; Zhang, W.C.; Du, G.H. Pinocembrin protects the neurovascular unit by reducing inflammation and extracellular proteolysis in MCAO rats. *J. Asian Nat. Prod. Res.* **2010**, *12*, 407–418. [CrossRef] [PubMed]

29. Saad, M.A.; Abdel Salam, R.M.; Kenawy, S.A.; Attia, A.S. Pinocembrin attenuates hippocampal inflammation, oxidative perturbations and apoptosis in a rat model of global cerebral ischemia reperfusion. *Pharmacol. Rep.* **2015**, *67*, 115–122. [CrossRef]

30. Tao, J.H.; Shen, C.; Sun, Y.C.; Chen, W.M.; Yan, G.F. Neuroprotective effects of pinocembrin on ischemia/reperfusion-induced brain injury by inhibiting autophagy. *Biomed. Pharmacother.* **2018**, *106*, 1003–1010. [CrossRef]

31. Guang, H.M.; Gao, M.; Zhu, S.Y.; He, X.L.; He, G.R.; Zhu, X.M.; Du, G.H. Effect of pinocembrin on Mitochondrial function in Rats with Acute focal Cerebral Ischemia. *Chin. Pharm. Bull.* **2012**, *28*, 24–29.

32. Guang, H.M.; Du, G.H. Protections of pinocembrin on brain mitochondria contribute to cognitive improvement in chronic cerebral hypoperfused rats. *Eur. J. Pharmacol.* **2006**, *542*, 77–83. [CrossRef] [PubMed]

33. Shi, L.L.; Qiang, G.F.; Gao, M.; Zhang, H.A.; Chen, B.N.; Yu, X.Y.; Xuan, Z.H.; Wang, Q.Y.; Du, G.H. Effect of pinocembrin on brain mitochondrial respiratory function. *Acta Pharm. Sin.* **2011**, *46*, 642–649.

34. Zhang, P.F.; Xu, S.T.; Zhu, Z.Y.; Xu, J.Y. Multi-target design strategies for the improved treatment of Alzheimer's disease. *Eur. J. Med. Chem.* **2019**, *176*, 228–247. [CrossRef] [PubMed]

35. Wang, T.; Xie, X.X.; Ji, M.; Wang, S.W.; Zha, J.; Zhou, W.W.; Yu, X.L.; Wei, C.; Ma, S.; Xi, Z.Y.; et al. Naturally occurring autoantibodies against Aβ oligomers exhibited more beneficial effects in the treatment of mouse model of Alzheimer's disease than intravenous immunoglobulin. *Neuropharmacology* **2016**, *105*, 561–576. [CrossRef] [PubMed]

36. Liu, R.; Wu, C.X.; Zhou, D.; Yang, F.; Tian, S.; Zhang, L.; Zhang, T.T.; Du, G.H. Pinocembrin protects against β-amyloid-induced toxicity in neurons through inhibiting receptor for advanced glycation end products (RAGE)-independent signaling pathways and regulating mitochondrion-mediated apoptosis. *BMC Med.* **2012**, *10*, 105–125. [CrossRef] [PubMed]

37. Liu, R.; Li, J.Z.; Song, J.K.; Zhou, D.; Huang, C.; Bai, X.Y.; Xie, T.; Zhang, X.; Li, Y.J.; Wu, C.X.; et al. Pinocembrin improves cognition and protects the neurovascular unit in Alzheimer related deficits. *Neurobiol. Aging* **2014**, *35*, 1275–1285. [CrossRef]

38. Liu, R.; Li, J.Z.; Song, J.K.; Sun, J.L.; Li, Y.J.; Zhou, S.B.; Zhang, T.T.; Du, G.H. Pinocembrin protects human brain microvascular endothelial cells against fibrillar amyloid-β (1-40) injury by suppressing the MAPK/NF-κB inflammatory pathways. *Biomed. Res. Int.* **2014**, *2014*, 1–14. [CrossRef]

39. Wang, Y.; Miao, Y.; Mir, A.Z.; Cheng, L.; Wang, L.; Zhao, L.; Cui, Q.; Zhao, W.; Wang, H. Inhibition of beta-amyloid-induced neurotoxicity by pinocembrin through Nrf2/HO-1 pathway in SH-SY5Y cells. *J. Neurol. Sci.* **2016**, *368*, 223–230. [CrossRef]

40. Pieczenik, S.R.; Neustadt, J. Mitochondrial dysfunction and molecular pathways of disease. *Exp. Mol. Pathol.* **2007**, *83*, 84–92. [CrossRef]

41. Wang, Y.M.; Gao, J.H.; Miao, Y.C.; Cui, Q.F.; Zhao, W.L.; Zhang, J.Y.; Wang, H.Q. Pinocembrin protects SH-SY5Y cells against MPP$^+$-induced neurotoxicity through the mitochondrial apoptotic pathway. *J. Mol. Neurosci.* **2014**, *53*, 537–545. [CrossRef] [PubMed]

42. Wang, H.Q.; Wang, Y.M.; Zhao, L.N.; Cui, Q.F.; Wang, Y.H.; Du, G.H. Pinocembrin attenuates MPP$^+$-induced neurotoxicity by the induction of heme oxygenase-1 through ERK1/2 pathway. *Neurosci. Lett.* **2016**, *612*, 104–109. [CrossRef] [PubMed]

43. Jin, X.H.; Liu, Q.; Jia, L.L.; Li, M.; Wang, X. Pinocembrin attenuates 6-OHDA-induced neuronal cell death through Nrf2/ARE pathway in SH-SY5Y cells. *Cell Mol. Neurobiol.* **2015**, *35*, 323–333. [CrossRef] [PubMed]

44. Pei, B.; Sun, J. Pinocembrin alleviates cognition deficits by inhibiting inflammation in diabetic mice. *J. Neuroimmunol.* **2018**, *314*, 42–49. [CrossRef] [PubMed]

45. Gu, X.Y.; Zhang, Q.; Du, Q.; Shen, H.; Zhu, Z.H. Pinocembrin attenuates allergic airway inflammation via inhibition of NF-κB pathway in mice. *Int. Immuno. Pharmacol.* **2017**, *53*, 90–95. [CrossRef] [PubMed]

46. Zhang, D.W.; Huang, B.; Xiong, C.J.; Yue, Z. Pinocembrin inhibits matrix metalloproteinase expression in chondrocytes. *IUBMB Life* **2015**, *67*, 36–41. [CrossRef]

47. Zhou, L.T.; Wang, K.J.; Li, L.; Li, H.; Geng, M. Pinocembrin inhibits lipopolysaccharide-induced inflammatory mediators' production in BV2 microglial cells through suppression of PI3K/Akt/NF-κB pathway. *Eur. J. Pharmacol.* **2015**, *761*, 211–216. [CrossRef]

48. Soromou, L.W.; Chu, X.; Jiang, L.; Wei, M.; Huo, M.; Chen, N.; Guan, S.; Yang, X.; Chen, C.; Feng, H.; et al. In vitro and in vivo protection provided by pinocembrin against lipopolysaccharide-induced inflammatory responses. *Int. Immuno. Pharmacol.* **2012**, *14*, 66–74. [CrossRef]

49. Shi, L.L.; Chen, B.N.; Gao, M.; Zhang, H.A.; Li, Y.J.; Wang, L.; Du, G.H. The characteristics of therapeutic effect of pinocembrin in transient global brain ischemia/reperfusion rats. *Life Sci.* **2011**, *88*, 521–528. [CrossRef]

50. Su, Q.; Sun, Y.H.; Ye, Z.L.; Yang, H.F.; Kong, B.H.; Li, L. Pinocembrin protects endothelial cells from oxidized LDL-induced injury. *Cytokine* **2018**, *111*, 475–480. [CrossRef]

51. Promsan, S.; Jaikumkao, K.; Pongchaidecha, A.; Chattipakorn, N.; Chatsudthipong, V.; Arjinajarn, P.; Pompimon, W.; Lungkaphin, A. Pinocembrin attenuates gentamicin-induced nephrotoxicity in rats. *Can. J. Physiol. Pharmacol.* **2016**, *94*, 808–818. [CrossRef] [PubMed]

52. Soromou, L.W.; Zhang, Y.; Cui, Y.; Wei, M.; Chen, N.; Yang, X.; Huo, M.; Baldé, A.; Guan, S.; Deng, X.M.; et al. Subinhibitory concentrations of pinocembrin exert anti-Staphylococcus aureus activity by reducing α-toxin expression. *J. Appl. Microbiol.* **2013**, *115*, 41–49. [CrossRef] [PubMed]

53. Gabaston, J.; Richard, T.; Cluzet, S.; Palos Pinto, A.; Dufour, M.C.; Corio-Costet, M.F.; Mérillon, J.M. Pinus pinaster Knot: A Source of Polyphenols against Plasmoparaviticola. *J. Agric. Food Chem.* **2017**, *65*, 8884–8891. [CrossRef] [PubMed]

54. Melaku, Y.; Worku, T.; Tadesse, Y.; Mekonnen, Y.; Schmidt, J.; Arnold, N.; Dagne, E. Antiplasmodial Compounds from Leaves of Dodonaea angustifolia. *Curr. Bioact. Compd.* **2017**, *13*, 268–273. [CrossRef] [PubMed]

55. Tundis, R.; Frattaruolo, L.; Carullo, G.; Armentano, B.; Badolato, M.; Loizzo, M.R.; Aiello, F.; Cappello, A.R. An ancient remedial repurposing: Synthesis of new pinocembrin fatty acid acyl derivatives as potential antimicrobial/anti-inflammatory agents. *Nat. Prod. Res.* **2019**, *33*, 162–168. [CrossRef] [PubMed]

56. Li, L.; Pang, X.B.; Chen, B.N.; Gao, L.; Wang, L.; Wang, S.B.; Wang, S.B.; Liu, D.P.; Du, G.H. Pinocembrin inhibits angiotensin II-induced vasoconstriction via suppression of the increase of $Ca^{2+}$ and ERK1/2 activation through blocking AT1 R in the rat aorta. *Biochem. Biophys. Res. Commun.* **2013**, *435*, 69–75. [CrossRef] [PubMed]

57. Li, L.; Yang, H.G.; Yuan, T.Y.; Zhao, Y.; Du, G.H. Rho kinase inhibition activity of pinocembrin in rat aortic rings contracted by angiotensin II. *Chin. J. Nat. Med.* **2013**, *11*, 258–263. [CrossRef]

58. Zhu, X.M.; Fang, L.H.; Li, Y.J.; Du, G.H. Endothelium-dependent and -independent relaxation induced by pinocembrin in rat aortic rings. *Vascul. Pharmacol.* **2007**, *46*, 160–165. [CrossRef]

59. Sun, Z.L.; Zhang, Y.Z.; Zhang, F.; Zhang, J.W.; Zheng, G.C.; Tan, L.; Wang, C.Z.; Zhou, L.D.; Zhang, Q.H.; Yuan, C.S. Quality assessment of Penthorum chinense Pursh through multicomponent qualification and fingerprint, chemometric, and antihepatocarcinoma analyses. *Food Funct.* **2018**, *9*, 3807–3814. [CrossRef]

60. Wang, Y.H.; Huang, M.Z.; Xuan, H.Z.; Yi, X.S. Different biological activities of pinocembrin and luteolin. *Apiculture China* **2017**, *68*, 13–15.

61. Zhang, P.; Xu, J.; Hu, W.; Yu, D.; Bai, X.L. Effects of Pinocembrin Pretreatment on Connexin 43 (Cx43) Protein Expression After Rat Myocardial Ischemia-Reperfusion and Cardiac Arrhythmia. *Med. Sci. Monit.* **2018**, *24*, 5008–5014. [CrossRef] [PubMed]

62. Granados-Pineda, J.; Uribe-Uribe, N.; García-López, P.; Ramos-Godinez, M.D.P.; Rivero-Cruz, J.F.; Pérez-Rojas, J.M. Effect of Pinocembrin Isolated from Mexican Brown Propolis on Diabetic Nephropathy. *Molecules* **2018**, *23*, 852. [CrossRef] [PubMed]

63. Liu, Y.; Liang, X.; Zhang, G.; Kong, L.; Peng, W.; Zhang, H. Galangin and Pinocembrin from Propolis Ameliorate Insulin Resistance in HepG2 Cells via Regulating Akt/mTOR Signaling. *Evid. Based Complement. Alternat. Med.* **2018**, *2018*, 1–10. [CrossRef]

64. Ajmala Shireen, P.; Abdul Mujeeb, V.M.; Muraleedharan, K. Theoretical insights on flavanones as antioxidants and UV filters: A TDDFT and NLMO study. *J. Photochem. Photobiol. B* **2017**, *170*, 286–294. [CrossRef] [PubMed]

65. Ying, J.; Du, G.H. A New Method for Quantitative Study of Drug Disposal and Its Application. *Chin. Pharm. Bull.* **2009**, *25*, 323.

66. Sayre, C.L.; Alrushaid, S.; Martinez, S.E.; Anderson, H.D.; Davies, N.M. Pre-Clinical Pharmacokinetic and Pharmacodynamic Characterization of Selected Chiral Flavonoids: Pinocembrin and Pinostrobin. *J. Pharm. Phar. Sci.* **2015**, *18*, 368–395. [CrossRef]

67. Sayre, C.L.; Takemoto, J.K.; Martinez, S.E.; Davies, N.M. Chiral analytical method development and application to pre-clinical pharmacokinetics of pinocembrin. *Biomed. Chromatogr.* **2013**, *27*, 681–684. [CrossRef]

68. Guo, W.W.; Qiu, F.; Chen, X.Q.; Ba, Y.Y.; Wang, X.; Wu, X. In-vivo absorption of pinocembrin-7-*O*-β-D-glucoside in rats and its in-vitro biotransformation. *Sci. Rep.* **2016**, *6*, 29340. [CrossRef]

69. Yan, B.; Cao, G.Y.; Yan, J.L.; Hu, X.; Peng, Y.Y.; Bian, Z.R.; Li, K.X. Determination of Pinocembrin in Human Urine by LC-MS / MS and Research on Urinary Excretion. *Chin. Pharm. J.* **2014**, *49*, 1540–1544.

70. Wang, D.M.; Liu, H.T.; Tong, Y.F.; Wu, S. Synthesis of two metabolites of pinocembrin. *Chin. J. New Drugs* **2013**, *22*, 1130–1132, 1165.

71. Yan, B.; Cao, G.Y.; Sun, T.H.; Zhao, X.; Hu, X.; Yan, J.L.; Peng, Y.Y.; Shi, A.X.; Li, Y.; Xue, W.; et al. Determination of pinocembrin in human plasma by solid-phase extraction and LC/MS/MS: Application to pharmacokinetic studies. *Biomed. Chromatogr.* **2014**, *28*, 1601–1606. [CrossRef]

72. Cao, G.Y.; Ying, P.Y.; Yan, B.; Xue, W.; Li, K.X.; Shi, A.X.; Sun, T.H.; Yan, J.L.; Hu, X. Pharmacokinetics, safety, and tolerability of single and multiple-doses of pinocembrin injection administered intravenously in healthy subjects. *J. Ethnopharmacol.* **2015**, *168*, 31–36. [CrossRef]

73. ŠarićMustapić, D.; Debeljak, Ž.; Maleš, Ž.; Bojić, M. The Inhibitory Effect of Flavonoid Aglycones on the Metabolic Activity of CYP3A4 Enzyme. *Molecules* **2018**, *23*, 2553. [CrossRef]
74. Navrátilová, L.; Ramos Mandíková, J.; Pávek, P.; Mladěnka, P.; Mladěnka, P.; Trejtnar, F. Honey flavonoids inhibit hOATP2B1 and hOATP1A2 transporters and hOATP-mediated rosuvastatin cell uptake in vitro. *Xenobiotica* **2018**, *48*, 745–755. [CrossRef]

*molecules*

**MDPI**

*Review*

# Could Polyphenols Help in the Control of Rheumatoid Arthritis?

**Siyun Sung [1], Doyoung Kwon [1], Eunsik Um [2] and Bonglee Kim [1,3,*]**

[1] College of Korean Medicine, Kyung Hee University, Seoul 02453, Korea; stellasung95@khu.ac.kr (S.S.); doyoung@khu.ac.kr (D.K.)

[2] Department of Clinical Korean Medicine, Graduate School, Kyung Hee University, Seoul 02453, Korea; flare0722@khu.ac.kr

[3] Department of Pathology, College of Korean Medicine, Graduate School, Kyung Hee University, Seoul 02453, Korea

* Correspondence: bongleekim@khu.ac.kr; Tel.: +82-2-961-9217

Received: 14 March 2019; Accepted: 20 April 2019; Published: 22 April 2019

check for updates

**Abstract:** Rheumatoid arthritis (RA) is a chronic, systemic, joint-invading, autoimmune inflammatory disease, which causes joint cartilage breakdown and bone damage, resulting in functional impairment and deformation of the joints. The percentage of RA patients has been rising and RA represents a substantial burden for patients around the world. Despite the development of many RA therapies, because of the side effects and low effectiveness of conventional drugs, patients still need and researchers are seeking new therapeutic alternatives. Polyphenols extracted from natural products are effective on several inflammatory diseases, including RA. In this review polyphenols are classified into four types: flavonoids, phenolic acids, stilbenes and others, among which mainly flavonoids are discussed. Researchers have reported that anti-RA efficacies of polyphenols are based mainly on three mechanisms: their anti-inflammatory, antioxidant and apoptotic properties. The main RA factors modified by polyphenols are mitogen-activated protein kinase (MAPK), interleukin-1β (IL-1β), IL-6, tumor necrosis factor-α (TNF-α), nuclear factor κ light chain enhancer of activated B cells (NF-κB) and c-Jun N-terminal kinases (JNK). Polyphenols could be potent alternative RA therapies and sources for novel drugs for RA by affecting its key mechanisms.

**Keywords:** rheumatoid arthritis; natural products; polyphenol; flavonoids; phenolic acid; stilbene

## 1. Introduction

Rheumatoid arthritis (RA) is a notorious chronic autoimmune inflammatory joint disease, which can cause cartilage and bone damage [1]. This disease is characterized by synovial inflammation, swelling, autoantibody production, cartilage and bone destruction, and systemic features such as cardiovascular, pulmonary, and skeletal disorders. It is associated with progressive disability, systemic complications, early death and socioeconomic costs [2]. As of 2015 is estimated that RA affects about 24.5 million people [3]. This number includes 0.5 to 1% of adults in the developed world, 5 to 50 per 100,000 patients newly added each year [1]. Although the critical damage caused by this disease is well known and thus widely studied, the mechanism(s), underlying cause and pathway(s) of RA are not well-known.

The number of therapeutic solutions available for treating RA has continuously grown in the past 30 years. These solutions include non-steroidal anti-inflammatory drugs, glucocorticoids, disease-modifying anti-rheumatic drugs (DMARDs) of synthetic origin (e.g., methotrexate and c-Jun N-terminal kinase (JNK) inhibitors) and of biological origin (ex. tumor necrosis factor (TNF) inhibitors, interleukin (IL)-6 inhibitor, and B cell-depleting drugs) [4]. Recently, medications that suppress the Janus kinase (JAK) pathways have shown noticeable effects as RA treatments, showing higher efficacy compared to the traditional ones. Tofacitinib and baricitinib, especially, are among the medications that show the most considerable effect and therefore have been extensively studied in clinical trial programs [4]. However, traditional DMARDs frequently present side-effects such as cytopenia, transaminase elevation, and poor tolerability. Another class of newly emerging solutions, the JAK inhibitors, also often cause gastrointestinal side-effects, lymphopenia, neutropenia, elevated cholesterol, and more infections [4]. On the basis of micro-environmental changes, severe synovial systematical reorganization and local fibroblast activation trigger synovial inflammation occur in RA [5]. The essential triggers of RA are unknown, but several genetic loci related to RA have been found [6]. These include major histocompatibility complex, class II, DR beta 1 (HLA-DRB1), Signal transducer and activator of transcription 4 (STAT4), protein tyrosine phosphatase (PTPN22), peptidyl arginine deiminase type I, IV (PAD14), and cytotoxic T-lymphocyte antigen 4 (CTLA4) [7]. Environmental factors such as smoking may stimulate the development of the disease by modifying genetic factors, but the specific mechanism(s) remain unknown [7].

Interactions of T cells, B cells, and related cytokines play key roles in developing RA symptoms such as synovitis, bone destruction, and cartilage degradation. The major cytokines that play a significant role in this process are TNFα, IL-6, IL-1, and IL-17 [8]. Like other auto-immune diseases, no perfect medications to treat RA have been developed. Recently many researchers have been trying to develop solutions for RA from natural products which have low toxicity and therefore assumed to have less side-effects. Polyphenols is one of the major classes of natural products that have been studied in this context. They are plant secondary metabolites that normally play a role in blocking ultraviolet radiation or pathogens. Numerous studies have shown that polyphenol-rich diets exert cardioprotective, anti-cancer, anti-diabetic and anti-aging effects [9]. Recognizing the strong anti-inflammatory effect of polyphenols and its potential role as a treatment for RA, we review herein the literature works that elucidate the effects of polyphenols on RA.

## 2. Polyphenols and Rheumatoid Arthritis

### 2.1. Phenolic Acids

Hydroxybenzoic and hydroxycinnamic acids are characteristic phenolic acids. Phenolic acids account for about a third of the polyphenolic compounds in our diet and are found in all plant material, but they are particularly abundant in acidic-tasting fruits. Caffeic acid, gallic acid, and ferulic acid are some common phenolic acids. Phenolic acids showing anti-RA effects are arranged in Table 1. When monocyte and macrophage cells from rat were pre-exposed for 24 h to ferulic acid, which is found in grains, vegetables, fruits and nuts, nuclear factor of activated T cells c1 (NFATc1), c-Fos, NF-κB, tartrate-resistant acid phosphatase (TRAP), matrix metalloproteinases (MMP)-9, Cathepsin activities were depressed [10]. The natural polyphenol N-feruloylserotonin (N-f-5HT), extracted from *Leuzea carthamoides*, had RA-inhibitory effects via suppressing c-reactive protein (CRP), 12/15-lipoxygenase (LOX), TNF-α, inducible nitric oxide synthase (iNOS), IL-1β in liver and spleen cells of arthritic rats. This study was conducted for 28 days, with 3 mg/kg of N-f-5HT [11]. In the study of Lee, mRNA transcription of TNF-α was significantly attenuated in a human mast cell line (HMC-1) treated with gallotanin derived from *Euphorbia* [12]. Chlorogenic acid (CGA), derived from *Gardenia jasminoides*, inhibited the phosphorylation of p38, Akt, extracellular signal-regulated kinase (ERK) and IkB, also suppressed the mRNA expression of nuclear factor activated T cells cl (NFATcl). Furthermore, lipopolysaccharide (LPS)-induced bone erosion was alleviated in vivo when bone marrow macrophages (BMMs) were exposed to 10, 25, 50 μg/mM of CGA for 4 days [13]. *p*-Coumaric Acid (CA), which can be extracted from *Gnetm cleistostachyum*, was used in two studies. Both of them used the same dose of 100 mg/kg of CA to treat an adjuvant-induced arthritis (AIA) rat model. One trial with a duration of 8 days presented degradation of TNF-α and circulating immune complexes (CIC) levels while inducing alleviation of immunoglobulin G (IgG) [14]. In the other 16 day trial, CA treatment also reduced

TNF-α activation, suggest an anti-RA effect via attenuation of cytokines, chemokines, osteoclastogenic factors, transcription factors, and mitogen-activated protein kinase (MAPK). In detail, the affected cytokines and chemokines are IL-1β, IL-6, monocyte chemoattractant protein (MCP)-1, the osteoclast factors are receptor activator of nuclear factor kappa-B ligand (RANKL), TRAP, the pro-inflammatory cytokines are IL-1b, IL-6, IL-17, the inflammatory enzymes are iNOS and cyclooxygenase (COX)-2, the transcription factors are NF-κB-p65, p-NF-κB-p65, NFATc-1, c-Fos, MAP kinases are JNK, p-JNK, ERK1/2. However, osteoprotegerin (OPG) elevation was shown [15].

**Table 1.** Rheumatoid arthritis-inhibiting phenolic acids.

| Compound | Source | Cell Line/Animal Model | Dose/Duration | Mechanism | Reference |
|---|---|---|---|---|---|
| Ferulic acid | Grains (rice, wheat and oats), vegetables, fruits, nuts | monocyte/macrophage cells/Rat | 25, 50, 100 μM/24 h | ↓ NFATc1, c-Fos, NF-κB, TRAP, MMP-9, Cathepsin | [10] |
| Natural polyphenol N-feruloylserotonin (N-f-5HT) | *Leuzea carthamoides* | AA | 3 mg/kg/28 days | ↓ CRP, LOX, TNF-α, iNOS, IL-1β | [11] |
| Gallotanins | *Euphorbia* | HMC-1/human | 10 mg/mL/30 min | ↓ TNF-α, IL-1β, IL-6, NF-κB | [12] |
| Kaempferol (3,5,7,4′-tetrahydroxy-flavone) | Gallic acid | RASFs/human | 100 μM/ 2 days | ↓ IL-1β, MMPs, COX, PGE2 | [16] |
| Chlorogenic acid (CGA) | *Gardenia jasminoides* | osteoclast/ BMMs | 10, 25, 50 μg/mM/4 days | ↓ NF-κB, P38, Akt, ERK | [13] |
| *p*-Coumaric Acid (CA) | *Gnetm cleistostachyum* | AIA | 100 mg/kg/8 days | ↓TNF-α, CIC ↑IgG | [14] |
| *p*-Coumaric Acid (CA) | *Gnetm cleistostachyum* | AIA | 100 mg/kg/16 days | ↓TNF-α, IL-1β, IL-6, MCP-1, RANKL, TRAP, IL-1β, IL-6, IL-17, iNOS, COX-2, NF-κB-p65, p-NF-κB-p65, NFATc-1, c-Fos, JNK, p-JNK, ERK1/2 ↑OPG | [15] |

AA, adjuvant arthritis; HMC-1, human mast cell line; RASFs, rheumatoid arthritis synovial fibroblasts; BMMs, bone marrow-derived macrophages; AIA, adjuvant induced *arthritis*; NFATc1, nuclear factor of activated T cells c1; NF-κB, nuclear factor κ light chain enhancer of activated B cells; TRAP, tartrate-resistant acid phosphatase; MMP-9, matrix metalloproteinases-9; CRP, c-reactive protein; LOX, 12/15-lipoxygenase; TNF-α, tumor necrosis factor-α; iNOS, inducible nitric oxide synthase; IL-1β, interleukin-1β; COX, cyclooxygenase; PGE2, prostaglandin E2; CIC, circulating immune complexes; IgG, immunoglobulin G; MCP-1, monocyte chemoattractant protein-1; RANKL, receptor activator of nuclear factor kappa-B ligand; TRAP, tartrate-resistant acid phosphatase; JNK, c-Jun N-terminal kinases; OPG, osteoprotegerin; ↑—up-regulation; ↓—down-regulation.

## 2.2. Stilbenes

Stilbenes have a 1,2-diphenylethylene nucleus that can be of two types: (*E*)-stilbenes which are the *trans* isomers and (*Z*)-stilbenes which are *cis* isomers [17]. Stilbenes are polyphenols with anti-inflammatory, cell death activation, and anti-oxidant effects. Among more than 400 natural stilbenes, the most popular one is resveratrol (RSV). RSV was reported as a new potential agent to suppress inflammation-induced arthritis (Table 2). RSV, which is originated from red grapes, showed anti-RA effect on FLSs of AA that was given with a dose of 5, 15, 45 mg/kg of the compound for 12 days, by inhibiting Beclin1, LC3A/B, manganese-dependent superoxide dismutase (MnSOD) and inducing MtROS [18]. A dose on FLSs in humans of 50 µg for 24 h also demonstrated an anti-RA effect via suppression of COX-2, prostaglandin E2 (PGE2), nicotinamide adenine dinucleotide phosphate (NADPH) oxidase, Akt, p38 MAPK, ERK1/2, reactive oxygen species (ROS), NF-κB [19]. On human synovial membrane in a test conducted with resveratrol at a dose of 6.25, 12.5, 25, 50 µM, resveratrol exerted the same effect by regulating IL-1β, MMP-3, p-Akt, MMP-3, PI3K-Akt [20]. In the randomized controlled clinical trial by Hani, 50 patients were given a 1 g RSV capsule for 3 months. This study suggested that taking RSV has significant clinical effect in RA. Also, RF positivity, SJC-28, TJC-28, CRP, erythrocyte sedimentation rate (ESR), uncarboxylated osteocalcin (ucOC), MMP-3, TNF-α, IL-6, disease Activity Score-28 for Rheumatoid Arthritis with ESR (DAS28-ESR) levels were alleviated [21]. Furthermore, RSV relieved RA symptoms by downregulating IgG1, IgG2a when a dose of 20 mg/kg was used. After treatment of draining lymph node (DLN) cells and Th17 cells of rat with 40 µM of RSV for 72 h, expressions of IL-17 and IFN-γ were decreased. With the same cell line, injection of 30 µM or 50 µM for 3 days led to suppression of TH-17, IL-17 [22]. Finally, RSV-exposed FLSs in AA showed a decline of Beclin1, LC3A/B, MnSOD and increase of mitochondrial (Mt) ROS [23].

## 2.3. Flavonoids

Flavonoids are a type of polyphenol which consist of two phenyl rings in a general 15-carbon skeleton structure. They can be classified into flavones, flavonols, flavanones, flavanonols, flavanols or catechins, anthocyanins, and chalcones [24]. Quercetin and epigallocatechin-3-gallate, a tea flavonoid, are some of the best known flavonoids. These compounds have beneficial effects such as anti-inflammatory and anti-cholinesterase activity and therefore are used to treat many diseases. For example, a flavonoid-rich diet was reported to be associated with a reduced risk of cardiovascular disease [25]. Citrus flavonoids can modulate lipid metabolism and thus can be used as a treatment of metabolic dysregulation [26]. The anti-inflammatory effects of flavonoids can also be applied to attenuating the symptoms of rheumatoid arthritis (Table 3). A-glucosylhesperidin is extracted from citrus fruits, and exerts anti-RA effects via downregulation of tumor necrosis factor α (TNFα) at a dose of 3 mg per 0.3 mL when it was administered on a collagen-induced arthritis (CIA) rat model 3 times a week for 31 days [27]. Anthocyanin from cherries showed anti-RA effects by inhibiting TNFα, prostaglandin E2 (PGE2), and malondialdehyde (MDA) and inducing superoxide dismutase (SOD) at doses of 10, 20, and 40 mg/kg when adjuvant induced arthritis (AIA) rats were treated for 14 days [28]. Also, cocoa polyphenol, which consists of epicatechin, catechin, flavonol glycosides, and procyanidin, downregulated vascular endothelial growth factor (VEGF), NF-kB, and activator protein (AP)-1 and increased formation of p-Akt, p-p70S6K, p-extracellular signal-regulated kinases (ERK), p-p90 kDa ribosomal S6 kinase (p90RSK), p-mitogen-activated protein kinase kinase 4 (MKK4), p-c-Jun N-terminal kinase (JNK), p- PI3K when a JB6 P+ mouse epidermal cell model was treated with doses of 10 and 20 µM /mL for 1 h [29]. Epigallocatechin-3-gallate (EGCG), a well-known compound from Camellia sinensis, exerted anti-RA effects on human rheumatoid arthritis synovial fibroblasts (RASF) by downregulating epithelial neutrophil-activating peptide (ENA)-78, RANTES, growth-regulated oncogene (GRO)-α, IL-1–induced MMP-2, chemokine-induced MMP-2 at doses of 10, 20, 30, 40 and 50 µM when administered for 12 h [30]. Doses of 125, 250, 500 nM of EGCG for 24 h also demonstrated anti-RA effects on human rheumatoid arthritis synovial fibroblasts (RASFs) via suppression of mitogen-activated protein kinase (MAPK), MMP-1, MMP-3, p-extracellular regulated kinases (ERK)1/2, p-JNK, p-p38, and AP-1 formation [31]. When EGCG was given to CIA rats for 3 weeks, at a dose of 20, 30, 40, and 50 mg/kg, it inhibited type II collagen (CII) antigen-specific IgG2a, IL-1β, IL-6, TNFα, IL-17, VEGF, nitrotyrosine, iNOS, c-Fos, nuclear factor of activated T cells c1 (NFATc1), cathepsin K (CTSK), MMP9, p-STAT3 727, IL-17, chemokine (C-C motif) ligand 6 (CCL6), aryl hydrocarbon receptor (AHR), IL-21, p-STAT3 705, p-ERK, receptor activator of nuclear factor κ B (RANK), tartrate-resistant acid phosphatase (TRAP), and calcitonin receptor (CTR), while it induced IL-10, TGF-β, suppressor of cytokine signaling 3 (SOCS3), Foxp3 [32]. Furthermore, IL-6, TNFα, and interferon (IFN)-γ were suppressed, but anti-CII specific IgG1 antibodies were activated when CIA rats were treated for 3 weeks with a dose of 10 mg per kg of the rats' weight [33]. EGCG at doses of 10 mg/kg for 5 days also showed anti-RA effects on pristane-induced arthritic (PIA) rats via inhibition of myeloperoxidase (MPO) [34]. When both human osteoclasts of peripheral blood monocytes and mice were treated for 15 days with a dose of 20 and 50 μM, CTR, carbonic anhydrase II, cathepsin K, α-v integrin, β-3 integrin, and NF-ATc1 were downregulated [35]. On osteoclast precursor cells and mature rat osteoclasts, 7 days of EGCG treatment with a dose of 10 and 100 μM restrained multinucleated osteoclast formation, MMP-9, and MMP-2, showing anti-RA effects [36]. Another flavonoid, fisetin from Rhus verniciflua Stokes, displayed anti-RA effects on human RA fibroblast-like synoviocytes (RAFLS) when they were treated with a dose of 0.1, 1, or 10 μg/mL for 72 h. FLS proliferation, TNFα, IL-6, IL-8, monocyte chemoattractant protein (MCP)-1, and VEGF were suppressed [37]. Under the same conditions described above, a flavonol-rich residual layer of the hexane fraction (RVHxR) derived from Rhus verniciflua Stokes, also inhibited FLS proliferation, TNFα, IL-6, IL-8, MCP-1, and VEGF and further inhibited p-ERK and p-JNK, while upregulated p-p38-MAPK [37]. Gallic acid, extracted from Cinnamomum zeylanicum L. bark, reduced RA symptoms by suppressing TNFα expression on adjuvant-induced arthritis (AIA) rats at a dose of 200 mg/kg for 12 and 21 days. When concanavalin (Con-A)-stimulated lymphocytes were treated at a dose of 40 μg/100 μL for 72 h, IL-2, IL-4, and IFN-γ were repressed [38]. One of the main flavonoids in soybean is genistein, which is reported to have anti-RA effects on CIA rat by reducing IFN-γ and T-bet, and the Th1/Th2 ratio, while it upregulates GATA-3 and IL-4. These effects were demonstrated in a dose of 1 mg/kg rat weight after treatment for 42 days [39]. Genistein also exerted the same effect via inhibition of FLS proliferation and MMP-9 when tested on human RAFLS for 24 h at a dose of 10 μg/mL [40]. Hesperidin ameliorated RA symptoms, inhibiting ELA, TBAR, and nitrite (NO), while inducing glutathione (GSH), SOD, and catalase when CIA rats were treated for 22 days with a dose of 160 mg/kg [41]. Kaempferol (3,4′,5,7-tetrahydroxyflavone), which is derived from diverse sources such as propolis and grapefruits, presented anti-RA effects on synovial tissues of patients with knee arthroplasty. When 10, 50, 100, and 200 μM of the flavonoid were applied on the tissue for 2 days, MAPK, NF-κB, and RASFs were inhibited. When a dose of 100 μM of the compound was applied on the tissue for 48 h, MAPK, NF-κB, MMP-1, MMP-3, COX-2, and PGE2 were inhibited [16]. Malvidin-3-O-β glucoside, extracted from red grape skin extract powder, relieved RA symptoms by downregulating TNF-α, IL1, macrophage inflammatory protein 1a (MIP1a), IL-8, IL-6, NO, and NOx when a dose of 1, 10, and 100 μM was applied to human peripheral blood monocyte-derived macrophages for 24 h. When peritoneal macrophages of rat were treated under the same conditions IL-1β, TNF-α, and IL-8 were suppressed [42]. Mangiferin (1,3,6,7-tetrahydroxyxanthone-C2-β-D-glucoside), derived from the family Thymelaeaceae, showed anti-RA effects via suppression of NF-κB, ERK1/2, IL-1β, IL-6, TNF-α, and RANKL when it was tested on CIA-induced DBA/1 rats at doses of 100 and 400 mg/kg for both 14 days and 27 days [43]. Morin (ML-morin) from various fruits and vegetables reduced RA symptoms by decreasing ROS, NO, iNOS, NF-κB-p65, TNF-α, IL-1-beta, IL-6, MCP-1, VEGF, RANKL, and STAT-3 formation in spleen and synovial macrophages of Wistar albino rats when they were treated with a dose of 10 mg/kg for 3 days [44]. Naringin, which can be extracted from grapes and citrus fruits, presented anti-RA effects on AIA rat by reducing TNFα, IL-1β, IL-6, and Bcl-2 formation while increasing Bax formation. The rats were treated with the compound at a dose of 20 mg/kg and of 40 mg/kg for 28 days, and both doses showed similar results [45]. Theaflavin-3,3′-digallate (TFDG), derived from Camellia sinensis exerted anti-RA effects via downregulation of multinucleated osteoclast formation, MMP-9 and MMP-2 of osteoclast precursor cells and mature osteoclast of rats treated with a dose of 10 and 100 μM for 7 days [36]. Thymoquinone (TQ) is extracted from Nigella sativa, and induces anti-RA effects by downregulating IL-6, IL-8, intercellular adhesion molecules (ICAM)-1, vascular cell adhesion protein (VCAM)-1, Cad-11, p38, and JNK in human RA synovium which was treated with 1, 2, 3, 4, and 5 μM of the compound for 2 h [46]. A similar effect was shown through a different mechanism when it was administered to CIA rats at 2.5 mg/kg for 5 days: TQ reduced IL-1β formation in CIA rat [47]. TQ also inhibited LPS-induced FLS proliferation, LPS-induced IL-1β, TNFα, MMP-13, COX-2, prostaglandin, $H_2O_2$-induced 4-hydroxynonenal (HNE), p-p38 -MKK, p-ERK, and p-NF-κB-p65 in human RAFLS treated with 1, 2, 5, and 10 μM of the compound for 1 h. When it was given to AIA rats at a dose of 5 mg/kg for 1 day, an anti-RA effect appeared via downregulation of HNE, IL-1β, and TNFα [48].

## 2.4. Other Compounds

Other polyphenols were also studied for their anti-RA mechanisms (Table 4). EVOO-polyphenol extract (PE), which is extracted from extra virgin oil (EVOO), exerts anti-RA effects via downregulation of TNF-α, IL-1β, IL-6, PEG2, p38, JNK, p65, and IκB-α at a dose of 100 and 200 mg/kg when collagen-induced arthritis (CIA) rats were treated for 13 days [49]. Hydroxytyrosol acetate (Hty-Ac), also from EVOO, showed anti-RA effects by inhibiting IgG1, IgG2a, COMP, MMP-3, TNF-Q, IFN-S, IL-1R, IL-6, IL-17A, nuclear factor (erythroid-derived 2)-like 2 (Nrf2), and heme oxygenase 1 (HO-1) at a dose of 0.5% when CIA rats were treated for 42 days [50]. Curcuminoid from turmeric rhizome or ginger rhizome, induced TNF-α, IL-1β, IL-6, IL-4, IL-10, SOD, CAT, and GSH, while suppressing lipid peroxidation (LPO), alanine transaminase (ALAT), and alkaline phosphatase (ALP) when it was given to AIA rats at a dose of 200 mg/kg/28 days [51]. Curcumin, also from the same origin, relieved RA symptoms of MH7A via downregulation of IL-1β, IL-6, NF-κB, ERK1/2, and AP-1, and upregulation of lactate dehydrogenase (LDH). The cells were treated with a dose of 12.5, 25, and 50 μM for 6 h [52]. Under the same conditions, RA symptoms in RAFLS were attenuated by repression of IL-1β, IL-6, NF-κB, ERK1/2, AP-1, and VEGF-A [38]. Curcumin, from rhizome of *Curcuma longa*, likewise demonstrated anti-RA effects via suppression of Bcl-2, caspase-3, caspase-9, ADP-ribose, and COX-2 of FLS when human FLS were treated for 24 h with a dose of 25, 50, 75, and 100 μM [53]. Curcumin oil-water nanoemulsions (CM-Ns) from the herb turmeric mitigated RA symptoms by downregulating NF-κB, TNF-α, and IL-1β in AIA rats which was treated with 50 mg/kg of CM-Ns for 24 h [54]. Emodin, extracted from *Rheum palmatum*, showed anti-RA effects on CIA rats that were given a dose of 10 mg/kg of the compound for 11 days, by inhibiting NF-κB, MMP, and M-CSF [55]. A dose of 5, 10, and 20 mg/kg on CIA rats for 21 days also demonstrated anti-RA effects via suppression of TNF-α, IL-6, and PGE2 [56]. On human synovial membrane which was administered emodin it the same effect by downregulating histone deacetylase (HDAC), HDAC1, VEGF, COX-2, COX-2, VEGF, hypoxia-inducible factor (HIF)-1a, MMP-1, MMP-13, NF-κB, and MAPK [57].

Table 2. Rheumatoid arthritis-inhibiting stilbenes

| Compound | Source | Cell Line/Animal Model | Dose/Duration | Mechanism | Reference |
|---|---|---|---|---|---|
| Resveratrol | Red grapes | FLSs/AA | 5, 15, 45 mg/kg/12 days | ↑ MtROS; ↓ Beclin1, LC3A/B, MnSOD | [18] |
| Resveratrol | Red grapes | FLSs/Human | 50 µg/24 h | ↓ COX-2, PGE2, NADPH oxidase, ROS, Akt, p38, MAPK, ERK1/2, NF-κB | [19] |
| Resveratrol | Red grapes | FLSs/Human | 6.25, 12.5, 25, 50 µM/1 h | ↓ IL-1β, MMP-3, P-Akt, PI3K-Akt | [20] |
| Resveratrol | Red grapes | Human * randomized controlled clinical trial | 1000 mg/day/3 month | ↓ RF, MMP-3, TNF-α, IL-6, | [21] |
| Resveratrol | Red grapes | 1) CIA 2) DLN cell/CIA 3) Th17 cell/CIA | 1) 20 mg/kg/10days 2) 40 µM/72 h 3) 30 µM, 50 µM/3 days | 1) ↓ IgG1, IgG2a 2) ↓ IL-17, IFN-γ 3) ↓ TH-17, IL-17 | [22] |
| Resveratrol | Red grapes | CFA induced rat | 10 mg/kg/day/7 days | ↓ RF, MMP-3, COMP, IgG, ANA, TNF-a, MPO, MDA; ↑ IL-10, GSH | [23] |

FLSs, fibroblast-like synoviocytes; CIA, collagen-induced arthritis; DLN, draining lymph node; CFA, cetylated fatty acids; MtROS, mitochondrial ROS; LC3, microtubule-associated protein 1a/1b-light chain 3; MnSOD, manganese-dependent superoxide dismutase; COX, cyclooxygenase; PGE2, prostaglandin E2; NADPH, nicotinamide adenine dinucleotide phosphate; ROS, reactive oxygen species; MAPK, mitogen-activated protein kinase; NF-κB, nuclear factor κ light chain enhancer of activated B cells; IL-1β, interleukin-1β; MMP-3, matrix metalloproteinases-3; PI3K, phosphoinositide 3-kinases; RF, rheumatoid factor; TNF-α, tumor necrosis factor-α; IgG, immunoglobulin G; IFN, interferon; COMP, cartilage oligomeric matrix protein; ANA, antinuclear antibodies; MPO, myeloperoxidase; MDA, malondialdehyde; GSH, glutathione; ↑—up-regulation, ↓—down-regulation.

**Table 3.** Rheumatoid arthritis-inhibiting flavonoids

| Compound | Source | Cell Line/Animal Model | Dose/Duration | Mechanism | Reference |
|---|---|---|---|---|---|
| A-glucosylhesperidin | Citrus fruit | CIA rat | 3 mg/0.3 mL/3 times a week, 31 days | ↓ TNFα | [27] |
| Anthocyanin | Cherries | AIA rat (Male Sprague Dawley) | 10, 20, 40 mg/kg/14 days | ↓TNFα, PGE2, MDA ↑SOD | [28] |
| Cocoa polyphenol (epicatechin, catechins, flavonol glycosides and procyanidin) | Cocoa | JB6 P+ mouse epidermal cells | 0, 10, 20 µM/mL/1 h | ↓ VEGF, NF-kB, AP-1 | [29] |
| Epigallocatechin-3-gallate (EGCG) | Green tea (*Camellia sinensis*) | RASFs | 10, 20, 30, 40, 50 µM/12 h | ↓ p-Akt, p-p70S6K, p- ERK, p- p90RSK, p-MKK4, p-JNK, p- PI3K ↓ENA-78, RANTES, GRO-alpha, MMP-2 | [30] |
| Epigallocatechin-3-gallate (EGCG) | Green tea | CIA rat (DBA/1I) | 20, 30, 40, 50 mg/kg/3 weeks | ↓IgG2a, IL-1β, IL-6, TNFα, TRAP, IL-17, VEGF, nitrotyrosine, iNOS, p-STAT3, c-Fos, NFATc1, CTSK, MMP9, p-STAT3 727, IL-17, CCL6, AHR, IL-21, p-STAT3 705, p-ERK, RANK, CTR ↑IL-10, TGF- β, SOCS3, Foxp3 | [32] |
| Epigallocatechin gallate | Green tea | PIA rats (Dark Agouti) | 10 mg/kg/5 days | ↓MPO | [34] |
| Epigallocatechin-3-gallate (EGCG) | Green tea (*Camellia sinensis*) | 1) Human osteoclasts of peripheral blood monocytes 2) DBA/1 mice | 20 µM, 50 µM/15 days | ↓ CTR, carbonic anhydrase II, cathepsin K, alpha-v integrin, β-3 integrin, NF-ATc1 | [35] |
| Epigallocatechin-3-gallate (EGCG) | Green tea | CIA rat (DBA/1I) | 10 mg/kg/3 weeks | ↓ IL-6, TNFα, IFN-γ ↑anti-CII specific IgG1 antibodies | [33] |
| Epigallocatechin-3-gallate (EGCG) | *Camellia sinensis* | Osteoclast precursors cells mature osteoclasts | 10, 100 µM/7 days | ↓Multinucleated osteoclast formation, MMP-9, MMP-2 | [36] |
| Epigallocatechin 3-gallate (EGCG) | Green tea (*Camellia sinensis*) | RASFs | 125, 250, 500 nM/24 h | ↓ MAPK, MMP-1, MMP-3, p-ERK1/2, p-JNK, p-p38, AP-1 | [31] |
| Fisetin | *Rhus verniciflua* Stokes | RA FLs | 0.1, 1, 10 µg/mL/72 h | ↓ TNFα, IL-6, IL-8, MCP-1, VEGF | [37] |
| Flavonol-rich residual layer of hexane fraction (RVHxR) | *Rhus verniciflua* Stokes | RA FLs | 0.1, 1, 10 µg/mL/72 h | ↓ TNFα, IL-6, IL-8, MCP-1, VEGF ↓p-ERK, p-JNK, ↑P-p38-MAPK | [37] |
| Gallic acid | *Cinnamomum zeylanicum* Bark | 1) AIA rat 2) Concanavalin (Con-A) stimulated lymphocyte | 1) 200 mg/kg/12 days, 21 days 2) 40 µg/100 µL/72 h | 1) ↓TNF-α 2) ↓ IL-2, IL-4, IFN-γ | [38] |
| Genistein | | CIA rats | 1 mL/kg/42 days | ↓IFN-γ, Th1/Th2, T-bet ↑GATA-3, IL-4 | [39] |
| Genistein | Soybean | RA FLS | 10 µg/mL/24 h | ↓ MMP-9 | [40] |
| Hesperidin | | CIA rat (Wistar rat) | 160 mg/kg/ 22 days | ↓ ELA, TBARS, nitrite ↑GSH, SOD, catalase | [41] |
| Malvidin-3-O-β-glucoside | Red grape skinExtract powder | 1) peripheral blood monocyte-derived macrophages 2) peritoneal macrophages 3) AIA rat | 1) 1, 10, 100 µM/24 h 2) 1, 10, 100 µM/24 h 3) 25 mg/kg/10 days | 1) ↓TNF-α, IL1, MIP1a, IL-8, IL-6, NO, NOx 2) ↓ IL-1β, TNF-α, IL-8 | [42] |

**Table 3.** *Cont.*

| Compound | Source | Cell Line/Animal Model | Dose/Duration | Mechanism | Reference |
|---|---|---|---|---|---|
| Mangiferin | Thymelaeaceae family (e.g., *Phaleria cumingii*) | CIA rat (DBA/1) | 100 and 400 mg/kg/14 days and 27 days | ↓ NF-κB, ERK1/2,IL-1β, IL-6, TNF-α, RANKL | [43] |
| Morin (ML–morin) | Fruits, vegetables, tea | Spleen and synovial macrophages | 10 mg/kg/3 days | ↓ ROS, NO, iNOS, NF-κB-p65, TNF-α, IL-1 β, IL-6, MCP-1, VEGF, RANKL, STAT-3 | [44] |
| Naringin | Grape, citrus fruit | AIA rat (Female Sprague-Dawley) | 1) 20 mg/kg/28 days<br>2) 40 mg/kg/28 days | ↓ TNFα, IL-1β, IL-6, Bcl-2<br>↑ Bax | [45] |
| Theaflavin-3,3′-digallate (TFDG) | *Camellia sinensis* | osteoclast precursors cells mature osteoclasts | 10, 100 μM/7 days | ↓ Multinucleated osteoclast formation, MMP-9, MMP-2 | [36] |
| Thymoquinone | *Nigella sativa* | RA synovium | 1, 2, 3, 4, 5 μM/2 h | ↓ IL-6, IL-8, ICAM-1, VCAM-1, Cad-11, p38, JNK | [46] |
| Thymoquinone | *Nigella sativa* | CIA rat (Sprague-Dawley Wistar rat) | 2.5 mg/kg/5 days<br>5 mg/kg/5 days | ↓ IL-1β | [47] |
| Thymoquinone | *Nigella sativa* | 1) RA FLS<br>2) AIA rat | 1) 0, 1, 2, 5, 10 μM/1 h<br>2) 5 mg/kg/1 day | 1) ↓LPS-induced IL-1β, TNFα, MMP-13, COX-2, prostaglandin, HNE, p-p38-MKK, p-ERK, p-NF-κB-p65<br>2) ↓ HNE, IL-1β, TNFα | [48] |

RA FLS, RA fibroblast-like synoviocytes; MKK4, mitogen-activated protein kinase kinase 4; ENA-78, Epithelial neutrophil- activating protein 78; RANTES, regulated on activation, normal t cell expressed and secreted; GRO, growth-regulated oncogene; VEGF, vascular endothelial growth factor; AHR, aryl hydrocarbon receptor; ICAM, intercellular adhesion molecules; VCAM, vascular cell adhesion protein; LPS, lipopolysaccharides; HNE, $H_2O_2$-induced 4-hydroxynonenal; ↑—up-regulation; ↓—down-regulation.

Table 4. Rheumatoid arthritis-inhibiting other polyphenols.

| Compound | Source | Cell Line/Animal Model | Dose/Duration | Mechanism | Reference |
|---|---|---|---|---|---|
| EVOO-polyphenol extract (PE) | EVOO | CIA in DBA-1/J | 100, 200 mg/kg/13 days | ↓TNF-α, IL-1β, IL-6, PEG2, p38, JNK, p65, IκB- α ↓IgG1, IgG2a, COMP, MMP-3, | [49] |
| Hydroxytyrosol acetate (Hty-Ac) | EVOO | CIA in DBA-1/J | 0.05%/42 days | TNF-Q, IFN-S, IL-1R, IL-6, IL-17A, Nrf2, HO-1 | [50] |
| Curcuminoid | 1. Turmeric rhizome 2. Ginger rhizome | AIA | 200 mg/kg/28 days | ↑TNF-α, IL-1β, IL-6, IL-4, IL-10, SOD, CAT, GSH ↓LPO, ALAT, ALP | [51] |
| Curcumin | Turmeric rhizome | 1) MH7A 2) RA-FLS | 12.5, 25, 50 μM/6 h | 1)　↓IL-1β, IL-6, NF-κB, ERK1/2, AP-1 ↑LDH  2)　↓IL-1β, IL-6, NF-κB, ERK1/2, AP-1, VEGF-A | [52] |
| Curcumin oil-water nanoemulsions (CM-Ns) | Herb turmeric | AIA | 50 mg/kg/14 days | ↓NF-κB, TNF-α, IL-1β | [54] |
| Curcumin | Rhizome of *Curcuma longa* | FLS/Patient | 0, 25, 50, 75, 100 μM/24 h | ↓Bcl-2, COX-2↑caspase-3, caspase-9 | [53] |
| Emodin | *Rheum palmatum* | CIA DBA/1 J | 10 mg/kg/11 days | ↓NF-κB, MMP, M-CSF | [55] |
| Emodin | *Rheum palmatum* | CIA | 5, 10, 20 mg/kg/21 days | ↓TNF-α, IL-6, PGE2 | [56] |
| Emodin | *Rheum palmatum* | Synovial membrane/Humans | 0.1, 1, 10 μM/24 h | ↓HDAC, HDAC1, VEGF, COX-2, HIF-1a, MMP-1, MMP-13, NF-κB, MAPK | [57] |

MH7A, rheumatoid synovial cell; Nrf2, nuclear factor (erythroid-derived 2)-like 2; HO-1, heme oxygenase 1; LPO, lipid peroxidation; ALAT, alanine transaminase; ALP, alkaline phosphatase; LDH, lactate dehydrogenase; M-CSF, macrophage colony-stimulating factor; HDAC, histone deacetylase; HIF, hypoxia-inducible factor; ↑—up-regulation; ↓—down-regulation.

## 3. Discussion

Rheumatoid arthritis (RA) is an autoimmune disease that induces chronic joint inflammation, which causes cartilage and bone damage [1]. Synovial inflammation, swelling, autoantibody production, cartilage and bone destruction, and systemic features such as cardiovascular, pulmonary, and skeletal disorders are the main symptoms of this disabling autoimmune disease. Currently, non-steroidal anti-inflammatory drugs (NSAIDS), glucocorticoids, DMARDs, immunosuppressants, and biologic agents have been used to treat this autoimmune disease. DMARDs, especially, has been acknowledged as an effective early intervention for RA, their efficacy being validated by several randomized trials [58,59]. However, each DMARD showed different toxicity that causes side-effects such as diarrhea and rashes [59], and therefore various studies have been conducted to find a better solution for RA treatment. We saw the potential of finding the solution in natural products, especially polyphenols.

Flavonoids demonstrate anti-RA effects through diverse mechanisms [24]. α-Glucosyl-hesperidin showed results in an animal model study, but it lacked a specific discussion on the mechanism of the effect [27]. A study on cocoa polyphenol extract (CPE) discusses in depth the effect of the compound on several different inflammatory routes such as VEGF regulation, PI3K-Akt, and MAPK pathways [29]. Epigallocatechin-3-gallate (EGCG), from green tea is the most widely studied polyphenol related to RA. Lee et al. conducted an in-depth study on this compound using CIA rats and found specific elements that regulate and are regulated by Th17 cells and p-STAT3. This study further has observed gene-level events, which showed that control of *Nrf2* gene may lead to anti-RA effects [32]. Yun et al., on the other hand, noted the mechanism of EGCG's effect from a different perspective, comprehensively focusing on MMP production via the MAPK and AP-1 pathways [31]. Morinobu et al. focused their study on the role of nuclear factor of activated T cells c1 (NF-ATc1) in EGCG's effect on osteoclasts [35]. On the other hand, a study by Leichsenring, et al. lacked a detailed discussion on the mechanism of EGCG [34]. A study of Oka, et al. on both EGCG and TFDG also gave an incomplete description of the mechanism of the compounds' anti-RA effects [36]. A study on a flavonol-rich residual layer of hexane fraction (RVHxR) gave a poor examination of the role of elements in the MAPK pathway of angiogenesis [37]. A study on genistein by Zhang et al. gave a limited discussion of the possible mechanism of the compound's effect on RA [40]. Umar et al. focused their studies on the effect of hesperidin on lipid peroxidation, which is another large category of RA pathogenic mechanisms. This research considered diverse lipid peroxidation factors, providing dense information about the effect of hesperidin in lipid peroxidation which causes RA [41]. Decendit et al. nicely designed a study on malvidin-3-*O*-b glucoside. The study included both animal model experiments, as well as animal and human cell experiments. The study also descriptively illustrated the malvidin-3-*O*-β-glucoside-related RA pathogenic pathway targeting macrophages [42]. Thymoquinone (TQ) was studied systematically in two studies. Vaillancourt, et al conducted an in-depth study on TQ, setting three stages of experiments, which included in vivo experiments on human RAFLS and a rat model and in vitro tests on an animal model. This study observed the effect of TQ on RA based on three different pathogenic pathways, which are lipid peroxidation, inflammation, and bone destruction. Interactions of elements that comprise each pathway are described in detail through an organized experiment process [48]. Umar et al, suggested a new point of view in studying the effects of polyphenols on RA pathogenesis. The study focuses on the role of apoptosis signal-regulating kinase 1 (ASK1) in the TNF-α signaling pathway and explains the role of its related factors in RA reduction [46]. In a study by Tekeoglu et al. three different experimental groups and a control group were used, but the results was unhelpful in explaining the molecular mechanism of the regulating effect of TQ [47].

Phenolic acids, plant metabolites that are widely spread throughout the plant kingdom, also possess anti-RA effects. A study on ferulic acid made profound observations on the effect of the compound in the RA pathogenic pathway, especially targeting the relation of RANKL, an osteogenic factor, and NF-κB signaling pathway [10]. Kwak et al. conducted an incomplete study on chlorogenic acid. Considering that RA's pathological pathway contains various immunological factors, only using osteoclasts as the experiment cell line is limiting [13]. Neog et al. conducted a thoughtful study on the effect of *p*-coumaric acid (CA), also focusing on the system related to RANK and its interaction with T cell immune system factors [15]. Another study of CA, designed by Pragasam et al., was conducted under similar experimental conditions as used by Neoget al., but was imperfect in elucidating the molecular mechanism of CA's anti-Ra effect [14].

Stilbenes are polyphenols that have two phenyl moieties connected by a two-carbon methylene bridge. Most of the studies on stilbenes that showed anti-RA effects were made on resveratrol. Three noticeable studies were made on resveratrol. Tsai et al. particularly noted resveratrol's role in regulating COX-2 and PGE2 interaction. This study is unique because it focuses on the effect of particulate matter (PM) from air pollution on RA, and sees how resveratrol affects the inflammatory pathways of RA caused by PM [19]. Wahba et al. reported the effect of this compound from three perspectives. They observed immunological changes, inflammatory systemic changes, and oxidative stress changes.

Choosing specific biomarkers for each part, this study specifically elucidated the role of resveratrol in each pathway [23]. A study by Xuzhu et al. observed three different levels of the object, which included the CIA animal model, DLN cells, and Th17 cells. Despite the effort to observe the result in diverse ways, this study failed to identify the specific mechanism of resveratrol's effect on RA regulation at a molecular level [22].

In addition to flavonoids, phenolic acids, and stilbenes there are several other polyphenols that are hard to classify. Among them, curcumin (CM)-related molecules and emodin are the most actively studied polyphenols. Ramadan et al. systemically examined the anti-inflammatory and anti-oxidant effects of curcuminoids shrewdly considering diverse factors related to the pathways [51]. Kloesch et al. comprehensively tested the effect of CM on various inflammation pathway factors, but the duration of this study was too short [52]. A study by Zheng, et al. suggested a new way to increase the bioavailability of CA by forming CM-loaded Ns (CM-Ns). They also skillfully designed their experiments with three different experimental groups and one control group. However, a study on the molecular mechanism of CM-Ns' anti-Ra effect was lacking [54]. Park et al. conducted an n in depth study on how emodin targets the apoptosis pathway, mainly focusing on Bax/Bcl-2 imbalance and activation of caspase-9 and caspase-3 [53]. A study of Ha, et al. on emodin thoroughly studied different aspects of the RA pathogenic pathway and tested the compound in vivo, which signified its role in inflammatory conditions [57]. Zhu et al. further conducted an animal model study on the effect of emodin on RA symptoms, but their explanation on the therapeutic mechanism in the pathogenic pathway was deficient [56].

Polyphenol inhibit RA progress mainly by acting on three pathways: the inflammatory pathway, the oxidative pathway, and the apoptotic pathway. The inflammatory pathway regulated by polyphenols is mainly via the MAPK pathway and through regulation of NFATC1 gene in osteoblasts. The key molecules related to these processes are MAPK, IL-1$\beta$, IL-6, TNF-$\alpha$, NF-$\kappa$B, JNK, ERK1/2, AP-1 and COX-2 (Figure 1).

Although they are not mentioned frequently among the studies, the oxidative and apoptotic pathways are also attributed a role in the reduction of RA symptoms by polyphenols (Figure 2). The key elements in the oxidative pathway which are controlled by polyphenols are mostly in the PI3-K/Akt pathway that produces HO-1 through transcription of the Nrf-2 gene. Other than this pathway, iNOS is frequently mentioned as the target of polyphenols. For the apoptotic pathway, only the pathway that involves Bcl-2 is indicated among many studies.

Studies on polyphenols' anti-RA effects have mainly focused on their influence on inflammation pathways. There are some studies that concentrate on the anti-oxidative and apoptotic effect of polyphenols which result in a reduction of RA symptoms, but those are few in number. Further studies are needed in clarify the molecular studies mechanism of polyphenols' anti-oxidative and apoptotic effects that regulate RA's pathogenic pathways.

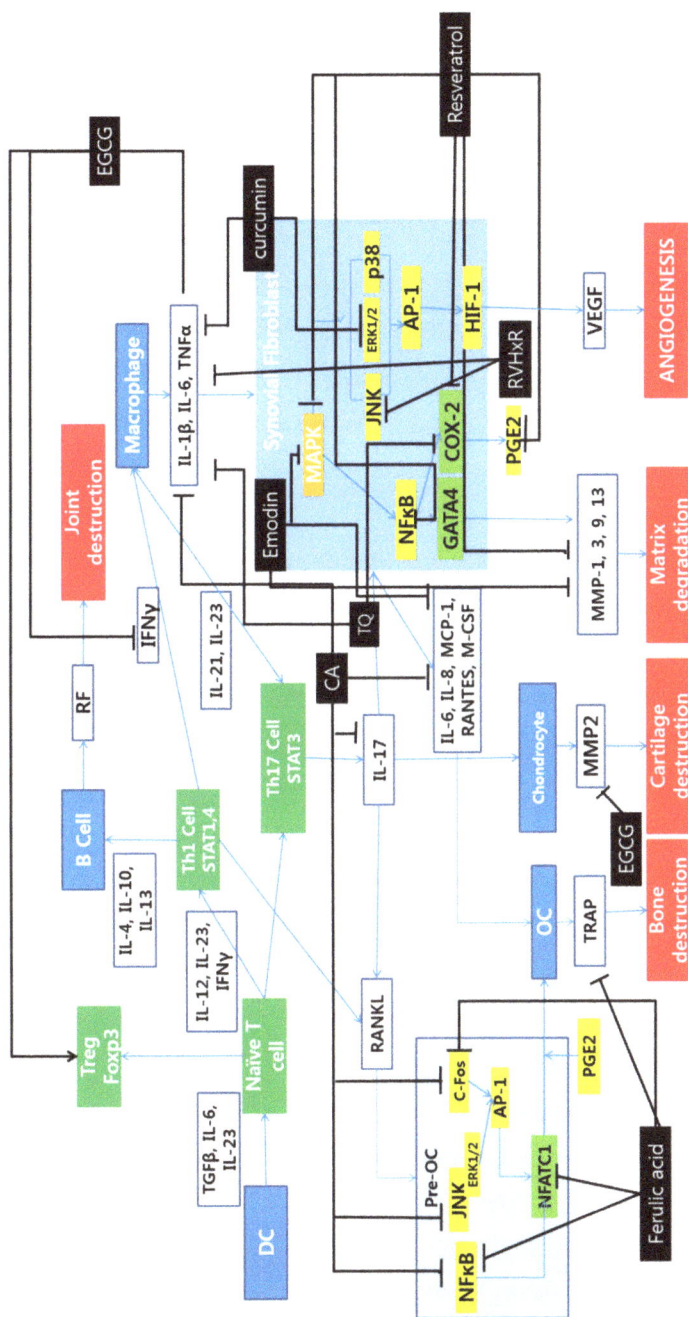

**Figure 1.** Schematic diagram of anti-inflammatory mechanisms of polyphenols. OC, osteoclasts; DC, dendritic cells; TQ, thymoquinone; CA, coumaric acid.

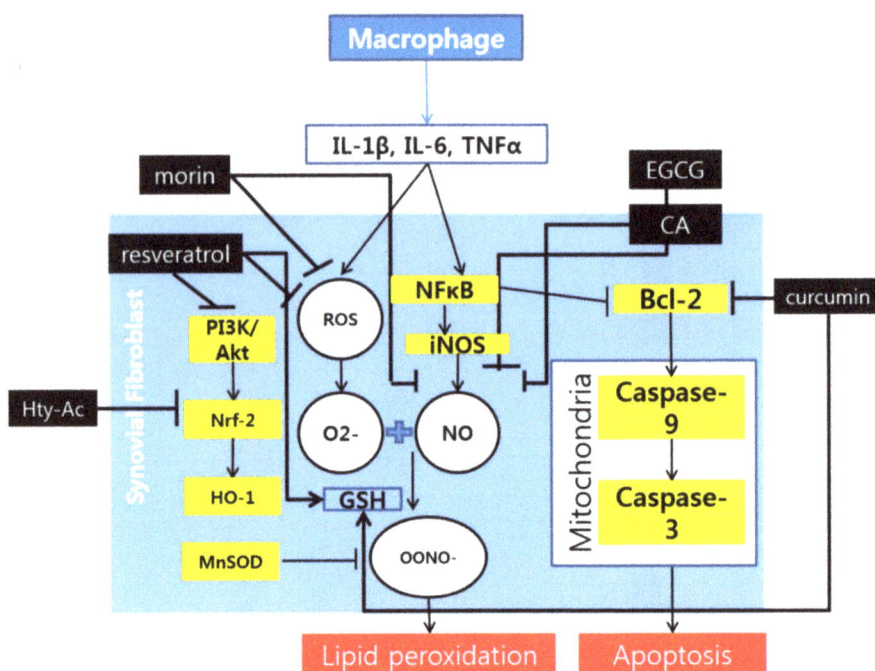

**Figure 2.** Schematic diagram of the anti-oxidative and apoptotic mechanisms of polyphenols.

In this review we have organized and summarized the role of each polyphenol compound in diverse pathogenic pathways of RA. This work will be significant in providing systematized information for developing natural-product-based RA therapeutic solutions.

## 4. Methods

Searches regarding the anti-RA effects of polyphenol were conducted on PubMed and Google Scholar in August of 2018. When searching for appropriate studies, we included "rheumatoid arthritis", and "polyphenol" as keywords. Only articles written in English, published from 2006 to 2018 were selected for further review. We selected studies which met the following criteria: (i) studies based on in vitro or in vivo experiments that demonstrate the anti-RA effects of polyphenols; (ii) studies that show statistically significant analysis data ($p < 0.05$); (iii) studies that were not shown to have errors by subsequent studies; (iv) studies written in English. For classifying the type of polyphenol, we used the method of Soto et al. [60].

## 5. Conclusions

Polyphenols reduce rheumatoid arthritis symptoms by regulating an extensive collection of RA-related molecules, including MAPK, IL-1β, IL-6, TNF-α, NF-κB, JNK, ERK1/2, AP-1 and COX-2. Studies on polyphenols' anti-RA effect were mainly focused on their influence on inflammation pathways. Further studies are needed for clarifying the molecular mechanism of polyphenol's anti-oxidative and apoptotic effects that also regulate RA's pathogenic pathways. Based on these preclinical data, clinical trials could be conducted.

**Author Contributions:** Investigation, writing—original draft preparation, S.S., D.K.; writing—review and editing, E.U., B.K; supervision, funding acquisition, B.K.

**Funding:** This Research was supported by the "2018 KIOM Undergraduate Research Program (C18054)" funded by Korea Institute of Oriental Medicine; (2018) URP program of College of Korean Medicine, Kyung Hee University; Basic Science Research Program through the National Research Foundation of Korea (NRF) funded by the Ministry of Education (NRF-2016R1D1A1B03933656).

**Conflicts of Interest:** The authors declare no conflict of interest.

## References

1. Smolen, J.S.; Aletaha, D.; McInnes, I.B. Rheumatoid arthritis. *Lancet* **2016**, *388*, 2023–2038. [CrossRef]
2. Firestein, G.S. Evolving concepts of rheumatoid arthritis. *Nature* **2003**, *423*, 356–361. [CrossRef] [PubMed]
3. Disease, G.B.D.; Injury, I.; Prevalence, C. Global, regional, and national incidence, prevalence, and years lived with disability for 310 diseases and injuries, 1990-2015: A systematic analysis for the Global Burden of Disease Study 2015. *Lancet* **2016**, *388*, 1545–1602. [CrossRef]
4. Burmester, G.R.; Pope, J.E. Novel treatment strategies in rheumatoid arthritis. *Lancet* **2017**, *389*, 2338–2348. [CrossRef]
5. McInnes, I.B.; Schett, G. The pathogenesis of rheumatoid arthritis. *N. Engl. J. Med.* **2011**, *365*, 2205–2219. [CrossRef] [PubMed]
6. McInnes, I.B.; Schett, G. Cytokines in the pathogenesis of rheumatoid arthritis. *Nat. Rev. Immunol.* **2007**, *7*, 429–442. [CrossRef] [PubMed]
7. Viatte, S.; Plant, D.; Raychaudhuri, S. Genetics and epigenetics of rheumatoid arthritis. *Nat. Rev. Rheumatol* **2013**, *9*, 141–153. [CrossRef]
8. Choy, E. Understanding the dynamics: Pathways involved in the pathogenesis of rheumatoid arthritis. *Rheumatology* **2012**, *51* (Suppl. 5), v3–v11. [CrossRef]
9. Pandey, K.B.; Rizvi, S.I. Plant polyphenols as dietary antioxidants in human health and disease. *Oxid. Med. Cell. Longev.* **2009**, *2*, 270–278. [CrossRef]
10. Doss, H.M.; Samarpita, S.; Ganesan, R.; Rasool, M. Ferulic acid, a dietary polyphenol suppresses osteoclast differentiation and bone erosion via the inhibition of RANKL dependent NF-kappaB signalling pathway. *Life Sci.* **2018**. [CrossRef] [PubMed]
11. Paskova, L.; Kuncirova, V.; Ponist, S.; Mihalova, D.; Nosal, R.; Harmatha, J.; Hradkova, I.; Cavojsky, T.; Bilka, F.; Siskova, K.; et al. Effect of N-Feruloylserotonin and Methotrexate on Severity of Experimental Arthritis and on Messenger RNA Expression of Key Proinflammatory Markers in Liver. *J. Immunol. Res.* **2016**, *2016*, 7509653. [CrossRef]
12. Lee, S.H.; Park, H.H.; Kim, J.E.; Kim, J.A.; Kim, Y.H.; Jun, C.D.; Kim, S.H. Allose gallates suppress expression of pro-inflammatory cytokines through attenuation of NF-kappaB in human mast cells. *Planta Med.* **2007**, *73*, 769–773. [CrossRef]
13. Kwak, S.C.; Lee, C.; Kim, J.Y.; Oh, H.M.; So, H.S.; Lee, M.S.; Rho, M.C.; Oh, J. Chlorogenic acid inhibits osteoclast differentiation and bone resorption by down-regulation of receptor activator of nuclear factor kappa-B ligand-induced nuclear factor of activated T cells c1 expression. *Biol. Pharm. Bull.* **2013**, *36*, 1779–1786. [CrossRef]
14. Pragasam, S.J.; Venkatesan, V.; Rasool, M. Immunomodulatory and anti-inflammatory effect of p-coumaric acid, a common dietary polyphenol on experimental inflammation in rats. *Inflammation* **2013**, *36*, 169–176. [CrossRef]
15. Neog, M.K.; Joshua Pragasam, S.; Krishnan, M.; Rasool, M. p-Coumaric acid, a dietary polyphenol ameliorates inflammation and curtails cartilage and bone erosion in the rheumatoid arthritis rat model. *Biofactors* **2017**, *43*, 698–717. [CrossRef] [PubMed]
16. Yoon, H.Y.; Lee, E.G.; Lee, H.; Cho, I.J.; Choi, Y.J.; Sung, M.S.; Yoo, H.G.; Yoo, W.H. Kaempferol inhibits IL-1beta-induced proliferation of rheumatoid arthritis synovial fibroblasts and the production of COX-2, PGE2 and MMPs. *Int. J. Mol. Med.* **2013**, *32*, 971–977. [CrossRef]
17. Sirerol, J.A.; Rodriguez, M.L.; Mena, S.; Asensi, M.A.; Estrela, J.M.; Ortega, A.L. Role of Natural Stilbenes in the Prevention of Cancer. *Oxid. Med. Cell. Longev.* **2016**, *2016*, 3128951. [CrossRef] [PubMed]
18. Zhang, J.; Song, X.; Cao, W.; Lu, J.; Wang, X.; Wang, G.; Wang, Z.; Chen, X. Autophagy and mitochondrial dysfunction in adjuvant-arthritis rats treatment with resveratrol. *Sci. Rep.* **2016**, *6*, 32928. [CrossRef] [PubMed]

19.  Tsai, M.H.; Hsu, L.F.; Lee, C.W.; Chiang, Y.C.; Lee, M.H.; How, J.M.; Wu, C.M.; Huang, C.L.; Lee, I.T. Resveratrol inhibits urban particulate matter-induced COX-2/PGE2 release in human fibroblast-like synoviocytes via the inhibition of activation of NADPH oxidase/ROS/NF-kappaB. *Int. J. Biochem. Cell Biol.* **2017**, *88*, 113–123. [CrossRef]
20.  Tian, J.; Chen, J.W.; Gao, J.S.; Li, L.; Xie, X. Resveratrol inhibits TNF-alpha-induced IL-1beta, MMP-3 production in human rheumatoid arthritis fibroblast-like synoviocytes via modulation of PI3kinase/Akt pathway. *Rheumatol. Int.* **2013**, *33*, 1829–1835. [CrossRef]
21.  Khojah, H.M.; Ahmed, S.; Abdel-Rahman, M.S.; Elhakeim, E.H. Resveratrol as an effective adjuvant therapy in the management of rheumatoid arthritis: A clinical study. *Clin. Rheumatol.* **2018**, *37*, 2035–2042. [CrossRef] [PubMed]
22.  Xuzhu, G.; Komai-Koma, M.; Leung, B.P.; Howe, H.S.; McSharry, C.; McInnes, I.B.; Xu, D. Resveratrol modulates murine collagen-induced arthritis by inhibiting Th17 and B-cell function. *Ann. Rheum. Dis.* **2012**, *71*, 129–135. [CrossRef] [PubMed]
23.  Wahba, M.G.; Messiha, B.A.; Abo-Saif, A.A. Protective effects of fenofibrate and resveratrol in an aggressive model of rheumatoid arthritis in rats. *Pharm. Biol.* **2016**, *54*, 1705–1715. [CrossRef]
24.  Panche, A.N.; Diwan, A.D.; Chandra, S.R. Flavonoids: An overview. *J. Nutr. Sci.* **2016**, *5*, e47. [CrossRef]
25.  Kim, K.; Vance, T.M.; Chun, O.K. Greater flavonoid intake is associated with improved CVD risk factors in US adults. *Br. J. Nutr.* **2016**, *115*, 1481–1488. [CrossRef] [PubMed]
26.  Mulvihill, E.E.; Burke, A.C.; Huff, M.W. Citrus Flavonoids as Regulators of Lipoprotein Metabolism and Atherosclerosis. *Annu. Rev. Nutr.* **2016**, *36*, 275–299. [CrossRef]
27.  Kometani, T.; Fukuda, T.; Kakuma, T.; Kawaguchi, K.; Tamura, W.; Kumazawa, Y.; Nagata, K. Effects of alpha-glucosylhesperidin, a bioactive food material, on collagen-induced arthritis in mice and rheumatoid arthritis in humans. *Immunopharmacol. Immunotoxicol.* **2008**, *30*, 117–134. [CrossRef]
28.  He, Y.H.; Zhou, J.; Wang, Y.S.; Xiao, C.; Tong, Y.; Tang, J.C.; Chan, A.S.; Lu, A.P. Anti-inflammatory and anti-oxidative effects of cherries on Freund's adjuvant-induced arthritis in rats. *Scand. J. Rheumatol.* **2006**, *35*, 356–358. [CrossRef]
29.  Kim, J.E.; Son, J.E.; Jung, S.K.; Kang, N.J.; Lee, C.Y.; Lee, K.W.; Lee, H.J. Cocoa polyphenols suppress TNF-alpha-induced vascular endothelial growth factor expression by inhibiting phosphoinositide 3-kinase (PI3K) and mitogen-activated protein kinase kinase-1 (MEK1) activities in mouse epidermal cells. *Br. J. Nutr.* **2010**, *104*, 957–964. [CrossRef]
30.  Ahmed, S.; Pakozdi, A.; Koch, A.E. Regulation of interleukin-1beta-induced chemokine production and matrix metalloproteinase 2 activation by epigallocatechin-3-gallate in rheumatoid arthritis synovial fibroblasts. *Arthritis Rheum.* **2006**, *54*, 2393–2401. [CrossRef]
31.  Yun, H.J.; Yoo, W.H.; Han, M.K.; Lee, Y.R.; Kim, J.S.; Lee, S.I. Epigallocatechin-3-gallate suppresses TNF-alpha -induced production of MMP-1 and -3 in rheumatoid arthritis synovial fibroblasts. *Rheumatol. Int.* **2008**, *29*, 23–29. [CrossRef] [PubMed]
32.  Lee, S.Y.; Jung, Y.O.; Ryu, J.G.; Oh, H.J.; Son, H.J.; Lee, S.H.; Kwon, J.E.; Kim, E.K.; Park, M.K.; Park, S.H.; et al. Epigallocatechin-3-gallate ameliorates autoimmune arthritis by reciprocal regulation of T helper-17 regulatory T cells and inhibition of osteoclastogenesis by inhibiting STAT3 signaling. *J. Leukoc. Biol.* **2016**, *100*, 559–568. [CrossRef] [PubMed]
33.  Min, S.Y.; Yan, M.; Kim, S.B.; Ravikumar, S.; Kwon, S.R.; Vanarsa, K.; Kim, H.Y.; Davis, L.S.; Mohan, C. Green Tea Epigallocatechin-3-Gallate Suppresses Autoimmune Arthritis Through Indoleamine-2,3-Dioxygenase Expressing Dendritic Cells and the Nuclear Factor, Erythroid 2-Like 2 Antioxidant Pathway. *J. Inflamm.* **2015**, *12*, 53. [CrossRef] [PubMed]
34.  Leichsenring, A.; Backer, I.; Furtmuller, P.G.; Obinger, C.; Lange, F.; Flemmig, J. Long-Term Effects of (−)-Epigallocatechin Gallate (EGCG) on Pristane-Induced Arthritis (PIA) in Female Dark Agouti Rats. *PLoS ONE* **2016**, *11*, e0152518. [CrossRef] [PubMed]
35.  Morinobu, A.; Biao, W.; Tanaka, S.; Horiuchi, M.; Jun, L.; Tsuji, G.; Sakai, Y.; Kurosaka, M.; Kumagai, S. (−)-Epigallocatechin-3-gallate suppresses osteoclast differentiation and ameliorates experimental arthritis in mice. *Arthritis Rheum.* **2008**, *58*, 2012–2018. [CrossRef]
36.  Oka, Y.; Iwai, S.; Amano, H.; Irie, Y.; Yatomi, K.; Ryu, K.; Yamada, S.; Inagaki, K.; Oguchi, K. Tea polyphenols inhibit rat osteoclast formation and differentiation. *J. Pharmacol. Sci.* **2012**, *118*, 55–64. [CrossRef]

37. Lee, J.D.; Huh, J.E.; Jeon, G.; Yang, H.R.; Woo, H.S.; Choi, D.Y.; Park, D.S. Flavonol-rich RVHxR from Rhus verniciflua Stokes and its major compound fisetin inhibits inflammation-related cytokines and angiogenic factor in rheumatoid arthritis fibroblast-like synovial cells and in vivo models. *Int. Immunopharmacol.* **2009**, *9*, 268–276. [CrossRef]

38. Rathi, B.; Bodhankar, S.; Mohan, V.; Thakurdesai, P. Ameliorative Effects of a Polyphenolic Fraction of Cinnamomum zeylanicum L. Bark in Animal Models of Inflammation and Arthritis. *Sci. Pharm.* **2013**, *81*, 567–589. [CrossRef] [PubMed]

39. Wang, J.; Zhang, Q.; Jin, S.; He, D.; Zhao, S.; Liu, S. Genistein modulate immune responses in collagen-induced rheumatoid arthritis model. *Maturitas* **2008**, *59*, 405–412. [CrossRef]

40. Zhang, Y.; Dong, J.; He, P.; Li, W.; Zhang, Q.; Li, N.; Sun, T. Genistein inhibit cytokines or growth factor-induced proliferation and transformation phenotype in fibroblast-like synoviocytes of rheumatoid arthritis. *Inflammation* **2012**, *35*, 377–387. [CrossRef]

41. Umar, S.; Kumar, A.; Sajad, M.; Zargan, J.; Ansari, M.; Ahmad, S.; Katiyar, C.K.; Khan, H.A. Hesperidin inhibits collagen-induced arthritis possibly through suppression of free radical load and reduction in neutrophil activation and infiltration. *Rheumatol. Int.* **2013**, *33*, 657–663. [CrossRef]

42. Decendit, A.; Mamani-Matsuda, M.; Aumont, V.; Waffo-Teguo, P.; Moynet, D.; Boniface, K.; Richard, E.; Krisa, S.; Rambert, J.; Merillon, J.M.; et al. Malvidin-3-O-beta glucoside, major grape anthocyanin, inhibits human macrophage-derived inflammatory mediators and decreases clinical scores in arthritic rats. *Biochem. Pharmacol.* **2013**, *86*, 1461–1467. [CrossRef]

43. Tsubaki, M.; Takeda, T.; Kino, T.; Itoh, T.; Imano, M.; Tanabe, G.; Muraoka, O.; Satou, T.; Nishida, S. Mangiferin suppresses CIA by suppressing the expression of TNF-alpha, IL-6, IL-1beta, and RANKL through inhibiting the activation of NF-kappaB and ERK1/2. *Am. J. Transl. Res.* **2015**, *7*, 1371–1381. [PubMed]

44. Sultana, F.; Neog, M.K.; Rasool, M. Targeted delivery of morin, a dietary bioflavanol encapsulated mannosylated liposomes to the macrophages of adjuvant-induced arthritis rats inhibits inflammatory immune response and osteoclastogenesis. *Eur. J. Pharm. Biopharm.* **2017**, *115*, 229–242. [CrossRef]

45. Zhu, L.; Wang, J.; Wei, T.; Gao, J.; He, H.; Chang, X.; Yan, T. Effects of Naringenin on inflammation in complete freund's adjuvant-induced arthritis by regulating Bax/Bcl-2 balance. *Inflammation* **2015**, *38*, 245–251. [CrossRef]

46. Umar, S.; Hedaya, O.; Singh, A.K.; Ahmed, S. Thymoquinone inhibits TNF-alpha-induced inflammation and cell adhesion in rheumatoid arthritis synovial fibroblasts by ASK1 regulation. *Toxicol. Appl. Pharmacol.* **2015**, *287*, 299–305. [CrossRef] [PubMed]

47. Tekeoglu, I.; Dogan, A.; Ediz, L.; Budancamanak, M.; Demirel, A. Effects of thymoquinone (volatile oil of black cumin) on rheumatoid arthritis in rat models. *Phytother. Res.* **2007**, *21*, 895–897. [CrossRef]

48. Vaillancourt, F.; Silva, P.; Shi, Q.; Fahmi, H.; Fernandes, J.C.; Benderdour, M. Elucidation of molecular mechanisms underlying the protective effects of thymoquinone against rheumatoid arthritis. *J. Cell Biochem.* **2011**, *112*, 107–117. [CrossRef]

49. Rosillo, M.A.; Alcaraz, M.J.; Sanchez-Hidalgo, M.; Fernandez-Bolanos, J.G.; Alarcon-de-la-Lastra, C.; Ferrandiz, M.L. Anti-inflammatory and joint protective effects of extra-virgin olive-oil polyphenol extract in experimental arthritis. *J. Nutr. Biochem.* **2014**, *25*, 1275–1281. [CrossRef]

50. Rosillo, M.A.; Sanchez-Hidalgo, M.; Gonzalez-Benjumea, A.; Fernandez-Bolanos, J.G.; Lubberts, E.; Alarcon-de-la-Lastra, C. Preventive effects of dietary hydroxytyrosol acetate, an extra virgin olive oil polyphenol in murine collagen-induced arthritis. *Mol. Nutr. Food Res.* **2015**, *59*, 2537–2546. [CrossRef]

51. Ramadan, G.; Al-Kahtani, M.A.; El-Sayed, W.M. Anti-inflammatory and anti-oxidant properties of Curcuma longa (turmeric) versus Zingiber officinale (ginger) rhizomes in rat adjuvant-induced arthritis. *Inflammation* **2011**, *34*, 291–301. [CrossRef] [PubMed]

52. Kloesch, B.; Becker, T.; Dietersdorfer, E.; Kiener, H.; Steiner, G. Anti-inflammatory and apoptotic effects of the polyphenol curcumin on human fibroblast-like synoviocytes. *Int. Immunopharmacol.* **2013**, *15*, 400–405. [CrossRef] [PubMed]

53. Park, C.; Moon, D.O.; Choi, I.W.; Choi, B.T.; Nam, T.J.; Rhu, C.H.; Kwon, T.K.; Lee, W.H.; Kim, G.Y.; Choi, Y.H. Curcumin induces apoptosis and inhibits prostaglandin E(2) production in synovial fibroblasts of patients with rheumatoid arthritis. *Int. J. Mol. Med.* **2007**, *20*, 365–372. [CrossRef] [PubMed]

54. Zheng, Z.; Sun, Y.; Liu, Z.; Zhang, M.; Li, C.; Cai, H. The effect of curcumin and its nanoformulation on adjuvant-induced arthritis in rats. *Drug Des. Dev. Ther.* **2015**, *9*, 4931–4942. [CrossRef] [PubMed]

55. Hwang, J.K.; Noh, E.M.; Moon, S.J.; Kim, J.M.; Kwon, K.B.; Park, B.H.; You, Y.O.; Hwang, B.M.; Kim, H.J.; Kim, B.S.; et al. Emodin suppresses inflammatory responses and joint destruction in collagen-induced arthritic mice. *Rheumatology* **2013**, *52*, 1583–1591. [CrossRef]

56. Zhu, X.; Zeng, K.; Qiu, Y.; Yan, F.; Lin, C. Therapeutic effect of emodin on collagen-induced arthritis in mice. *Inflammation* **2013**, *36*, 1253–1259. [CrossRef]

57. Ha, M.K.; Song, Y.H.; Jeong, S.J.; Lee, H.J.; Jung, J.H.; Kim, B.; Song, H.S.; Huh, J.E.; Kim, S.H. Emodin inhibits proinflammatory responses and inactivates histone deacetylase 1 in hypoxic rheumatoid synoviocytes. *Biol. Pharm. Bull.* **2011**, *34*, 1432–1437. [CrossRef]

58. Fleury, G.; Mania, S.; Hannouche, D.; Gabay, C. The perioperative use of synthetic and biological disease-modifying antirheumatic drugs in patients with rheumatoid arthritis. *Swiss Med. Wkly.* **2017**, *147*, w14563. [CrossRef]

59. Cho, S.-K.; Bae, S.-C. Pharmacologic treatment of rheumatoid arthritis. *J. Korean Med. Assoc.* **2017**, *60*. [CrossRef]

60. Soto, M.; Falqué, E.; Domínguez, H. Relevance of Natural Phenolics from Grape and Derivative Products in the Formulation of Cosmetics. *Cosmetics* **2015**, *2*, 259–276. [CrossRef]

*molecules*       MDPI

*Review*

# Recent Advances in the Discovery and Biosynthetic Study of Eukaryotic RiPP Natural Products

**Shangwen Luo and Shi-Hui Dong ***

State Key Laboratory of Applied Organic Chemistry, College of Chemistry and Chemical Engineering, Lanzhou University, Lanzhou 730000, China; luosw@lzu.edu.cn
* Correspondence: dongsh@lzu.edu.cn; Tel.: +86-931-8912500; Fax: +86-931-8915557

Academic Editors: Pavel B. Drasar and Vladimir A. Khripach
Received: 22 March 2019; Accepted: 18 April 2019; Published: 18 April 2019

**Abstract:** Natural products have played indispensable roles in drug development and biomedical research. Ribosomally synthesized and post-translationally modified peptides (RiPPs) are a group of fast-expanding natural products attribute to genome mining efforts in recent years. Most RiPP natural products were discovered from bacteria, yet many eukaryotic cyclic peptides turned out to be of RiPP origin. This review article presents recent advances in the discovery of eukaryotic RiPP natural products, the elucidation of their biosynthetic pathways, and the molecular basis for their biosynthetic enzyme catalysis.

**Keywords:** natural product; RiPP; ribosomally synthesized; post-translationally modified peptides

## 1. Introduction

### 1.1. Common Features of RiPP Biosynthesis

Ribosomally synthesized and post-translationally modified peptides (RiPPs) are ribosomally synthesized and post-translationally modified peptide natural products. As their name indicates, all RiPP natural products are encoded by structural genes and are initially synthesized as precursor peptides by ribosome (Figure 1). In most RiPPs, the precursor peptide consists of a sequence-conserved amino N-terminal leader peptide and a hypervariable core sequence. Many eukaryotic precursor peptides, as described in this review, have a carboxyl C-terminal recognition sequence that is important for excision and cyclization [1]. In general, the precursor peptide is first synthesized by the ribosome. Then, the core peptide is subjected to post-translational modifications, many of which are guided by leader peptides and recognition sequences. Finally, the leader sequences and recognition sequences are removed by proteolysis to generate mature peptides. Notably, some post-translational modifications are leader/recognition sequence independent, catalyzed after removal of the flanking sequences.

Due to the fact that RiPP core peptides are directly translated from open reading frames (ORFs) in the genomes of the producing organisms, genome mining algorithms and toolkits were developed to correlate the mature RiPP with its corresponding biosynthetic gene cluster (BGC) and to use that information to search for more homologous BGCs. On the other hand, by analyzing the sequence of a putative homologous BGC, the sequence and even the raw structure of its corresponding mature RiPP can be predicted as well [2,3].

### 1.2. Designation of RiPP Families

Nisin, produced by *Lactococcus lactis*, is one of the longest known RiPPs first reported in the 1920s and has been used as a food preservative since the 1960s [1]. Its structure is characterized by the presence of lanthionine residues, giving the name lanthipeptides (for lanthionine-containing peptides) to this family of RiPPs. It was not until recently that the biosynthetic mechanisms for lanthionine residues in nisin were revealed [4–6]. The family of lanthipeptides are further divided into four classes according to different domain organizations of the key biosynthetic enzymes that install the thioether crosslinks in their characteristic lanthionine residues [1].

Historically, RiPP families have been either defined based on the producing organisms, such as microcins from Gram-negative bacteria, or their bioactivities, such as bacteriocins that exhibit antibacterial activities. A consensus was reached in 2013 within the scientific community to designate RiPPs based on their structural and biosynthetic commonality [1]. Accordingly, a variety of RiPP families were defined, such as linaridins, proteusins, linear azol(in)e-containing peptides (LAPs), cyanobactins, thiopeptides, bottromycins, lasso peptides, microviridins, and sactipeptides.

### 1.3. Engineering Potential of RiPP Antibiotics

Many RiPPs natively display potent antibiotic activities, such as lasso peptides, lanthipeptides, and thiopeptides. Biosynthetic studies of those antibiotic RiPPs showed that their biosynthetic pathways are modular. More importantly, many of RiPP biosynthetic enzymes are promiscuous and can tolerate alternative substrates. This plasticity in RiPP biosynthesis gives rise to engineering efforts in making new-to-nature compounds with higher potency and better bioavailability. As more RiPP biosynthetic pathways are revealed and more RiPP biosynthetic enzymes are thoroughly investigated, the therapeutic potential of RiPPs will be significantly increased by those engineering efforts [7].

Bacterial RiPPs have been more extensively studied than eukaryotic RiPPs in the past. However, eukaryotic RiPPs are equally important in providing novel chemical scaffolds and valuable enzymatic transformations. To this end, this review article briefly introduces characteristics of various RiPP natural products from eukaryotic organisms, including the discovery of novel eukaryotic RiPPs and their structural characteristics, emphasizes on recent advances in eukaryotic RiPP biosynthetic studies, and, finally, discusses the unique features of eukaryotic RiPP biosynthesis comparing to bacterial RiPP biosynthesis.

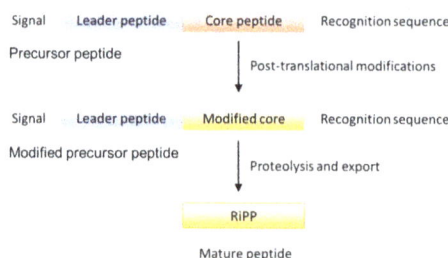

**Figure 1.** General Ribosomally synthesized and post-translationally modified peptide (RiPP) natural products biosynthetic pathway. Adapted from Reference [1].

## 2. Fungal RiPP

### 2.1. RiPP from Basidiomycetes

#### 2.1.1. Amatoxins and Phallotoxins

Mushrooms in the genus *Amanita* account for most of fatal mushroom poisonings [8]. Their toxicity is caused by a group of bicyclic peptides named amatoxins. Amatoxins are also biosynthesized by mushrooms in other unrelated genera, such as *Galerina*, *Lepiota*, and *Conocybe*. They can cause liver failure and death by inhibiting ribonucleic acid (RNA) polymerase II [9]. Amanita mushrooms are also responsible for producing a structurally related group of toxins, called phallotoxins, which are orally inactive but toxic when injected. Phallotoxins act by stabilizing F-actin, and have been utilized to stain the cytoskeleton (Figure 2) [9].

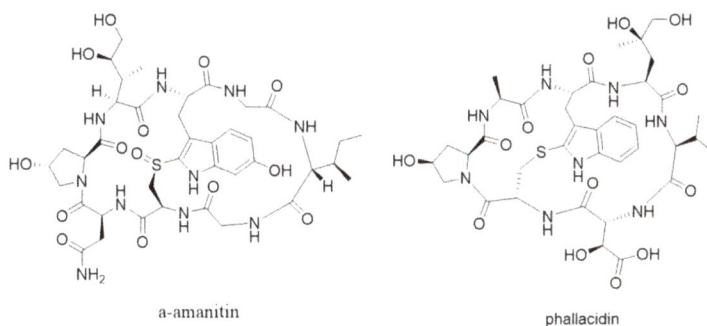

Figure 2. Structures of α-amanitin and phallacidin.

Genome survey sequencing revealed that amatoxins and phallotoxins are biosynthesized via ribosomal pathways [8,10]. It was also shown that genes encoding precursor peptides for amatoxins and phallotoxins are prevalent in toxic mushrooms and form a large family, known as the MSDIN family for the first five conserved amino acid residues in the precursor peptides [11,12]. Members of the MSDIN family are characterized by a hypervariable core region flanked by conserved leader and recognition sequences. Moreover, the core is flanked by invariant proline residues that act as proteolytic targets by a prolyl oligopeptidase (POP), named POPB (Figure 3).

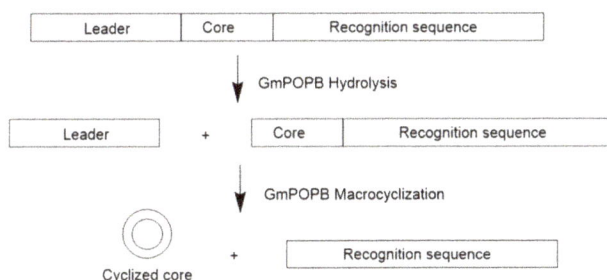

**Figure 3.** Schematic of prolyl oligopeptidase (POPB) catalyzed proteolysis and cyclization. GmPOPB is the POPB from *Galerina marginata* species. Adapted from Reference [13].

POPB is a member of the POP family of serine proteases. It differs from conventional POP (such as POPA that is also present in *Amanita* mushrooms) in that it catalyzes two nonprocessive reactions: Hydrolysis of leader peptide following the proline residue, and transpeptidation to form macrocycle of the core peptide [13]. This two-step mechanism of POPB catalysis was also supported by kinetic and structural studies (Figure 4). The enzyme first hydrolyzes N-terminal leader by the removal of 10 residues from a 35-residue precursor. The resulting 25 amino-acid peptide is conformationally trapped and forced to be released. After dissociation from the enzyme, the 25-mer is conformationally rearranged and rebounded by the enzyme. This process is possibly directed by the C-terminal follower peptide. Finally, the follower peptide is removed and the core peptide is macrocyclized in the active site of the same enzyme [14]. Due to its unusual two-step mechanism, high substrate tolerance in the core region, and satisfying kinetic efficiency, POPB has been exploited as a general catalyst for peptide macrocyclization [15].

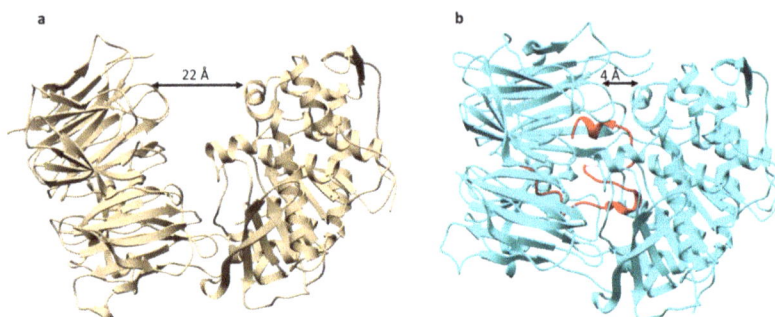

**Figure 4.** Overall structures of prolyl oligopeptidase POPB. (**a**) Structure of POPB from *Galerina marginata* species unbound to substrate (apoGmPOPB) in tan (PDB 5N4F), (**b**) S577A mutant of GmPOPB bound to 35-mer peptide (PDB 5N4C), S577A mutant in cyan, 35-mer peptide in red.

### 2.1.2. Borosins

Omphalotin A was isolated from the basidiomycete *Omphalotus olearius* with potent and selective nematotoxic activity. The structure of omphalotin A is characterized by a peptidic macrocycle with nine *N*-methylations on the amide backbone (Figure 5). It was postulated that omphalotin A was biosynthesized by a nonribosomal peptide synthetase (NRPS) pathway, because backbone *N*-methylation had never been observed for RiPP pathways. It was not until 2017 that a RiPP biosynthetic gene cluster, *oph*, was confirmed to be responsible for producing omphalotin A by two groups in parallel [16,17]. Even more surprisingly, the precursor peptide of omphalotin A is not present as a stand-alone substrate for post-translational modifications, but rather fused to the C-terminal of a protein with sequence homology to *S*-adenosylmethionine (SAM)-dependent methyltransferases [18,19]. The gene encoding this fusion protein was named *ophA*. Additional experiments showed that OphA autocatalytically methylates its own C terminus in a sequential manner from N to C terminus, followed by cleavage and cyclization by the prolyl oligopeptidase OphP to form omphalotin [16]. Van der Velden et al. [17] proposed the name "borosins" after the ancient mythological symbol Ouroboros for this new family of RiPPs.

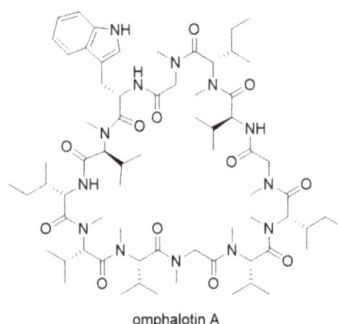

omphalotin A

**Figure 5.** Structure of omphalotin A.

The molecular mechanism of OphA automethylation was proposed based on structural studies by two groups in parallel (Figure 6) [20,21]. OphA acts by forming a homodimer, with each monomer resembling the appearance of a ring. In the co-complex structure, the C-terminal core peptide in monomer A sits into the methyltransferase active-site of monomer B (and vice versa), giving the dimer the appearance of two interlocked rings [20]. This structural arrangement results in substrate proximity and suggests an acid-base catalysis mechanism. The amide nitrogen is first deprotonated with the help of a basic amino acid residue (possibly arginine). The resulting negative charge is stabilized by tyrosine residues in close proximity. Finally, the proximity of a reactive SAM molecule promotes the alkylation reaction [21].

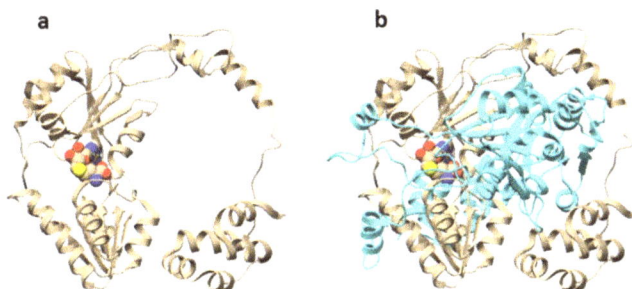

**Figure 6.** Overall structure of dbOphMA (PDB 6MJF). dbOphMA is OphA homolog from *Dendrothele bispora* species. (**a**) dbOphMA monomer bound to *S*-adenosyl homocystein (SAH), (**b**) interlocking organization of dbOphMA dimer.

## 2.2. RiPP from Ascomycetes

### 2.2.1. Dikaritins

Ustiloxins

Ustiloxins are the first example of natural products that are biosynthesized by RiPP pathways in filamentous fungi. The study of ustiloxin biosynthetic pathway represents the first example of complete RiPP gene cluster characterization in fungi [22]. Ustiloxin B was originally discovered from plant pathogenic fungus *Ustilaginoidea virens* with phytotoxic activity by inhibiting microtubule assembly [23]. The structure of ustiloxin B consist of a Tyr-Ala-Ile-Gly (YAIG) tetrapeptide and contains unusual norvaline modification on the hydroxylated tyrosine residue (Figure 7). The biosynthetic origin of ustiloxin B remained unknown until the genome mining method MIDDAS-M was developed for the detection of natural product biosynthetic gene clusters in fungi [24,25]. MIDDAS-M is the abbreviation of motif-independent de novo detection algorithm for secondary metabolite biosynthetic gene clusters. By scoring transcription levels of all putative gene clusters in *Aspergillus flavus* under different culture conditions, the MIDDAS-M method identified the gene cluster for ustiloxin B, named the *ust* cluster, which was later confirmed by knockout studies.

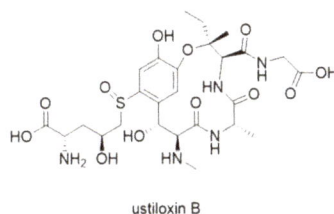

ustiloxin B

**Figure 7.** Structure of ustiloxin B.

In depth sequence analysis of the cluster revealed that the precursor peptide UstA contains 16-fold repeats of the YAIG core, and that each repeated core is flanked by conserved ED and KR motifs which are likely necessary for recognition by post-translationally modifying enzymes (Figure 8). Moreover, the N-terminal of UstA is a signal peptide-like sequence that also contains a KR motif. The overall organization of UstA, and the fact that KR motif is a known recognition site for Kex2 protease, which is a type of universal serine proteases, strongly suggest that ustiloxin B is a RiPP and that *ustA* encodes the precursor peptide [22]. Accordingly, the biosynthetic gene cluster of ustiloxin B in its original host, *U. virens*, was also characterized and confirmed to be a RiPP cluster [26].

MKLILTLLVSGLCALAAPAA**KR**DGVED<u>YAIGID</u>**KR**--
--NSVED<u>YAIGID</u>**KR**--
--NSVED<u>YAIGID</u>**KR**--
--NSVED<u>YAIGID</u>**KR**--
--NSVED<u>YAIGID</u>**KR**--
--NTVED<u>YAIGID</u>**KR**--
--NSVED<u>YAIGID</u>**KR**--
--NTVED<u>YAIGID</u>**KR**--
--NSVED<u>YAIGID</u>**KR**--
--NSVED<u>YAIGID</u>**KR**--
--GGSVED<u>YAIGID</u>**KR**--
--NSVED<u>YAIGID</u>**KR**--
--GSVED<u>YAIGID</u>**KR**--
--GTVED<u>YAIGID</u>**KR**--
--GGSVED<u>YAIGID</u>**KR**HGGH

**Figure 8.** Sequence of UstA from *Aspergillus flavus*. KR motifs to be recognized by Kex2 protease are marked in bold. Repeated YAIG core sequences are underlined. Adapted from Reference [22].

The entire biosynthetic pathway of ustiloxin B has been studied in detail by gene inactivation, heterologous expression, and in vitro biosynthetic enzyme functional reconstitution (Figure 9a) [27]. Gene disruption studies revealed that the three genes *ustQYaYb* are essential to give the first intermediate **2**. UstQ is a tyrosinase homolog. UstYa/UstYb are mutual homologs containing the DUF3328 motif and have no homology with functionally known enzymes. Heterologous expression of *ustQYaYb* in *Aspergillus oryzae* gave **2** as the sole product. Based on these results, it was speculated that the UstA precursor is first digested into 16 trideca-/tetradecapeptides by Kex2 proteases before cyclization by UstQYaYb. However, the proteases that removes the N- and C- terminal sequences flanking the core are still unknown. UstM is a methyl transferase. Introduction of *ustM* into *ustQYaYb* transformants generated **3**. UstF1/UstF2 are Class B bifunctional flavoprotein monooxygenases (FMO). Purified maltose binding protein (MBP)-tagged UstF1/UstF2 showed yellow color and strong absorption at 450 nm, indicating binding of flavin adenine dinucleotide (FAD). Incubating **4** with UstF1 in the presence of nicotinamide adenine dinucleotide phosphate (NADPH) resulted in **5**, which was transformed into an entgegen/zusammen (*E/Z*) mixture of **6** after incubating with UstF2 in the presence of NADPH. Treatment of **6** with 0.1% trifluoroacetic acid (TFA) afforded **8**, a hydrate form of **7**. *ustD* gene showed homology with pyridoxal 5′-phosphate (PLP)-dependent enzyme. Incubating **8** with MBP-tagged UstD in the presence of PLP and aspartic acid generated ustiloxin B. The reaction mechanism of UstD was studied by incubating the enzyme with PLP and aspartic acid and treating the reaction mixture with dansyl chloride. The above experiment generated dansylated alanine, indicating that UstD catalyzes decarboxylation of aspartate to form an enamine, which acts as a nucleophile and reacts with **7** to give **1** (Figure 9b). Given that *ustYa/ustYb* are located near the precursor peptide gene *ustA*, combined queries of homologs of *ustYa/ustYb* and *ustA* identified 94 homologous clusters in *Aspergilli* genome sequences.

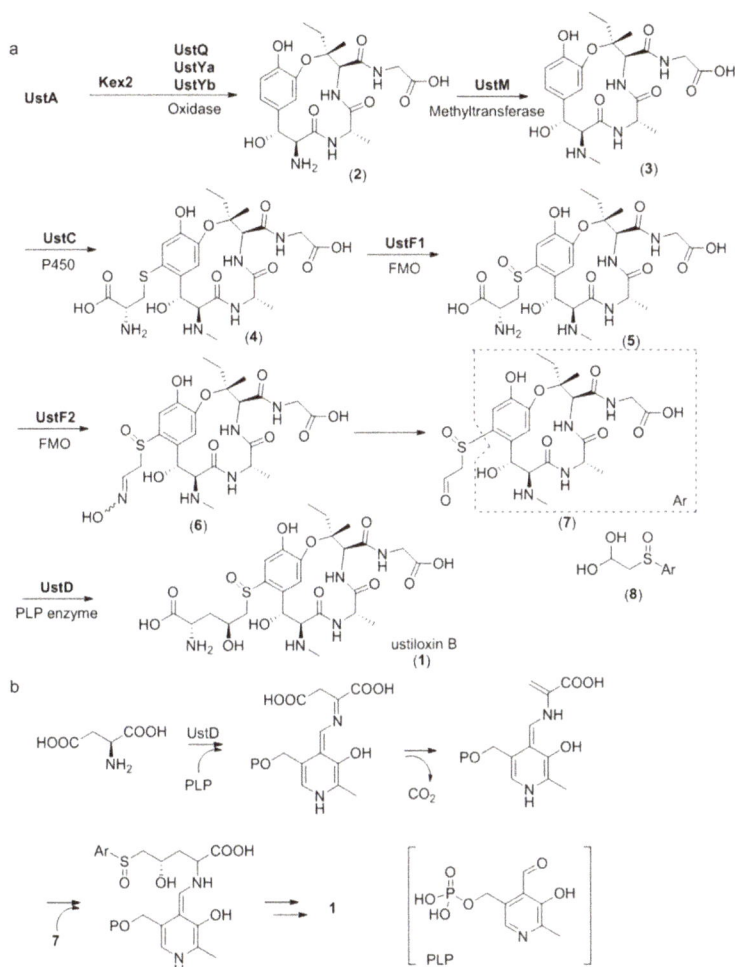

**Figure 9.** Proposed biosynthetic pathway for ustiloxin B. (**a**) Overall biosynthetic scheme, (**b**) Proposed mechanism of UstD catalyzed reaction. P450 = cytochrome P450 monooxygenase, FMO = flavoprotein monooxygenase, PLP = pyridoxal 5′-phosphate. Adapted from Reference [27].

Asperipins

Guided by the finding of 94 precursor peptide gene candidates by querying *ustYa/ustYb* and *ustA* in combination, a new cyclic peptide asperipin-2a was isolated from *Aspergillus flavus* [28]. Although asperipin-2a has high homology to ustiloxins in their gene clusters, they have distinct structural characteristics. Asperipin-2a has a hexa-peptidic core sequence of FYYTGY, forming a bicyclic structure connected by ether linkages between tyrosine side chains and β-carbons (Figure 10). The putative biosynthetic gene cluster for asperipin-2a is only composed of four genes: A precursor peptide gene *aprA*, a *ustYa/ustYb* homolog *aprY*, a transporter *aprT*, and an isoflavone reductase *aprR*. Heterologous expression of asperipin-2a gene cluster in *Aspergillus oryzae* showed that *aprY* is essential for biosynthesizing asperipin-2a, and indicated a sequential oxidative macrocyclization function for AprY [29].

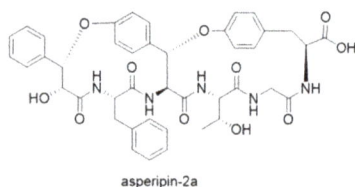

asperipin-2a

**Figure 10.** Structure of asperipin-2a.

Phomopsins

Phomopsins are a group of cyclic hexapeptide produced by the plant pathogenic fungus *Phomopsis leptostromiformis*. The structures of phomopsins are characterized by a 13-member macrocyclic ring formed by ether linkage between tyrosine and isoleucine (Figure 11). They are potent antimitotic compounds that target the vinca domain of tubulin, causing liver disease in livestock fed on infected plants [30]. A RiPP gene cluster was confirmed to be responsible for producing phomopsins by analyzing the genome sequence of *P. leptostromiformis* ATCC 26115 [31]. Similar to ustiloxin B precursor peptide gene *ustA*, *phomA*, the precursor peptide gene for phomopsins is also arranged in the same pattern. The N-terminal of PhomA is a signal-peptide like leader sequence, followed by eight repeats of core peptide flanked by conserved KR motifs. Knockout studies showed that the tyrosinase PhomQ is essential for phomopsin biosynthesis and is likely involved in forming the cyclic scaffold. In vitro enzymatic assays revealed that the methyltransferase PhomM installs methyl groups onto the N-terminal $\alpha$-amino group. A search for PhomA, PhomQ, and PhomM homologous proteins in the National Center for Biotechnology Information (NCBI) database resulted in the identification of 27 similar gene clusters, suggesting the presence of a family of fungal RiPP natural products. Because these compounds appear to associate with strains of the subkingdom Dikarya, the name "dikaritins" was proposed for this new family of peptides. A global sequence similarity network was constructed for all of the putative proteins from the identified gene clusters, showing that these gene clusters contain a set of highly conserved proteins, including PhomA homologs, PhomQ homologs, PhomR-like zinc finger transcription-regulating proteins, and S41 family peptidases. Noteworthy is the presence of DUF3328 proteins in all of the gene clusters, such as UstYa/UstYb and AprY, whose role in dikaritin biosynthesis remains to be elucidated [31].

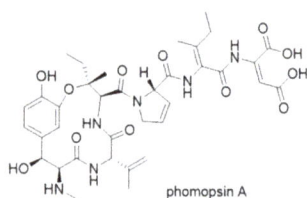

phomopsin A

**Figure 11.** Structure of phomopsin A.

## 2.2.2. Epichloëcyclins

Epichloëcyclins were discovered from grass endophytic fungi belonging to the genus *Epichloë*. MS/MS analyses indicated their structure characteristics to be a hepta-peptidic ring formed by oxidative cyclization on the tyrosine residue, and methylations on the lysine residue. Detailed structures of epichloëcyclins remain to be determined. The precursor peptide gene for epichloëcyclins was identified from fungal transcripts in endophyte-infected grasses and designated *gigA* (grass induced gene). GigA is composed of a signal sequence at its N-terminal, followed by four repeats of sequences containing the core peptide and conserved motifs, such as KR recognition site for Kex2 protease. Epichloëcyclins are the first example of RiPP natural products found in mutualistic symbiotic fungus, suggesting a possible bioactive role [32]. Whether epichloëcyclins belongs to the family of dikaritins or forms its own family of RiPP remains to be determined until the full biosynthetic gene cluster of epichloëcyclins can be identified.

## 3. Plant RiPP

### 3.1. Cyclotides

Cyclotides are plant derived RiPPs that are characterized by a head-to-tail cyclic peptide backbone and a signature cyclic cystine knot (CCK) motif [33]. They were discovered from plants of the Rubiaceae, Violaceae, Cucurbitaceae, and Fabaceae families [34–38]. Due to their insecticidal activities, cyclotides were thought to be plant defense agents. The broad range of other biological activities, such as antiviral, antimicrobial, and cytotoxic activities made cyclotides attractive for pharmaceutical applications [1]. The precursors of cyclotides can be present as dedicated proteins, similar to other RiPPs from bacteria and fungi. However, it was found that cyclotide precursors in *Clitoria ternatea* (Fabaceae family) are embedded within an albumin precursor, indicating RiPPs might be much more common than has been thought [36,37]. Cyclotide precursors consists of an endoplasmic reticulum (ER) domain, a pro-region (PRO), an N-terminal region (NTR), and one or more copies of the core sequence. The protease that removes the leader remains to be characterized. Butelase 1, a Asx-specific peptide ligase from cyclotide producing *C. ternatea*, was characterized to be responsible for cyclotide backbone cyclization [39]. Butelase 1 has high sequence homology with asparaginyl endopeptidase (AEP), and indeed showed AEP activity. However, it recognizes the C-terminal Asn/Asp-His-Val (D/NHV) sequence and is capable of cyclizing various peptides of plant and animal origin with high catalytic efficiencies (Figure 12) [39]. A recent structural study revealed that the active site of butelase 1 has only subtle differences from conventional AEPs, suggesting its efficient macrocyclization activity may be attributed to its peptide binding region (Figure 13) [40]. A co-crystal structure of butelase 1 with its peptide substrate will help us understand the mechanism of macrocyclization. Due to its high promiscuity and fast kinetics, butelase 1 has been applied in protein labelling [41,42], chemoenzymatic synthesis of bacteriocins [43], generating cyclic peptides with non-native amino acids [44], decorating *E. coli* cell surfaces [45], making peptide dendrimers [46], and preparing C-to-C fusion proteins [47].

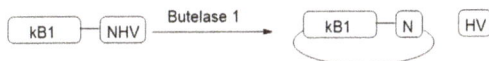

**Figure 12.** Schematic of butelase 1 catalyzed cyclization. kB1 containing 29 amino acid residues is the core peptide of plant cyclotide kalata B1. Asp-His-Val (NHV) is the C-terminal sequence recognized by butelase 1. Adapted from Reference [39].

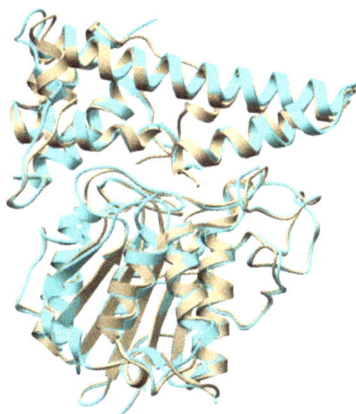

**Figure 13.** Crystal structure of butelase 1 (tan, PDB 6DHI) overlaid with structure of conventional AEP from *Oldenlandia affinis* (cyan, PDB 5H0I).

### 3.2. Orbitides

Orbitides refer to N-to-C cyclized plant peptides that do not contain disulfides. They are produced by at least nine plant families: Annonaceae, Caryophyllaceae, Euphorbiaceae, Lamiaceae, Linaceae, Phytolaccaceae, Rutaceae, Schizandraceae, and Verbenaceae [1]. Similar to cyclotides, orbitide precursor peptides also contain multiple copies of core sequences, resulting in a single precursor to be processed to multiple cyclic peptides. The biosynthetic pathway of orbitide segetalin A has been studied in detail (Figure 14). The 32-amino acid precursor presegetalin A1 is first processed by the serine protease OLP1 to remove the N-terminal 15 residues. Then, the peptide cyclase PCY1 cleaves the C-terminal 13 amino acids, with concomitant macrocyclization of the remaining six residues to form segetalin A [48]. PCY1 is identified as a member of the S9A protease family that includes POP enzymes. Kinetic analysis showed that PCY1 has similar $k_{cat}$ values, and five~10-fold higher $K_M$ values comparing to butelase 1 involved in cyclotide macrocyclization. Crystal structures of PCY1 revealed its transamidation and cyclization mechanisms (Figure 15). Upon binding of the follower peptide, PCY1 is maintained in a closed state that precludes solvent from the active site, potentially limiting the competing hydrolysis reaction. A key residue His659 sits on a mobile loop, which contributes to two roles: Activating Ser nucleophile to form acyl-enzyme intermediate, and deprotonating the $\alpha$-amine of the substrate for transamidation [49]. Using the obtained knowledge of PCY1 reaction molecular basis, a three residue C-terminal extension (F/I-Q-A/T) was designed to replace the native long recognition tail FQALDVQNASAPV, permitting PCY1 to work on synthetic substrates [50].

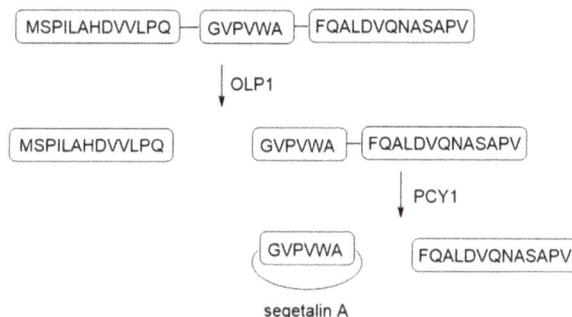

**Figure 14.** Schematic of segetalin A biosynthetic pathway. Letters in the figure represents single letter notations for amino acids. OLP1 is a serine protease. PCY1 is a peptide cyclase. Adapted from Reference [49].

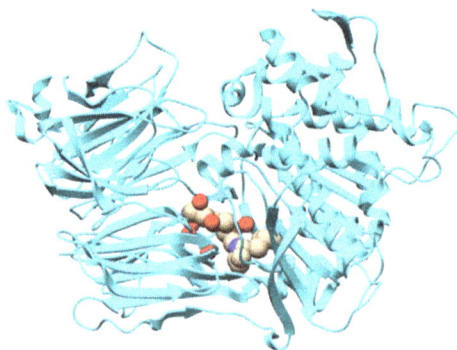

**Figure 15.** Overall structure of peptide cyclase PCY1 with follower peptide (PDB 5UW3).

## 4. Animal RiPP

Marine snails, such as cone snails, are known to produce a variety of ribosomally synthesized and post-translationally modified peptide venoms. Those venomes are produced by predatory cone snails, injected into the prey, and lead to paralysis. The best characterized marine snail peptides are conopeptides, also known as conotoxins. It was estimated that more than 500 species of predatory marine *Conus* snails are capable of producing conotoxins [51,52]. Those *Conus* species can produce as many as 70,000 structurally diverse conotoxins [53,54]. Many conotoxins act by targeting ion channels, thus have been widely used as basic research tools in neuroscience. A number of conotoxins showed therapeutic potential due to their unparalleled potency and selectivity against a wide range of receptors and ion channels [55]. For example, Ziconotide, a calcium channel agonist isolated from *Conus magus*, was approved by the United States Food and Drug Administration (US FDA) in 2004 for the treatment of chronic pain. A detailed review of conopeptides discovery and biosynthesis was made in 2013, and more structures of conopeptides have been characterized since then [1,56–58].

The precursor peptide sequences of conotoxins were studied by analyzing the transcriptomes of cone snails. Those studies revealed that conotoxin precursor transcript sequences consist of three regions: An ER signal peptide, a mature peptide region, and pre-/postpropeptide regions [59,60]. The ER signal peptide sequence is highly conserved, whereas the mature peptide region is highly diverse [60]. Types of conotoxin post-translational modifications include disulfide-bond formation, proline hydroxylation, *O*-glycosylation on serine or threonine residues, and glutamate $\gamma$-carboxylation [61,62]. A web-based ConoServer database (conoserver.org) was established to record known structures of conopeptides, classifications, post-translational modifications, and their general statistics. Due to that conopeptide biosynthetic genes are not organized in clusters, biosynthetic studies of animal RiPPs are extremely challenging. The details of conotoxin biosynthetic pathways and the mechanism and molecular basis for their post-translational modifications still remain largely unexplored [55].

## 5. Discussion

Eukaryotic RiPP pathways have some special features comparing to bacterial RiPP pathways. As described above, many fungal and plant RiPPs have N-terminal recognition sequences in their precursor peptides. Moreover, C-terminal signal sequences are also common in eukaryotic RiPP precursors. For example, in the case of cyclotides (Section 3.1), an ER signal sequence is present in their precursor peptides [63]. In addition, the core region of eukaryotic precursor peptides often has several repeats of the core sequence, flanked by conserved motifs. Although this manner is also found in cyanobactin biosynthesis, whose cores are present as repetitive cassettes, it is not common in other bacterial RiPP pathways [64]. Even more surprisingly, some eukaryotic RiPP precursors are not encoded as stand-alone genes, but rather as fusion or chimeric proteins. For example, omphalotin A precursor is fused to the C-terminal of a post-translationally modifying enzyme methyltransferase (Section 2.1.2), and some cyclotide precursors are embedded within an albumin precursor (Section 3.1). The presence of precursor peptides fused to other structural genes underscores the possibility that eukaryotic RiPP natural products are much more common than have been found. Interestingly, the biosynthesis of many eukaryotic RiPPs involves an N-C macrocyclization catalyzed by proteases. For example, amatoxins are macrocyclized by POPB enzymes, cyclotides are formed by a head-to-tail cyclization catalyzed by butelase 1, and the N-to-C cyclization of orbitides are catalyzed by PCY1 proteases [65]. The resulting macrocyclic peptides are more resistant to protease degradation in physiological environments, thus having more potential to be developed into novel therapeutics.

RiPP natural products are promising candidates for developing novel therapeutics, as many RiPPs have shown significant biological activities and great engineering potential [7]. The studies of eukaryotic RiPP biosynthesis are relatively more challenging than their bacterial counterparts, mainly due to the more complex genomic context. This challenge can be compromised by the development of optimal computational tools for mining eukaryotic genomes [2,66,67]. As more RiPP natural products are discovered, and more RiPP biosynthetic pathways are revealed, this group of fast expanding natural

*Molecules* **2019**, *24*, 1541

products will continue to provide compounds for industrial applications, and to inspire engineering efforts on enzymatic machineries.

**Funding:** This study was funded by the start-up funding from both Lanzhou University and State Key Laboratory of Applied Organic Chemistry.

**Acknowledgments:** The authors are thankful to State Key Laboratory of Applied Organic Chemistry, College of Chemistry and Chemical Engineering, Lanzhou University for help in conducting this study.

**Conflicts of Interest:** The authors declare no conflict of interest.

## References

1. Arnison, P.G.; Bibb, M.J.; Bierbaum, G.; Bowers, A.A.; Bugni, T.S.; Bulaj, G.; Camarero, J.A.; Campopiano, D.J.; Challis, G.L.; Clardy, J.; et al. Ribosomally synthesized and post-translationally modified peptide natural products: overview and recommendations for a universal nomenclature. *Nat. Prod. Rep.* **2013**, *30*, 108–160. [CrossRef] [PubMed]

2. Blin, K.; Wolf, T.; Chevrette, M.G.; Lu, X.W.; Schwalen, C.J.; Kautsar, S.A.; Duran, H.G.S.; Santos, E.L.C.D.L.; Kim, H.U.; Nave, M.; Dickschat, J.S.; et al. antiSMASH 4.0-improvements in chemistry prediction and gene cluster boundary identification. *Nucleic Acids Res.* **2017**, *45*, W36–W41. [CrossRef] [PubMed]

3. Tietz, J.I.; Schwalen, C.J.; Patel, P.S.; Maxson, T.; Blair, P.M.; Tai, H.C.; Zakai, U.I.; Mitchell, D.A. A new genome-mining tool redefines the lasso peptide biosynthetic landscape. *Nat. Chem. Biol.* **2017**, *13*, 470. [CrossRef]

4. Li, B.; Yu, J.P.J.; Brunzelle, J.S.; Moll, G.N.; van der Donk, W.A.; Nair, S.K. Structure and mechanism of the lantibiotic cyclase involved in nisin biosynthesis. *Science* **2006**, *311*, 1467. [CrossRef] [PubMed]

5. Ortega, M.A.; Hao, Y.; Zhang, Q.; Walker, M.C.; van der Donk, W.A.; Nair, S.K. Structure and mechanism of the tRNA-dependent lantibiotic dehydratase NisB. *Nature* **2015**, *517*, 509. [CrossRef] [PubMed]

6. Ortega, M.A.; Hao, Y.; Walker, M.C.; Donadio, S.; Sosio, M.; Nair, S.K.; van der Donk, W.A. Structure and tRNA Specificity of MibB, a Lantibiotic Dehydratase from Actinobacteria Involved in NAI-107 Biosynthesis. *Cell Chem. Biol.* **2016**, *23*, 370–380. [CrossRef] [PubMed]

7. Hudson, G.A.; Mitchell, D.A. RiPP antibiotics: biosynthesis and engineering potential. *Curr. Opin. Microbiol.* **2018**, *45*, 61–69. [CrossRef] [PubMed]

8. Hallen, H.E.; Luo, H.; Scott-Craig, J.S.; Walton, J.D. Gene family encoding the major toxins of lethal Amanita mushrooms. *Proc. Natl. Acad. Sci. USA* **2007**, *104*, 19097–19101. [CrossRef] [PubMed]

9. Walton, J.D.; Hallen-Adams, H.E.; Luo, H. Ribosomal Biosynthesis of the Cyclic Peptide Toxins of Amanita Mushrooms. *Biopolymers* **2010**, *94*, 659–664. [PubMed]

10. Luo, H.; Len-Adams, H.E.H.; Scott-Craig, J.S.; Walton, J.D. Ribosomal biosynthesis of alpha-amanitin in Galerina marginata. *Fungal Genet. Biol.* **2012**, *49*, 123–129. [CrossRef] [PubMed]

11. Pulman, J.A.; Childs, K.L.; Sgambelluri, R.M.; Walton, J.D. Expansion and diversification of the MSDIN family of cyclic peptide genes in the poisonous agarics Amanita phalloides and A-bisporigera. *BMC Genom.* **2016**, *17*, 1038. [CrossRef] [PubMed]

12. Luo, H.; Cai, Q.; Luli, Y.; Li, X.; Sinha, R.; Hallen-Adams, H.E.; Yang, Z.L. The MSDIN family in amanitin-producing mushrooms and evolution of the prolyl oligopeptidase genes. *IMA Fungus* **2018**, *9*, 225–242. [CrossRef] [PubMed]

13. Luo, H.; Hong, S.Y.; Sgambelluri, R.M.; Angelos, E.; Li, X.; Walton, J.D. Peptide Macrocyclization Catalyzed by a Prolyl Oligopeptidase Involved in alpha-Amanitin Biosynthesis. *Chem. Biol.* **2014**, *21*, 1610–1617. [CrossRef] [PubMed]

14. Czekster, C.M.; Ludewig, H.; McMahon, S.A.; Naismith, J.H. Characterization of a dual function macrocyclase enables design and use of efficient macrocyclization substrates. *Nat. Commun.* **2017**, *8*, 1045. [CrossRef]

15. Sgambelluri, R.M.; Smith, M.O.; Walton, J.D. Versatility of Prolyl Oligopeptidase B in Peptide Macrocyclization. *Acs Synth. Biol.* **2018**, *7*, 145–152. [CrossRef]

16. Ramm, S.; Krawczyk, B.; Muhlenweg, A.; Poch, A.; Mosker, E.; Sussmuth, R.D. A Self-Sacrificing N-Methyltransferase Is the Precursor of the Fungal Natural Product Omphalotin. *Angew. Chem. Int. Edit.* **2017**, *56*, 9994–9997. [CrossRef] [PubMed]

17. Van der Velden, N.S.; Kalin, N.; Helf, M.J.; Piel, J.; Freeman, M.F.; Kunzler, M. Autocatalytic backbone N-methylation in a family of ribosomal peptide natural products (vol 13, pg 833, 2017). *Nat. Chem. Biol.* **2017**, *13*, 833. [CrossRef] [PubMed]

18. Bowers, A.A. Methylating mushrooms. *Nat. Chem. Biol.* **2017**, *13*, 821–822. [CrossRef]

19. Aldemir, H.; Gulder, T.A.M. Expanding the Structural Space of Ribosomal Peptides: Autocatalytic N-Methylation in Omphalotin Biosynthesis. *Angew. Chem. Int. Edit.* **2017**, *56*, 13570–13572. [CrossRef]

20. Song, H.; van der Velden, N.S.; Shiran, S.L.; Bleiziffer, P.; Zach, C.; Sieber, R.; Imani, A.S.; Krausbeck, F.; Aebi, M.; Freeman, M.F.; et al. A molecular mechanism for the enzymatic methylation of nitrogen atoms within peptide bonds. *Sci Adv.* **2018**, *4*, eaat2720. [CrossRef] [PubMed]

21. Ongpipattanakul, C.; Nair, S.K. Molecular Basis for Autocatalytic Backbone N-Methylation in RiPP Natural Product Biosynthesis. *Acs Chem. Biol.* **2018**, *13*, 2989–2999. [CrossRef] [PubMed]

22. Umemura, M.; Nagano, N.; Koike, H.; Kawano, J.; Ishii, T.; Miyamura, Y.; Kikuchi, M.; Tamano, K.; Yu, J.J.; Shin-ya, K.; et al. Characterization of the biosynthetic gene cluster for the ribosomally synthesized cyclic peptide ustiloxin B in Aspergillus flavus. *Fungal Genet. Biol.* **2014**, *68*, 23–30. [CrossRef] [PubMed]

23. Koiso, Y.; Li, Y.; Iwasaki, S.; Hanaoka, K.; Kobayashi, T.; Sonoda, R.; Fujita, Y.; Yaegashi, H.; Sato, Z. Ustiloxins, Antimitotic Cyclic-Peptides from False Smut Balls on Rice Panicles Caused by Ustilaginoidea Virens. *J. Antibiot.* **1994**, *47*, 765–773. [CrossRef]

24. Umemura, M.; Koike, H.; Nagano, N.; Ishii, T.; Kawano, J.; Yamane, N.; Kozone, I.; Horimoto, K.; Shin-ya, K.; Asai, K.; et al. MIDDAS-M: Motif-Independent De Novo Detection of Secondary Metabolite Gene Clusters through the Integration of Genome Sequencing and Transcriptome Data. *PloS ONE* **2013**, *8*, e84028. [CrossRef] [PubMed]

25. Umemura, M.; Koike, H.; Machida, M. Motif-independent de novo detection of secondary metabolite gene clusters - toward identification from filamentous fungi. *Front. Microbiol.* **2015**, *6*, 371. [CrossRef]

26. Tsukui, T.; Nagano, N.; Umemura, M.; Kumagai, T.; Terai, G.; Machida, M.; Asai, K. Ustiloxins, fungal cyclic peptides, are ribosomally synthesized in Ustilaginoidea virens. *Bioinformatics* **2015**, *31*, 981–985. [CrossRef]

27. Ye, Y.; Minami, A.; Igarashi, Y.; Izumikawa, M.; Umemura, M.; Nagano, N.; Machida, M.; Kawahara, T.; Shin-ya, K.; Gomi, K.; et al. Unveiling the Biosynthetic Pathway of the Ribosomally Synthesized and Post-translationally Modified Peptide Ustiloxin B in Filamentous Fungi. *Angew. Chem. Int. Edit.* **2016**, *55*, 8072–8075. [CrossRef] [PubMed]

28. Nagano, N.; Umemura, M.; Izumikawa, M.; Kawano, J.; Ishii, T.; Kikuchi, M.; Tomii, K.; Kumagai, T.; Yoshimi, A.; Machida, M.; et al. Class of cyclic ribosomal peptide synthetic genes in filamentous fungi. *Fungal Genet. Biol.* **2016**, *86*, 58–70. [CrossRef]

29. Ye, Y.; Ozaki, T.; Umemura, M.; Liu, C.W.; Minami, A.; Oikawa, H. Heterologous production of asperipin-2a: proposal for sequential oxidative macrocyclization by a fungi-specific DUF3328 oxidase. *Org. Biomol. Chem.* **2019**, *17*, 39–43. [CrossRef]

30. Cormier, A.; Marchand, M.; Ravelli, R.B.G.; Knossow, M.; Gigant, B. Structural insight into the inhibition of tubulin by vinca domain peptide ligands. *EMBO Rep.* **2008**, *9*, 1101–1106. [CrossRef]

31. Ding, W.; Liu, W.Q.; Jia, Y.L.; Li, Y.Z.; van der Donk, W.A.; Zhang, Q. Biosynthetic investigation of phomopsins reveals a widespread pathway for ribosomal natural products in Ascomycetes. *Proc. Natl. Acad. Sci. USA* **2016**, *113*, 3521–3526. [CrossRef] [PubMed]

32. Johnson, R.D.; Lane, G.A.; Koulman, A.; Cao, M.S.; Fraser, K.; Fleetwood, D.J.; Voisey, C.R.; Dyer, J.M.; Pratt, J.; Christensen, M.; et al. A novel family of cyclic oligopeptides derived from ribosomal peptide synthesis of an in planta-induced gene, gigA, in Epichloe endophytes of grasses. *Fungal Genet. Biol.* **2015**, *85*, 14–24. [CrossRef] [PubMed]

33. Craik, D.J.; Daly, N.L.; Bond, T.; Waine, C. Plant cyclotides: A unique family of cyclic and knotted proteins that defines the cyclic cystine knot structural motif. *J. Mol. Biol.* **1999**, *294*, 1327–1336. [CrossRef]

34. Saether, O.; Craik, D.J.; Campbell, I.D.; Sletten, K.; Juul, J.; Norman, D.G. Elucidation of the Primary and 3-Dimensional Structure of the Uterotonic Polypeptide Kalata B1. *Biochemistry* **1995**, *34*, 4147–4158. [CrossRef]

35. Ireland, D.C.; Colgravel, M.L.; Nguyencong, P.; Daly, N.L.; Craik, D.J. Discovery and characterization of a linear cyclotide from Viola odorata: Implications for the processing of circular proteins. *J. Mol. Biol.* **2006**, *357*, 1522–1535. [CrossRef] [PubMed]

36. Giang, K.T.N.; Zhang, S.; Ngan, T.K.N.; Phuong, Q.T.N.; Chiu, M.S.; Hardjojo, A.; Tam, J.P. Discovery and Characterization of Novel Cyclotides Originated from Chimeric Precursors Consisting of Albumin-1 Chain a and Cyclotide Domains in the Fabaceae Family. *J. Biol. Chem.* **2011**, *286*, 24275–24287.

37. Poth, A.G.; Colgrave, M.L.; Lyons, R.E.; Daly, N.L.; Craik, D.J. Discovery of an unusual biosynthetic origin for circular proteins in legumes. *Proc. Natl. Acad. Sci. USA* **2011**, *108*, 10127–10132. [CrossRef]

38. Giang, K.T.N.; Lian, Y.L.; Pang, E.W.H.; Phuong, Q.T.N.; Tran, T.D.; Tam, J.P. Discovery of Linear Cyclotides in Monocot Plant Panicum laxum of Poaceae Family Provides New Insights into Evolution and Distribution of Cyclotides in Plants. *J. Biol. Chem.* **2013**, *288*, 3370–3380.

39. Nguyen, G.K.T.; Wang, S.J.; Qiu, Y.B.; Hemu, X.; Lian, Y.L.; Tam, J.P. Butelase 1 is an Asx-specific ligase enabling peptide macrocyclization and synthesis. *Nat. Chem. Biol.* **2014**, *10*, 732–738. [CrossRef]

40. James, A.M.; Haywood, J.; Leroux, J.; Ignasiak, K.; Elliott, A.G.; Schmidberger, J.W.; Fisher, M.F.; Nonis, S.G.; Fenske, R.; Bond, C.S.; et al. The macrocyclizing protease butelase 1 remains auto-catalytic and reveals the structural basis for ligase activity. *Plant J.* **2019**. [CrossRef] [PubMed]

41. Cao, Y.; Nguyen, G.K.T.; Tam, J.P.; Liu, C.F. Butelase-mediated synthesis of protein thioesters and its application for tandem chemoenzymatic ligation. *Chem. Commun.* **2015**, *51*, 17289–17292. [CrossRef] [PubMed]

42. Nguyen, G.K.T.; Cao, Y.; Wang, W.; Liu, C.F.; Tam, J.P. Site-Specific N-Terminal Labeling of Peptides and Proteins using Butelase 1 and Thiodepsipeptide. *Angew. Chem. Int. Edit.* **2015**, *54*, 15694–15698. [CrossRef] [PubMed]

43. Hemu, X.; Qiu, Y.B.; Nguyen, G.K.T.; Tam, J.P. Total Synthesis of Circular Bacteriocins by Butelase 1. *J. Am. Chem. Soc.* **2016**, *138*, 6968–6971. [CrossRef] [PubMed]

44. Nguyen, G.K.T.; Hemu, X.; Quek, J.P.; Tam, J.P. Butelase-Mediated Macrocyclization of d-Amino-Acid-Containing Peptides. *Angew. Chem. Int. Edit.* **2016**, *55*, 12802–12806. [CrossRef] [PubMed]

45. Bi, X.B.; Yin, J.; Nguyen, G.K.T.; Rao, C.; Halim, N.B.A.; Hemu, X.; Tam, J.P.; Liu, C.F. Enzymatic Engineering of Live Bacterial Cell Surfaces Using Butelase 1. *Angew. Chem. Int. Edit.* **2017**, *56*, 7822–7825. [CrossRef] [PubMed]

46. Cao, Y.; Nguyen, G.K.T.; Chuah, S.; Tam, J.P.; Liu, C.F. Butelase-Mediated Ligation as an Efficient Bioconjugation Method for the Synthesis of Peptide Dendrimers. *Bioconjugate Chem.* **2016**, *27*, 2592–2596. [CrossRef] [PubMed]

47. Harmand, T.J.; Bousbaine, D.; Chan, A.; Zhang, X.H.; Liu, D.R.; Tam, J.P.; Ploegh, H.L. One-Pot Dual Labeling of IgG 1 and Preparation of C-to-C Fusion Proteins Through a Combination of Sortase A and Butelase 1. *Bioconjugate Chem.* **2018**, *29*, 3245–3249. [CrossRef]

48. Barber, C.J.S.; Pujara, P.T.; Reed, D.W.; Chiwocha, S.; Zhang, H.X.; Covello, P.S. The Two-step Biosynthesis of Cyclic Peptides from Linear Precursors in a Member of the Plant Family Caryophyllaceae Involves Cyclization by a Serine Protease-like Enzyme. *J. Biol. Chem.* **2013**, *288*, 12500–12510. [CrossRef]

49. Chekan, J.R.; Estrada, P.; Covello, P.S.; Nair, S.K. Characterization of the macrocyclase involved in the biosynthesis of RiPP cyclic peptides in plants. *Proc. Natl. Acad. Sci. USA* **2017**, *114*, 6551–6556. [CrossRef] [PubMed]

50. Ludewig, H.; Czekster, C.M.; Oueis, E.; Munday, E.S.; Arshad, M.; Synowsky, S.A.; Bent, A.F.; Naismith, J.H. Characterization of the Fast and Promiscuous Macrocyclase from Plant PCY1 Enables the Use of Simple Substrates. *ACS Chem. Biol.* **2018**, *13*, 801–811. [CrossRef]

51. Olivera, B.M.; Gray, W.R.; Zeikus, R.; Mcintosh, J.M.; Varga, J.; Rivier, J.; Desantos, V.; Cruz, L.J. Peptide Neurotoxins from Fish-Hunting Cone Snails. *Science* **1985**, *230*, 1338–1343. [CrossRef] [PubMed]

52. Olivera, B.M.; Rivier, J.; Clark, C.; Ramilo, C.A.; Corpuz, G.P.; Abogadie, F.C.; Mena, E.E.; Woodward, S.R.; Hillyard, D.R.; Cruz, L.J. Diversity of Conus Neuropeptides. *Science* **1990**, *249*, 257–263. [CrossRef] [PubMed]

53. Livett, B.G.; Gayler, K.R.; Khalil, Z. Drugs from the sea: Conopeptides as potential therapeutics. *Curr. Med. Chem.* **2004**, *11*, 1715–1723. [CrossRef] [PubMed]

54. Davis, J.; Jones, A.; Lewis, R.J. Remarkable inter- and intra-species complexity of conotoxins revealed by LC/MS. *Peptides* **2009**, *30*, 1222–1227. [CrossRef] [PubMed]

55. Akondi, K.B.; Muttenthaler, M.; Dutertre, S.; Kaas, Q.; Craik, D.J.; Lewis, R.J.; Alewood, P.F. Discovery, Synthesis, and Structure Activity Relationships of Conotoxins. *Chem. Rev.* **2014**, *114*, 5815–5847. [CrossRef] [PubMed]

56. Lebbe, E.K.M.; Ghequire, M.G.K.; Peigneur, S.; Mille, B.G.; Devi, P.; Ravichandran, S.; Waelkens, E.; D'Souza, L.; De Mot, R.; Tytgat, J. Novel Conopeptides of Largely Unexplored Indo Pacific Conus sp. *Mar. Drugs* **2016**, *14*, 199. [CrossRef]

57. Reimers, C.; Lee, C.H.; Kalbacher, H.; Tian, Y.; Hung, C.H.; Schmidt, A.; Prokop, L.; Kauferstein, S.; Mebs, D.; Chen, C.C.; et al. Identification of a cono-RFamide from the venom of Conus textile that targets ASIC3 and enhances muscle pain. *Proc. Natl. Acad. Sci. USA* **2017**, *114*, E3507–E3515. [CrossRef]

58. Turner, A.H.; Craik, D.J.; Kaas, Q.; Schroeder, C.I. Bioactive Compounds Isolated from Neglected Predatory Marine Gastropods. *Mar. Drugs* **2018**, *16*, 118. [CrossRef] [PubMed]

59. Woodward, S.R.; Cruz, L.J.; Olivera, B.M.; Hillyard, D.R. Constant and Hypervariable Regions in Conotoxin Propeptides. *EMBO J.* **1990**, *9*, 1015–1020. [CrossRef] [PubMed]

60. Moller, C.; Melaun, C.; Castillo, C.; Diaz, M.E.; Renzelman, C.M.; Estrada, O.; Kuch, U.; Lokey, S.; Mari, F. Functional Hypervariability and Gene Diversity of Cardioactive Neuropeptides. *J. Biol. Chem.* **2010**, *285*, 40673–40680. [CrossRef]

61. Bandyopadhyay, P.K.; Colledge, C.J.; Walker, C.S.; Zhou, L.M.; Hillyard, D.R.; Olivera, B.M. Conantokin-G precursor and its role in gamma-carboxylation by a vitamin K-dependent carboxylase from a Conus snail (vol 273, pg 5447, 1998). *J. Biol. Chem.* **1998**, *273*, 14658–14658.

62. Bandyopadhyay, P.K.; Garrett, J.E.; Shetty, R.P.; Keate, T.; Walker, C.S.; Olivera, B.M. gamma-glutamyl carboxylation: An extracellular posttranslational modification that antedates the divergence of molluscs, arthropods, and chordates. *Proc. Natl. Acad. Sci. USA* **2002**, *99*, 1264–1269. [CrossRef]

63. Shafee, T.; Harris, K.; Anderson, M. Chapter Eight - Biosynthesis of Cyclotides. In *Advances in Botanical Research*; Craik, D.J., Ed.; Academic Press: Cambridge, MA, USA, 2015; Volume 76, pp. 227–269.

64. Gu, W.; Dong, S.-H.; Sarkar, S.; Nair, S.K.; Schmidt, E.W. Chapter Four - The Biochemistry and Structural Biology of Cyanobactin Pathways: Enabling Combinatorial Biosynthesis. In *Methods in Enzymology*; Moore, B.S., Ed.; Academic Press: Cambridge, MA, USA, 2018; Volume 604, pp. 113–163.

65. Ongpipattanakul, C.; Nair, S.K. Biosynthetic Proteases That Catalyze the Macrocyclization of Ribosomally Synthesized Linear Peptides. *Biochemistry* **2018**, *57*, 3201–3209. [CrossRef] [PubMed]

66. Medema, M.H.; Fischbach, M.A. Computational approaches to natural product discovery. *Nat. Chem. Biol.* **2015**, *11*, 639–648. [CrossRef] [PubMed]

67. Van der Lee, T.A.J.; Medema, M.H. Computational strategies for genome-based natural product discovery and engineering in fungi. *Fungal Genet. Biol.* **2016**, *89*, 29–36. [CrossRef] [PubMed]

MDPI

St. Alban-Anlage 66

4052 Basel

Switzerland

Tel. +41 61 683 77 34

Fax +41 61 302 89 18

www.mdpi.com

*Molecules* Editorial Office

E-mail: molecules@mdpi.com

www.mdpi.com/journal/molecules

www.ingramcontent.com/pod-product-compliance
Lightning Source LLC
Chambersburg PA
CBHW051838210326
41597CB00033B/5695